D1259736

Profiles of Rafinesque

Miniature of Rafinesque, attributed to William Birch.
Courtesy of Transylvania University Library.

Profiles of Rafinesque

EDITED

BY

Charles Boewe

THE UNIVERSITY OF TENNESSEE PRESS
KNOXVILLE

Library of Congress Cataloging-in-Publication Data

Profiles of Rafinesque / edited by Charles Boewe.—1st ed.
p. cm.
Includes bibliographical references.
ISBN 1-57233-225-5 (cl.: alk. paper)
1. Rafinesque, C. S. (Constantine Samuel), 1783–1840.
2. Naturalists—United States—Biography.
3. Naturalists—France—Biography.
I. Boewe, Charles E., 1924–
QH31.R13 P76 2003
508'.092—dc21   2003004035

I well remember that when I came to America . . . Linnaeus was here, as in England, the nec plus ultra of Zoology and Botany, while I who already belonged to the French school . . . was deemed a rash youth and innovator. . . . I have lived to see my youthful rashness become science, and . . . I may live yet to see my mature insanity . . . become wisdom. . . .

C. S. RAFINESQUE, 1832

CONTENTS

Contents

III. The Philologist

Contents

IV. The Writer

V. The Legend

ILLUSTRATIONS

Illustrations

ACKNOWLEDGMENTS

Acknowledgment is made for texts reprinted by the kind permission of the following:

Transylvania University for "The Life and Work of Rafinesque."

Marseilles, la Revue Culturelle de la Ville, for "Genealogical Note on the Rafinesque Family."

Verhandelingen der Koninklijke Nederlandse Akademie van Wetenschappen for "Rafinesque Portraits."

Bartonia, Journal of the Philadelphia Botanical Club, for "Rafinesque among the Field Naturalists," "Opinions of Rafinesque Expressed by His American Botanical Contemporaries," and "Rafinesque's Sentimental Botany."

Tryonia, Department of Malacology, Academy of Natural Sciences of Philadelphia, for "Rafinesque on Classification."

Archivio Botanico for "Rafinesque: A Concrete Case."

Library of the Arnold Arboretum for "Rebuttal (from *Index Rafinesquianus*)."

Kentucky Review for "The Fall from Grace of that 'Base Wretch' Rafinesque."

Anthropological Linguistics for "Rafinesque's Linguistic Activity."

American Museum of Natural History for "Unraveling the *Walam Olum.*"

Kluwer Academic Publishers for "Indian Languages and the Prix Volney."

Center for Maya Research for "The Beginning of Maya Hieroglyphic Study."

New York Botanical Garden Press for "The Medicine and Medicinal Plants of Rafinesque."

Historical Society of Pennsylvania for "Who's Buried in Rafinesque's Tomb?"

Evolution of the Rafinesque Biography

CHARLES BOEWE

I

IN THE CENTURY between the publication of Linnaeus's *Species plantarum* (1753), a book that established the need for binomial nomenclature in botany, and the publication of Darwin's *Origin of Species* (1859), a book that provided the phylogenetic basis for the classification of plants, no naturalist of the period—save perhaps Audubon—has received more attention from posterity than has C. S. Rafinesque. Unlike his friend John James Audubon, who has attracted the attention of biographers having a wide range of interests, Rafinesque has been written about mostly by fellow naturalists who came across him in the course of their own scientific work. Because he proposed and published Latin technical names for so many plants and a lesser number for animals, biologists encounter his name in their professional literature almost as often as that of Linnaeus. As a result, what they have written about him usually is colored by primary interests lying elsewhere than in the art of biography.

The first of these biologists was Asa Gray (1810–1888), the second was Samuel Steman Haldeman (1812–1880), both of whom had been personally acquainted with their subject. Not yet America's leading botanist that he later would become, Gray summarized Rafinesque's botanical career in 1841 in a twenty-page obituary article for the *American Journal of Science*, at that time America's leading journal in its field. The next year and in the same place, Haldeman performed a similar function for Rafinesque's zoological career in a ten-page article, a length proportional to the lesser importance of Rafinesque's zoological writings. Even within the tradition of the obligatory obituaries scientists write about one another, which conventionally are bland, objective but comprehensive, and reveal the strengths of character of the subject while skipping over all but his cognizable weaknesses, these two articles are markedly flawed. Gray, writing with the hidden agenda of destroying the reputation of his subject's work so that no one would ever again bother to consult it, condemned Rafinesque with ambiguous approbation. Haldeman, to some extent more charitable toward his subject and like him having more varied interests than Gray, nevertheless so underscored his subject's credulity and gullibility that the uninformed would hesitate to accept Rafinesque's word on anything not reinforced by parallel evidence from another source. Though he ended his career as a professor of comparative philology, Haldeman also had taught both zoology and geology, but in his article he said nothing about Rafinesque's contributions in either geology or philology. Of Rafinesque's manuscript lectures, some of which Haldeman owned, he remarked that those on zoological subjects "indicate but little talent."

Haldeman would have been obliged to look into one branch of Rafinesque's writings to prepare his own *Monograph on the Freshwater Univalve Mollusca of the United States* (1840–45), a subject in which Rafinesque had been a pioneer. It was freshwater mollusks that also attracted the attention of Rafinesque's next biographer, Richard Ellsworth Call (1856–1917). Like Rafinesque, Call was a jack-of-all-trades. After working for the U.S. Geological Survey in Arkansas and Utah, he went to Kentucky to teach chemistry and physics at Louisville's Manual Training High School and, while living in that city, earned an M.D. degree but never practiced medicine. There, on the banks of the Ohio River, he took up the study of its fishes, a subject in which Rafinesque also had been a trailblazer. With the historical collections of the founder of Louisville's Filson Club Historical Society to draw on, he wrote what has long been considered the only "book-length" biography of Rafinesque before he moved on to other jobs, first in Indiana (where he published in 1900 a *Descriptive Illustrated Catalogue of the Mollusca* of that state), then in New York, where he held a series of science-related jobs—none of them lasting very long, as a result of what his own obituary called his "usual lack of mental equipoise."

Aside from its front matter and considerable bibliography, Call's *The Life and Writings of Rafinesque* is only 132 pages long. The author himself called it a "brochure," and it is reported that he read out the entire manuscript at one long sitting of the Filson Club. Issued in 1895 as Filson Club Publication number ten,

Introduction

the book was bulked up by the use of large type, thick paper, and extra wide margins. In his preface Call remarks that he began his study of Rafinesque

> in an attempt to clear up certain matters connected with the synonymy of a large and important group of fresh-water mollusks—the *Unionidae*. A number of very remarkable facts connected with the personality of its subject were thus incidentally learned. As the collection of data proceeded the facts gathered seemed of sufficient importance to group them for presentation to the literary and scientific world, in the hope that a better and more intelligent understanding of this eccentric naturalist might result.

In this short passage Call touched most of the bases that have continued to characterize writing about Rafinesque. The biographical sketch is incidental to a scientific problem; the details of Rafinesque's life are a remarkable discovery; these details are "data" (as though Rafinesque himself might be thought of as a mollusk needing to be classified); the man was an eccentric; perhaps for this reason he has not been understood, and the present writer is going to set matters right. All this is expressed in the passive voice which, in Call's time and until long afterwards, was considered the fitting inflection for most verbs in any prose smelling of science.

A better ichthyologist than Call already had seen fit to expose his own incompetence as a biographer of Rafinesque after having wrestled with the puzzle of identifying the latter's Ohio River fishes. This was David Starr Jordan (1851–1931), who was, first, president of Indiana University and later of Stanford University. In 1886 Jordan published a ten-page sketch of the life of Rafinesque in *Popular Science Monthly* that had at least as many serious errors as pages; he reprinted an augmented version of it two years later in a book called *Science Sketches*, where it was picked up and reprinted again in William Jay Youmans's *Pioneers of Science in America* (1896), a book still considered a reliable reference work. Jordan never would have stated as "facts" about fishes the kind of unexamined imaginary conclusions he penned about Rafinesque. Among these are the fable that, at age twelve, Rafinesque "published his first scientific paper"; that somehow, through his writings, Rafinesque "became, in a measure, one of the ancestors of Mormonism"; and, of course, that he died "in miserable lodgings in an unfriendly garret" where the landlord wanted to sell his corpse to a medical school. And surely, as a university president, Jordan would have expelled any student who took liberties with the words of others as he did when ostensibly quoting extracts from Rafinesque's autobiography.

The next "biographer" of note was Thomas Jefferson Fitzpatrick (1868–1952), whose bibliographical work is discussed here in chapter 18. In a classic case of the tail wagging the dog, Fitzpatrick's *Rafinesque: A Sketch of His Life with Bibliography* often was reviewed as a biography when it appeared in 1911, but the fifty-page sketch that opens the book was intended only to introduce the 170 pages of serious bibliography that follow. A botany professor by profession, Fitzpatrick shared his contemporaries' fondness for the passive voice; this proclivity for meekly backing into sentences tricked him into many a howling dangling participle such as: "Securing a horse a journey was made. . . ." Because most of what Fitzpatrick put into his sketch was only a clumsy paraphrase of Rafinesque's own autobiography, there is little reason to read the imitation when one can go direct to the original—even though Rafinesque himself was no phrasemaker.

One other biographical effort merits mention. *Rafinesque in Lexington, 1819–1826*, which contains only eighty-four small pages of text, was written on commission by Huntley Dupre (1892–1968), a professor at the University of Kentucky, for publication in 1945 by the Bur Press, an art press in Lexington, Kentucky. More distinguished for its typography than for the biography it contains, the book says a good bit about Lexington in the second decade of the nineteenth century, and less about Rafinesque. This content is not lacking in value, however, for it is material that sets the scene for his life in Kentucky and is drawn in large part from contemporaneous sources, including Lexington newspapers and the unpublished Crosby Papers. Then in the possession of the great-granddaughter of Horace Holley, these papers have since been deposited at the University of Louisville, where they are now open to the public. They help to explain the naturalist's troubled relationship with the president of Transylvania University, an issue Dupre skated over all too briefly.

All the authors mentioned so far who wrote on the life of Rafinesque—with the exception of Asa Gray—based their narratives on *A Life of Travels* (1836) and usually quoted from it extensively, so it is time to explain how that so-called autobiography came about and why it has been both helpful and misleading. None of these biographers went very far beyond the confines of its narrow scope. Most of them had little appreciation for the need always to check autobiographical statements against other, independent sources, nor did they have much interest even in seeking out such sources. Both Call and Fitzpatrick had access to letters written by Rafinesque, but neither thought it worthwhile to try to put flesh on the bare bones of their subject's own narrative by reference to his correspondence. Call did seek testimony about Rafinesque's appearance from people still alive who remembered him; Fitzpatrick uncritically reprinted swatches of commentary by several who thought they knew something *about* him. Call printed a transcript of Rafinesque's will without finding any reason to comment on its biographical significance; Fitzpatrick did comment, but only on what was obvious.

Introduction

All of these writers were unavoidably ignorant about the kind of document they were dealing with in *A Life of Travels*. In his prefatory remarks, Rafinesque had written that his little book was "a translation from a sketch of my travels written in French in 1833 and sent to France, to my only sister," which led naturally to the later assumption that it was a kind of extended personal letter to the author's family, none of whom he had seen in person for nearly three decades—hence, a narrative not really intended for the public at all. Rafinesque also said that the book of fewer than 150 pages "was abridged from my memoirs, and journals of travels, begun to be written since 1796." We know that whatever he had begun as a record of his life in 1796 had been lost in the shipwreck on his return to the United States in 1815, so all but eighteen years of *A Life of Travels* had to be from memory. Slight as the privately published book was, the author nevertheless suggested that it "may be deemed an introduction to my works," but he emphasized at the same time that "it is not a mere biographical sketch." So, then, what is it?

The answer developed as a result of a chain of fortuitous Franco-American incidents. While working at Temple University in 1975 as a visiting scientist, the embryologist Georges Reynaud, in his spare time a genealogist, collected information about Rafinesque from the rich Philadelphia repositories. When he returned home, to the Université de Provence, he discovered that Rafinesque was virtually unknown in France. Resolving to dispel this fog of ignorance, Reynaud set out to translate *A Life of Travels* back into its original French, but before the job was finished he learned that the 1833 original manuscript still existed. It was in Paris, in the hands of a surgeon, Dr. Jean Rafinesque (for medicine had become almost the hereditary profession of the Rafinesques since Constantine's time), who was happy to put the manuscript at Dr. Reynaud's disposal.

Try as he might, Reynaud could not find a French publisher interested in bringing out the none-too-elegant writing of this unknown Frenchman who had drifted so far away from his *patrie* that he wrote such uncouth Franglais as "stagecoach" when he should have written "diligence" and "railroad" instead of "chemin de fer." Much as Rafinesque himself had to popularize natural science in Kentucky before he could teach it, Reynaud set out to popularize Rafinesque in France so he could publish the autobiography there. In the Marseilles municipal magazine he published in 1978 a substantial article, "Un grand naturaliste méconnu" (A Great Unknown Naturalist) about Rafinesque and, by persistent campaigning, finally got a brief mention of him admitted in the *Grand dictionnaire encyclopédique Larousse,* but only in 1984. He continues to pester the municipal authorities to name a street for the Rafinesque family, preferably one in the vicinity of the suburban bastide of the naturalist's grandmother, where Constantine spent many happy hours.

Charles Boewe

While seeking information at the Historical Society of Pennsylvania in Philadelphia about Rafinesque's grave and the alleged transporting of his bones to Lexington, I heard about Reynaud's earlier research there and wrote to him in Marseilles for a copy of his article. With it I received an account of his recovery of the autobiographical manuscript, his problems trying to get it published, and—not incidentally—the good news that Dr. Jean Rafinesque, the naturalist's great-great grandnephew, had a cache of family letters I had been unaware of, which he would be happy to let me use. Reynaud had all but given up hope of publishing the auto-biography. But when I took up the torch and brashly set out to find an American publisher, I ran into a similar string of obdurate negatives. Those publishers who had as much as heard of Rafinesque could not understand why anyone would be silly enough to want to read in French what was already available in English (for a type facsimile had been printed here in 1944), nor were they as impressed as I was that Reynaud had found 624 variants between the two texts. He and I resolved to go ahead and prepare an edition anyway. Having enlisted the additional help of Prof. Beverly Seaton (one of whose essays appears here as chapter 17), we three met in Lexington at Transylvania University's 1983 conference celebrating the bicentennial of C. S. Rafinesque's birth and planned our respective roles. Reynaud had transcribed the manuscript and now agreed to help identify persons named in it, especially the Europeans; Seaton, whose French is better than mine, agreed to copyread and also assist in identifying persons; and I drafted the introduction and notes (all in English), helped to identify persons, and promised to continue seeking publication. The final product, for which Seaton providentially found a small grant that made it possible for us to retain a professional typist comfortable with the French language, was finally scrutinized and approved by the three of us.

And, yes, it did get published, as *Précis ou abrégé des voyages, travaux, et recherches de C. S. Rafinesque: The Original Version of A Life of Travels*, in 1987 by the Verhan-delingen der Koninklijke Nederlandse Akademie van Wetenschappen Afd. Natu-urkunde in Amsterdam, where, ever since Erasmus, the cosmopolitan Dutch have been able to deal with any language, any subject, any time. In view of the skepti-cism we had faced already in the two countries where publication had seemed most likely, I feared the project might still fall foul of the academy's required "peer review" process. But the late Frans Stafleu, a Rafinesque enthusiast who had taken the project under his wing, calmed my apprehension. A distinguished member of the Royal Dutch Academy of Sciences himself, he assured me that as sponsor of the project he would be one of three referees and would have the right to select one of the two remaining ones. A two-thirds majority decision would prevail—which, in my opinion, is an agreeably democratic way to conduct a peer review.

Introduction

Since 1987, then, it has been known that this "abridgment"—rightly called by its author a précis of the story of his life—is merely the skeleton of the personal memoir he hoped to publish but never accomplished. The French manuscript had been addressed to the Société de Géographie, for whose *Bulletin* Rafinesque already had submitted a handful of essays, and it was sent to his sister Georgette in Bordeaux for her onward transmission to Paris. But Georgette was pregnant when it arrived, and she died in childbirth, early in 1834, before she could carry out her brother's wishes. After her death, the manuscript was preserved and passed down through the family. From Antoine, the brother of Constantine and Georgette, it passed to Antoine's son Jules, and so on down the male line finally to Jean. In the United States, when Jules's uncle Constantine saw that his manuscript was never going to be printed in France, he added 3,500 words to bring the narrative up to date, rearranged the material somewhat, translated it all into English, and had it printed in Philadelphia in 1836 at his own expense.

Knowing these details, we should be dismayed no longer that *A Life of Travels* concentrates so doggedly on where the author went and when and leaves out nearly all the particulars we would like to know about his human relationships. It was written for an audience of French geographers, whose *Bulletin* was filled with first-person accounts of intrepid French explorers. Moreover, Rafinesque's intention is further confirmed by his direction that the manuscript should be handed over to the *Annales des Voyages* if it was not found suitable for the *Bulletin*. Since both publications specialized in eyewitness accounts, so to speak, this is why Rafinesque apologized in English for what he called the "egotic form" of his narrative. Had he changed it to third person, "this would have caused another objection" from the publication media he had in mind.

Even taken together, *A Life of Travels* and *Précis ou abrégé* make a wobbly platform on which to erect a biography, but there also are a few additional biographical scraps from Rafinesque's pen that have turned up in recent years. A travel journal now at the University of Kansas gives some idea of the sort of memoir Rafinesque might have written. As a sample, we reproduced in the introduction to *Précis ou abrégé* its entries from April 9 to May 18, 1826, which cover Rafinesque's leave-taking from Transylvania and his progress through the state of Ohio on his way back to Philadelphia, where he spent the rest of his life. These snippets of telegraphic brevity must have been intended to recall to his mind the people, places, and events the author meant to expand on in his fuller memoir. Among those of human interest are his farewell visits to friends in the neighborhood of Lexington (among them the venerable Isaac Shelby, Revolutionary War hero and Kentucky's first governor), his attendance at a lecture on

Charles Boewe

astronomy in Cincinnati and his visit there with Daniel Drake (who later returned the courtesy by lambasting a book of Rafinesque's he admitted he had never read), his opinion on why the Owenite community at Yellow Springs had failed and his visit to natural curiosities in its vicinity, his stopping to play with children at a spring, his attendance at the meeting of a court where a man was being tried "for burning a child," tavern scenes to be remembered only by such cryptic entries as "Singer" and "Beer," etc. Brief as the entries are, they are engaging enough to make us wish the author had been able to complete his memoir as planned.

Another provocative document, one that surfaced after the publication of *Précis ou abrégé,* gives reason to believe there might still exist a more substantial autobiographical memoir from the pen of its author. Readers of *A Life of Travels* will recall that on his way west in 1818, Rafinesque contracted with the Pittsburgh booksellers Cramer & Spear to publish his "Travels in America." Compensation for their breach of contract is one of the outstanding claims he mentioned in his will. It now appears that such a book either actually was written or at least was thought through in outline, for the University of Kentucky has a three-page manuscript, foolscap size, bought in recent years from a Cincinnati rare book dealer, that is the kind of analytical table of contents books in the nineteenth century often had. In Rafinesque's distinctive handwriting, it summarizes the contents of eight chapters.

Many of the pithy entries refer to people and events that are well known, but we certainly would like to read the extended text of these items in Chapter I: "University of Pennsylvania—Death of Dr. [Benjamin Smith] Barton—I propose a new faculty of Natural Sciences—It is established—I wished the chairs of Botany or Zoology—I am offered that of Comparative Anatomy which I refuse." And after listing the names of new friends in Philadelphia, any one of whom would have warranted a paragraph or two in an extended memoir, Rafinesque returns to New York City.

Chapter II covers travels in the state of New York. When we remember David Starr Jordan's cocksure conclusion that if Rafinesque "ever loved any man or woman, except as a possible patron and therefore aid to his schemes of travel, he himself gives no record of it," the following entries—the first of several about personal relationships—come as a welcome contradiction. "A false friend—Emmeline—Strange Occurrences" where the word "Strange" is written above a canceled "Shocking." If Emmeline in New York was the false friend, their relationship must have gone sour about the same time that the mother of his children, Josephine Vaccaro, was marrying her comic actor in Palermo.

Introduction

Chapter III covers the "great western tour" of 1818, where this string of topics appears: "Transylvania University—Reorganization—I am offered the Chair of Chemistry—I prefer Botany and Natural History—A bad choice it appears."

Chapter IV opens with "A Winter in Philadelphia—Society" and intrigues us with "Anecdotes—Mary." Will we ever know more about Mary, and what she was to Constantine?

Chapter V is titled "Lexington" and promises to give us "State of Society—Manners—Death of Mr. Clifford." A later section would tell us about "Presid^t Holley," then take up "Literary Cabals—Intrigues." Is there a connection? Next come "Silvia & Laura—Female friends—anecdotes—Strange events—Hopes and Results." Imagination runs riot piecing together how these five topics might be related to one another.

Chapter VI deals with travels around Kentucky, and the clouds darken in its concluding entries: "Characters—Selfishness—A singular character—Sketches of human life—Despondency—Highlife—Slavery—Illiberality—Trustees—Inquisitions—Secret History—Presumption."

Chapter VII, "Labours," touches mostly on matters we know something about already, but it is of interest that here Rafinesque notes: "I begin my Memoirs—I write my travels," and he also alludes to a hitherto unknown literary project, "I collect materials for a Picture of Kentucky."

Chapter VIII is the "Conclusion," which is transcribed below in its entirety because it is so characteristic of Rafinesque's self image:

> My zeal for science—My labours not yet well known—My own character—Religious opinions—My parents were Protestants—Admiration of the Quakers &c—Discoveries in all branches of Sciences even in Geometry and Mental Philosophy—My analysis of the Human Will, manuscript of mine—I hope to do much for science if I live—Botanical Garden—I mean to render it like the Athenian Gardens in time—My enemies [*sic*]—I do not hate them altho' they hate me—They have prevented many of my productions from appearing—Strange circumstances—I am compelled to send some of my works to be printed 5000 miles off—Intimate friends—Since the death of J. D. Clifford I have not met *Un ami de coeur*—Opinions on me—Many false suppositions and mistakes—Easily rectified—I care not for the opinion of fools—My labours are better known in Europe than here and better appreciated—I was born too early for this new country—But *Nihil desperandum* & *Numquam otiosus* are my mottoes.

Charles Boewe

II

In the spirit of *nihil desperandum,* then, this book has been assembled in the belief that it will in some measure supply the lack of the authoritative biography C. S. Rafinesque deserves but has not yet received. The opening essay, "The Life and Work of Rafinesque," by Francis W. Pennell, is the longest and is intended to be the linchpin for the whole enterprise. This chapter comes from the keynote address Pennell gave in 1940 at Transylvania's symposium marking the centennial of Rafinesque's death. It was not printed until two years later, and the printed text implies that in the interim Pennell added additional material not in his original manuscript (there is a break in the footnoting, for instance). As transcribed here from the 1942 printed text, some deletions have been made and numerous additions have been supplied in the notes to bring the contents into congruence with the current knowledge of Rafinesque that has been attained over the past two generations.

Although Pennell's essay has been known all this while to Rafinesque scholars, it was unavailable to general readers because of the unusual circumstances of its publication. Along with five lesser and largely negligible pieces, it was printed in number seven of volume fifteen of the *Transylvania College Bulletin.* At that time, in 1942, special reduced postage rates applied to the mailing of periodicals; by issuing the required minimum number of issues each year it was possible for colleges to circulate their annual catalogs under the periodical rate and still have issues left over for whatever else they wished to publish during that year.

The *Bulletin's* No. 7 for 1942 is subtitled "Rafinesque Memorial Papers" and additionally contains an inconsequential account of "The Modern Naturalist," by Alfred E. Emerson, who came down from the University of Chicago, presumably to establish the 1940 context for viewing Rafinesque's pre-1840 contributions. H. B. Haag came from the Medical College of Virginia to speak on "Rafinesque's Interests—Medical Plants," but must have grown so impatient over the two-year delay in publication that meanwhile he printed his full text in *Science* magazine (cited here in Flannery's article), leaving only an abridged and insignificant text for the *Bulletin.* William M. Clay came over from the University of Louisville to talk on "Herpetology and Rafinesque"; as printed, his paper enlarged on "the advances in herpetology during the past one hundred years" and had little to say about Rafinesque's prior contribution. William E. Ricker, from Indiana University, was aware that the former president of his university, David Starr Jordan, had had something to do with reviving a few of Rafinesque's fish names, but in his "Research in the Biology of Fishes" he didn't even bother to insert Rafinesque's

Introduction

name in the title. No wonder a footnote says his "paper has been condensed." And W. D. Funkhouser, from the nearby University of Kentucky, speaking on "Rafinesque—Archaeologist" had the bad taste to declare at this celebratory party that "Rafinesque's archaeological work can not now be taken seriously"—largely, one gathers, because Rafinesque, unlike the pot-hunters of his time, chose not to desecrate archaeological sites by digging into them. Finally, the college librarian, Elizabeth Norton, provided two lists: the first was of the Rafinesque publications in her safekeeping; the second of the library's Rafinesque manuscripts. Both of these are now out of date, but Norton's successor wanted to have me lynched when I pointed out, forty years later, that four of the "Rafinesque manuscripts" could not possibly have come from his hand.

I mention these shortcomings of the "Rafinesque Memorial Papers" so readers of this book will not feel cheated in the suspicion of important material being omitted, so they will know what to expect if they ever do look into the full symposium report, and to hint at the way Pennell's contribution stands as a beacon towering over the flickering candles below it. I don't know what honoraria, if any, the other speakers received, but from his papers archived at the Academy of Natural Sciences of Philadelphia, it appears that Pennell got a round-trip train ticket, a hundred dollars—and probably meals and a bed at the home of one of the faculty.

Francis Whittier Pennell came from an old Quaker family, and he was educated at the University of Pennsylvania. An associate curator at the New York Botanical Garden from 1914 until 1922, he then returned to Philadelphia as curator of plants at the Academy of Natural Sciences. He remained there the rest of his life, except for several worldwide field excursions in search of material on the plant group he made his specialty, the Scrophulariaceae, the figwort family, which contains such attractive ornamentals as the lady's-slipper, snapdragon, and foxglove. It was his study of the literature on this plant family that caused his earliest interest in Rafinesque, and that resulted in technical papers on Rafinesque's overlooked names for Scrophulariaceae plants. Something about the association of plant names with the people who classified the plants they stand for often causes taxonomists to branch off into the history of botany, and in this Pennell was no exception. After publishing on Rafinesque's colleague "Elias Durand and His Association with the Academy of Natural Sciences of Philadelphia" in 1936, it was only natural that Pennell should continue with Rafinesque himself. After the major 1940 symposium effort, he continued to write shorter pieces about Rafinesque as new evidence turned up, often publishing it in *Bartonia,* the annual Proceeding of the Philadelphia Botanical Club, which he edited and published from the academy. During the period of the Transylvania address, he also was publishing in wider scope on "The Botanical Collectors of the Philadelphia Local Area," in two parts in *Bartonia*, 1940–43, which, despite its parochial title, is the best forty pages available on some of the important botanists of their time. Francis Pennell probably was the most qualified person in America to try to sum up the accomplishments of the life of Rafinesque at Transylvania's centennial program.

As published in 1942, Pennell's sixty-page article is a biography superior to Call's 1895 monograph. Like his predecessors, Pennell had to erect his study on the scaffolding of Rafinesque's *A Life of Travels,* but in addition he gave a judicious evaluation of the accomplishments of his subject as viewed from the state of knowledge of Pennell's own time, and he made up many of the deficiencies of the autobiography by drawing on letters found both in his own institution and elsewhere in Philadelphia. Among these, the letters written by Emilia Rafinesque (the naturalist's daughter) and Georgette Rafinesque Lanthois (his sister) introduce for the first time in biographical writing about the naturalist a warm, human dimension of his life. If Emilia's letters sound as emotional as the libretto for a Verdi opera, she had good cause to write that way. Forced to go on the stage as a teenager, seduced, impregnated, and abandoned by her British lover, she poured out her heart to the father she had hardly known. Georgette, even if skeptical about her niece's avowal of reformed virtue, is tender toward the brother who had wandered so long in foreign lands. Of course, in French his sister *tu*-ed him; and Pennell's decision to translate this pronoun with "thee" and "thou" probably grated less on his Quaker ear than it does on mine.

Strangely missing from Pennell's narrative is any mention of the long exchange of letters between Rafinesque and John Torrey. Torrey apparently did not keep file copies of his correspondence, and if Rafinesque preserved the originals they were destroyed in the general debacle made of his estate. But most of Rafinesque's letters to Torrey are at the New York Botanical Garden, where Pennell himself had worked for seven years, and they cover the final period when, back in Philadelphia, there was no longer any need for Rafinesque to correspond with Zaccheus Collins. Moreover, Collins died a decade before Rafinesque, but Torrey outlived him.

It is more understandable that in 1940 Pennell was unaware of the valuable collection of letters by Rafinesque to A. P. de Candolle preserved at the Conservatoire Botanique de Genève in Switzerland because the first publication describing them appeared only in 1952. I find it incredible, however, that he thought his audience would not be interested in Rafinesque's decade in Sicily—the place where Rafinesque published his first scientific books, where he worked for a time

Introduction

for the American consul, where he made so much money by trade that he could retire by age twenty-five, where he traveled all over a little-known part of the world, and, above all, where he fathered two children. This decade, too, is partially documented in Rafinesque correspondence, this time with William Swainson. Though these letters are at the Linnean Society of London, a printed calendar of them had been available since 1900.

With nothing to go on but *A Life of Travels*, all the biographers reproduced Rafinesque's own misleading information about his parentage, including his reversal of his father's *prenoms* and his ambiguous reference to his mother's birthplace. It remained for Georges Reynaud to straighten out these and other genealogical matters of the Rafinesque family in an appendix to his 1978 article. This "Notice généalogique sur la famille Rafinesque" has remained little known to American researchers, so it is a pleasure to present it here, in English translation, with several regrettable typographical errors of the original now corrected. Reynaud, who is bilingual, helped me over a number of rough spots through an e-mail interchange and corrected, expanded, and improved my translation until we both were satisfied with it. In addition to what it tells us about the naturalist and his immediate family, Reynaud's research throws light on the rise of the Rafinesques for more than a century before Constantine's birth.

Never published before, my "Last Days of Rafinesque" is included in the forlorn hope of squelching the sentimental mythologizing about the impoverished, neglected genius dying alone and unloved in his miserable garret and his frail corpse being hastily shoveled into a pauper's grave in a potter's field. Some fables are too compelling to be deflected by mere truth, though, and this may well be one of them.

Anyone who has come this far with Rafinesque wants to know what he looked like. There are portraits in both painted and verbal media—all more or less equivocal. Such information as is available about them is summed up here in a short piece that first appeared in *Précis ou abrégé* but has not become widely known, since misattributions continue to be made in popular articles and on the even more popular Internet.

III

Pennell gave the first and, what continues to be, the most judicious and well-rounded evaluation of Rafinesque as a naturalist. His appreciation of Rafinesque's joy of discovery was my own springboard for a paper read at a symposium on "Early Botany in the Trans-Allegheny Region" at Ohio State University in 1987.

Charles Boewe

Under the title "Rafinesque among the Field Naturalists," the paper was printed in *Bartonia*, the journal Pennell had edited throughout much of his career though now under different editorship, of course.

The work of most naturalists in the first half of the nineteenth century focused on collecting, preserving, classifying, and naming new plants and animals. Not every naturalist performed all these functions; some collected and preserved, leaving it to others to classify and name. Rafinesque not only did perform all four functions but, additionally, he set himself up as a theorist of classification and considered it his mission in life to continue the work of Linnaeus as lawgiver for classification. Hence, to understand his place in the history of natural history it is necessary to grapple with his theories and his practice against the background of his reading and in comparison with the practice of others of his generation.

No one was better equipped to undertake such a study than the late Arthur Cain. Already a noted authority on Linnaeus and his works, Cain also had a wide-ranging knowledge of the books that provided the furniture of Rafinesque's mind and of the practices of naturalists contemporary with him, especially the French, to whom Rafinesque looked for models. Like Reynaud, Cain came to Rafinesque studies as the result of a visit to the United States from his home base at the University of Liverpool. A specialist on terrestrial gastropods, including snails and slugs, he was working at the Academy of Natural Sciences of Philadelphia as a visiting scientist when he was shown a series of unionacean mussel shells arranged according to Rafinesque's classification scheme for them. In other words, he approached Rafinesque studies by exactly the same route as R. E. Call, Rafinesque's first biographer, as a result of pondering the relationship of *Unionidae* mussel shells to each other. Having worked extensively on the *Cepaea* genus of land snails, which exhibits immense variation in shell color and ornamentation, Cain immediately understood the problem of classifying a series of bivalve specimens that differed markedly from each other at the extremes when lined up, yet almost imperceptibly changed from individual to individual within that lineup. He wrote that Rafinesque's "arrangement was so far from mad that I was immediately intrigued." When he began reading, in three languages, the academy's notable collection of Rafinesque's own publications on the theory and practice of classification and nomenclature, as well as the secondary literature about him, "it became obvious that neither the contemptuous neglect of European, nor the enthusiastic praises of some American, authors really met the justice of the case." The outcome of this conviction was his *Constantine Samuel Rafinesque Schmaltz on Classification: A Translation of Early Works by Rafinesque with Introduction and Notes*, published in 1990 by the academy. Our excerpt here comes from the introduction to this book.

Introduction

It may be of interest to explain why, in his title, Cain called the naturalist both Rafinesque and Rafinesque Schmaltz. As the naturalist himself tells us, he found it "prudent" to include his matronym in Sicily because just across the Strait of Messina the French army was poised to invade, and he wanted to avoid being identified with the French. Since in Europe his books published during that decade are those best known, their author has continued to be called Rafinesque-Schmaltz there. After his return to the United States in 1815, he never again used the hyphenated form, and here he always has been called Rafinesque.

Before turning to the next subject, I have to mention for the historical record the contretemps that almost caused Cain's book never to see the light of day. The Academy of Natural Sciences naturally sent his manuscript out for peer review. One "peer" reviewer contemptuously refused even to discuss the manuscript, and the other risked apoplexy writing the savage report he turned in. His diatribe was even more intemperate than Croizat's vicious attack reprinted here; but, put more mildly, its recommendation was, let sleeping dogs lie, because "any reasonable scientist knows these publications of Rafinesque were not very valuable at the date of their publication and they are even less so now." Whatever this anonymous person's stature may be among slugs and snails, he was no peer of Arthur Cain, for in his smug linguistic ignorance he even lambasted the manuscript's British spelling conventions as "mistakes," which he also was unable to distinguish from Cain's artfully contrived English equivalents of the typos in the original French and Italian texts. I like to believe I helped to get Cain's work into print by pointing out that failures are as worthy of being studied in the history of science as are the successes and may be even more instructive. The episode also is a reminder that if Rafinesque's own nineteenth-century peers, whose negative opinions are well exhibited in Ronald Stuckey's essay, had had veto power over *their* colleagues' writings you would not be holding this book in your hands today.

Zoologists have written less about Rafinesque than have their botanical counterparts, perhaps because his zeal for devising new names has wrought less havoc in their profession. However, those names he did propose for animals display the same characteristics as his botanical nomenclature, and his classification of animals not infrequently was carried out second- or third-hand on the published writings of others, a practice that threw botanists into a paroxysm of wrath when his *Florula Ludoviciana* was published. Most of what has been written about Rafinesque's zoology concerns his fishes and bivalve mollusks, both of which he observed in large part direct from specimens. Little has been said about his mammals, so it came as a provocative challenge when I was asked to give a paper on the subject at the Fifth International Theriological Congress in Rome in 1989. The essay, "Rafinesque's Mammalian Taxonomy" is here published for the first time.

Ronald L. Stuckey's article on what Rafinesque's contemporaries said about him also was written for a conference. Stuckey, a professor of botany at Ohio State University, in some ways inherited the mantle of Francis Pennell and became this generation's botanist-interpreter of Rafinesque's work. As early as 1971 he made a notable study of the Rafinesque's specimens of vascular plants that are still to be found in the herbarium of the Academy of Natural Sciences of Philadelphia. Later he described how Rafinesque obtained some of those plants through public auction, wrote on Rafinesque's collecting practices in Kentucky and elsewhere in the Ohio Valley, and helped to update the secondary bibliography by identifying recent publications about Rafinesque. His article reprinted here from *Bartonia* was one of two worthy of preservation among those read at the 1983 bicentennial celebration of Rafinesque's birth held at Transylvania University. (Seaton's, also reprinted here, was the other.) The study of personal correspondence, such as Stuckey's, takes on added significance when it is remembered that during much of the nineteenth century the lack of scientific journals was so great that the exchange of letters between colleagues was an important means of professional communication among naturalists.

The scene shifts markedly when we come to the next contributor. Born in Turin in 1894 to French parents, Leon Croizat came to New York in 1923 as a refugee from fascism. A talented artist, he sold watercolors but could not make a living from his art. He came to know E. D. Merrill at the New York Botanical Garden, and in 1936, when Merrill was appointed director of the Arnold Arboretum, he offered Croizat a position there as a technical assistant. Fluent in eight languages, Croizat ransacked botanical literature from antiquity to the present and wrote and illustrated more than three hundred manuscript booklets of notes, which provided the database from which he mined his published articles and books. Much more like Rafinesque in personality than he himself would have liked to admit, he is mostly remembered for what many botanists consider his wrong-headed criticism of Darwin's concept of natural selection.

The job in Massachusetts lasted but a decade, for after Merrill was eased out of his directorship in 1946 Croizat was dismissed; he himself believed it was because he had been critical of his superiors. With no further employment prospects in the United States, he drifted south to Venezuela, and there, after several temporary positions, ended up as director of a provincial botanical garden, where he died in 1982. Knowing this much about his life, we can understand that in his article he is as much belaboring his old boss—despite having been befriended by him—as he is trying to exorcise the ghost of Rafinesque, which he believes has continued to haunt the science of botany.

Elmer Drew Merrill had a varied career also. Working for the U.S. government in the Philippines from 1902 to 1923, he became an authority on the plants of China and the Philippine archipelago. Later he was dean of the University of California College of Agriculture and director of the California Botanical Garden, then chief executive officer of the New York Botanical Garden, and finally director of Harvard's Arnold Arboretum. While in the last post he began his project to search out and index all the Latin plant names ever published by Rafinesque and to relate them to currently accepted names. In preparation for this Herculean task, he had begun writing about Rafinesque in 1942 and during that decade issued by photo-offset five volumes of out-of-print Rafinesque texts needed for the index, which was itself published by the Arnold Arboretum in 1949. It is from the introduction to his *Index Rafinesquianus* that our excerpt here, supplied with a made-up title, has been taken. More about Merrill as bibliographer will be found in the essay on "The Fugitive Publications of Rafinesque."

Finally, this section closes with an article of my own, first published in the *Kentucky Review,* that uncovers the start of the campaign to destroy the reputation of C. S. Rafinesque.

IV

When we come to a consideration of Rafinesque's philological studies Pennell is of no help, nor are any of the other nineteenth- and early-twentieth-century commentators. Rafinesque's excursions into prehistory, archaeology, ethnology, and linguistics were not as whimsical as they may have seemed to later readers, and all were strands of the same fabric. In his time it was still possible to include human beings in a generous and far-ranging concept of natural history. In 1820, in Lexington, Rafinesque did in fact give a public lecture "On the Natural History of Mankind." Never concerned with linguistics as a distinct science in its own right, he pursued linguistic studies as an adjunct of ethnology (a word he was using before most of his American contemporaries had heard of it). His investigation of archaeological remains and Indian languages would afford access, he believed, into the prehistory of the indigenous people of the New World and uncover traces of their migration from the Old.

His linguistic studies have been entirely neglected by most of those who have had anything to say about Rafinesque's work, because linguistics falls outside their field of interest. Only S. S. Haldeman had the competence to deal with linguistics,

Charles Boewe

and the fact that he chose not to do so hints that his intention was by no means to give a full picture of Rafinesque's accomplishments. Then, too, so little of what Rafinesque had to say in this area ever was published that scholars like Vilen Belyi lament the loss of manuscripts once said to exist.

Deeply versed in the history of linguistics in Europe, Belyi had begun an examination of the origin of linguistic study in the United States when he came upon fascinating references to C. S. Rafinesque. They were in an unlikely source, the *American Antiquities and Discoveries in the West,* a book cobbled together in 1838 by Josiah Priest, an Albany coach trimmer, saddler, and harness maker turned author, who quoted (not always with attribution) great swatches of Rafinesque's published prose, as often as not in order to controvert it. Working from the Vinnitsa Technical University in the Ukraine, Belyi could get this book and a few other printed texts on loan from the Lenin State Library in Moscow. They were enough to whet his appetite for more, especially for manuscripts he found mentioned but had no way to get his hands on. In 1983 he sent a query about materials in this country to Transylvania's librarian, who turned his letter over to me since I had just completed my revision of the 1911 Fitzpatrick bibliography. Although I had not found some of the manuscripts Belyi hoped to see, I knew of others that had not come to his attention and had copies of these shipped to him as well as printed materials from sources more reliable than Priest's book.

When it proved impossible under the time constraints for Belyi to present his study at Transylvania's Bicentennial Celebration, he sent it to Germany for publication in the *Zeitschrift für Phonetik, Sprachwissenschaft und Kommunikationsforschung.* There it languished until publication was achieved in 1991. Meanwhile, however, the breakup of the USSR deprived Belyi of further use of the Lenin State Library and, as far as this essay was concerned, such was the attendant confusion in the Ukraine and other new republics that if proofs of his article ever were sent to him from Germany he did not receive them. Perhaps because the text was in English the *Zeitschrift's* editor let serious errors pass uncorrected, and some sentences were little more than gibberish. Nevertheless, in Bloomington, Indiana, the editors of *Anthropological Linguistics* recognized merit in the blemished essay and offered Belyi the opportunity to reprint a corrected version. No longer able to check his citations against sources in the library in Moscow, Belyi was in a quandary. Fortunately, it proved possible for me check the article's references and to "see it through the press" as the old-fashioned expression has it, with the result that the improved version now reprinted here finally did appear in 1997 in *Anthropological Linguistics.* Despite a small degree of repetition with the three essays that follow Belyi's, I have let stand all that he says because his depth of understanding of the history of linguistics provides a context for the specificities of the following three.

I am tempted now to repeat what Pennell said about Rafinesque's lawsuit against the estate of Zaccheus Collins: "We come to an incident that I wish we could purge from the career of Rafinesque." This is the affair of the *Walam Olum*, whose publication brought Rafinesque what little renown he has enjoyed among scholars of American prehistory. The issues involving it, far from settled yet, are disclosed here in an essay by David Oestreicher and further elaborated in one of my own.

At the outset it must be acknowledged that the Oestreicher essay reprinted from the magazine *Natural History* is but a précis of two much longer, fully documented articles by him in the 1994 and 1995 issues of the annual *Bulletin of the Archaeological Society of New Jersey*. In turn, these articles are essentially an abstract of his 1995 547-page doctoral dissertation at Rutgers University. Massive documentation backs up the issues that are most succinctly stated in the more accessible article given here. Since Oestreicher's research and my own are so intertwined, their divergent results may be introduced in joint fashion.

As he says, Oestreicher had been studying Lenape folklore and the language of the Lenape people for nearly two decades when he undertook his analysis of the *Walam Olum* document. I had been interested in its provenance even longer, when, in 1984, a letter from Dr. Joan Leopold told me of her discovery in Paris of a 256-page manuscript along with a 14-page supplement to it, both in French, in a hand that had been identified as Rafinesque's. She invited me to write a chapter about these documents for her forthcoming book on the Prix Volney, for her discovery was the manuscript of one of the only two submissions for that competition's 1835 contest.

By 1986 my chapter was finished, but the book, now projected as three volumes, experienced the first of numerous delays. Two years later I prepared photocopies of my typescript and sent them to a number of anthropologists I knew to be interested in the "Walam Olum problem," on which the Paris manuscript appeared to shed new light—especially on the kind of material the pictographs were inscribed on, for it has been called everything from birch-bark strips to maple shingles. Apparently my "preprint" was in turn copied and further distributed somewhat like a chain letter during the next five years. It was cited in at least three different places as a "forthcoming" article. It actually did come forth in print in 1999, though now substantially enlarged and, I believe, improved because meanwhile Leopold—a professional in linguistics, as I am not—had made numerous suggestions for changes.

Charles Boewe

It was a copy of the 1986 preliminary version that caused David Oestreicher to seek my assistance in 1994. Having arranged for him to see a copy of the Prix Volney manuscript and providing copies of letters I had saved from correspondence with several of the authorities he mentions, including C. A. Weslager and John Witthoft, as well as copies of relevant unpublished Rafinesque letters, I share with Oestreicher some of the onus for revealing this blot on Rafinesque's character. But I hasten to declare that I claim none of the credit for uncovering the spurious nature of the *Walam Olum*. This disclosure resulted entirely from Oestreicher's brilliant detective work, and its accomplishment has made three generations of scholars look foolish for pursuing such chimeras as the enigmatic "Dr. Ward" and the equally enigmatic "John Burns" when the solution to the mystery lay in a careful scrutiny of texts published before 1836 and available to all earlier researchers. As a result of Oestreicher's analysis, the *Walam Olum* can no longer be thought of as the "American Iliad"; it is not the great creation myth, flood legend, and migration record of the Lenape people; it is not even a fair sample of the Lenape language.

David Oestreicher calls it a hoax. Where there is a hoax there must be a hoaxer; so proceeding to pile Pelion on Ossa, Oestreicher has hauled in every shortcoming ever attributed to Rafinesque to imply that it was he who devised the hoax in order a) to shore up his earlier speculation about the peopling of the New World, b) to get rich, and c) to cadge a pension out of the French government. But such ad hominem arguments seldom are convincing, and it is here that Oestreicher and I must part company.

Of course there *was* a hoaxer, but it is even more likely that gullible, credulous C. S. Rafinesque was his victim. Very likely he also was the unintended victim of whoever invented the "Charles LeRaye," from whose alleged account of his captivity by the Sioux the naturalist derived his information about the mule deer (see p. 148). Rafinesque himself was capable of much foolishness; he often was his own worst enemy, and, like all of us, he could sometimes be forgetful. But nowhere in my fifty years of acquaintance with all his known published and unpublished writings have I ever found him deceitful. In this regard, I agree with Merrill's psychiatrist, Dr. J. M. Woodall of Boston, who concluded that there is "no evidence for dishonesty or that he was motivated by a desire to deceive." Not only would a hoax of this magnitude be a form of deceit and require a conscious wish to deceive, it also would require, I think, a grotesque, sardonic sense of humor. The perpetrator of such a hoax wants to sneer and smirk at somebody— maybe jeer at posterity—but C. S. Rafinesque came as close to being humorless as may be possible for the mammal that laughs. In all those thousands of pages of

Introduction

writing there occurs only one feeble pun: when Benjamin Silliman barred his contributions to the *American Journal of Science,* Rafinesque called him a "Silly-man."

Now who would want to play such a trick on our inoffensive naturalist? I
cannot name a particular culprit, but their numbers were legion. Something about
Rafinesque's personality—perhaps the exceptional combination of egotism, certitude, naiveté, and paranoia in his makeup—challenged pranksters to tweak his
nose. At Transylvania a group of writers calling themselves Zolius & Co., who
probably were students, published a pedantic parody of the kind of writing found
in the *Western Review,* where they singled out Rafinesque among all its contributors
for special derision, calling him, among other snide epithets, "the stone on which
our wits were sharpened." From sources in the Transylvania library they also could
have contrived the pictographs of the *Walam Olum.* The Cincinnati scribblers especially enjoyed pulling Rafinesque's leg; the "glory, jest, and riddle of the age," as
Thomas Peirce called "Professor Muscleshellorum" in one of his *Odes of Horace in
Cincinnati.* An anonymous reviewer in the *Cincinnati Literary Gazette* of Rafinesque's
Ancient Annals of Kentucky pounced on its epigraph that the author had used more
than once, the Latin equivalent of "Never Idle," to conclude: "Nunquam otiosus,
Un-no-wewu uh-hau-youn-quohk neen don khow-wot, qua don mkhea-wowk-nihk
khow-wot; un-no-wew sook-te-pooh-tuh don aum-wau-weh soo-kut queh-now wim
neh wse-khi. Ow-waun dum wke-sih nooh tom-mon-nuh wpon-non-nuh-kau-wau-
con-nun?" Is that a code, or is it mere babble? At any rate, the mind that composed
it could just as easily have composed: "Tellenchen Kittapaki nillawi / Wémoltin
gutikuni nillawi / Akomen wapanaki nillawi / Ponskan-ponskan wémìwi Olini"—
which is, according to Rafinesque, a verse from the *Walam Olum* narrating "the passage to America." In the same magazine "A Brief Desultory and Imperfect
Vocabulary of the Comanchee Language" was dedicated to Rafinesque by a person signing himself only "B.," because, he said, "to affix my proper name would be
introducing a stranger, who has not pretensions to the literary celebrity and deep-
drawn lore that render the name Rafinesque a sufficient guarantee for any 'historical details,' without extorting the mortifying confession that they are borrowed
from Clavigero, Humboldt, or the more recent Bonnycastle." Need we say more
about the willingness in the 1820s to twit Rafinesque's pretensions to scholarship?

Rafinesque never claimed that the *Walam Olum* came to him from the
hand of a Native American. Presumably it was some unnamed Caucasian who
first obtained the pictographs, "as a reward for a medical cure," and only later
"were obtained from another individual the songs annexed thereto in the original language." He does not even say that it was he who brought the two forms
together. For all we know, he may be repeating the story told him by the con

Charles Boewe

man who artfully led him down a garden path. It seems reasonable to conclude that in Kentucky he did obtain some tangible artifact he believed to be an Indian document and that he spent more than a decade of misdirected ingenuity trying to make sense of it. Perhaps the most famous hoax perpetrated on him in Kentucky by a person well known to posterity helps to explain how this could happen.

It was his Francophone friend John James Audubon who delighted in getting Rafinesque hopelessly lost in a canebrake to so befuddle his brain that he not only believed Audubon's fraudulent account of a red-headed swallow but published a scientific description of this implausible bird. It is not impossible that Audubon painted for him such a mythical creature as well, as he may also of the *Notrema fissurella*, Rafinesque's "trivalve" mollusk that malacologists have chortled over ever since. But it was with phony fish that Audubon outdid himself. David Starr Jordan found five of them in *Ichthyologia Ohiensis*, the most notorious being "the wonder of the Ohio," for which we know Audubon thoughtfully provided both a description and a picture. This Devil-Jack Diamond-Fish, which Rafinesque described professionally and named *Litholepis adamantinus* (see his field sketch in fig. 19.1), was alleged to measure up to ten feet long and to weigh as much as four hundred pounds. Its scales were so hard they deflected bullets, and the backwoodsmen could start their campfires by striking one of its scales against steel to make sparks. (Doubters may examine the representation of an "actual" scale in Rafinesque's drawing.) "It lies sometimes asleep or motionless on the surface of the water," Rafinesque says, "and may be mistaken for a log." Audubon even convinced him that he had seen the monster itself, "but only at a distance." If an experienced ichthyologist could be persuaded to believe a log seen at a distance in the water was in fact a large fish, what might the same mind make of some inscrutable syllables when hot on the trail of Indian languages known at best through the faltering orthography of Caucasian informants? Or of esoteric pictographs which, as Oestreicher points out, resemble graphic signs ranging from Egyptian hieroglyphics to Mayan symbols? Why would a trickster need to invent afresh what could more plausibly be adapted from systems already known?

Confident that Rafinesque himself honestly believed his *Walam Olum* was a genuine Indian document, however maladroitly he may have handled it, I have retained what is said about it in the abridged version of my contribution to Leopold's book that is reprinted here. The contested issue of *Walam Olum* aside, this essay develops chronologically the stages of Rafinesque's linguistics investigations that led to his total accomplishment as detailed by Vilen Belyi. Belyi concludes with admirable savoir-faire that the fraudulence of one document—whoever may have been responsible for the deception—should not be permitted

Introduction

to diminish our appreciation of Rafinesque's overall pioneering achievement in linguistics.

No such controversy attends Rafinesque's Maya discoveries. His role was first noticed in 1964 by Günter Zimmermann, but it remained for George Stuart in the article reprinted here to ferret out the full details along with the texts used by Rafinesque. This essay rounds out much of the story of Rafinesque as linguistics scholar, though still needed is a study of the 236-page *Genius and Spirit of the Hebrew Bible* (1838), where Rafinesque elaborates a new way to transliterate Hebrew and gives a Cabalistic interpretation of the "hidden" meaning of Hebrew roots. Such an inquiry might tell us little about the Pentateuch, but it ought to illustrate a good deal about Rafinesque's lifelong fascination with secret codes, one of which he designed and in which he tried unsuccessfully to interest the State Department. It was, he believed, "unbreakable."

V

Among all the books and articles he produced, the two-volume *Medical Flora* was Rafinesque's most successful. Like his ill-starred *Walam Olum*, it has attracted considerable attention in recent years, especially from people interested in alternative medicine, including that derived from Native American lore. Worthy of mention in this regard is the 1970 *American Indian Medicine*, by Virgil J. Vogel, a book that uses Rafinesque's two volumes as its principal source. Similar Rafinesque material figures prominently in "The Impact of the Materia Medica of the North American Indians on Professional Practice," by David L. Cowen, published by the Internationale Gesellschaft für Geschichte de Pharmazie in 1984 in Band 53 of its *Veröffentlichungen*.

Michael A. Flannery was first to take up, one by one, the ninety-nine plants treated in greatest detail by Rafinesque and investigate the extent to which each enjoys either current use or has been an accepted medicament of the past. He was well positioned for the task, for when he wrote this article he was director of the Lloyd Library, a notable research library in Cincinnati assembled by the wealthy pharmaceutical manufacturer John Uri Lloyd, whose biography Flannery also wrote. Lloyd had specialized in formulating the medicines used by so-called Eclectic physicians, who preferred botanical remedies to the harsh chemical potions used by most conventional physicians of the time, so his library was richly stocked with both ancient and modern texts having anything to do with phytomedicinal knowledge.

For anyone wishing to pursue the subject further, it should be noted that Flannery did not analyze the last hundred pages of volume 2, where Rafinesque deals also with 488 plants, some covered more thoroughly in earlier pages, others taken up afresh, and still others having no alleged medicinal value at all. This arrangement is typical of Rafinesque's method of composition—which was to add one thing after another, often ordered by no other unifying principle than that of the alphabet, until he had filled up the space available to him. He seems seldom to have revised. He expected to include corrections as well as oversights in the "supplement" that appears at the end of most of his books. Other times his enthusiasm for a particular topic led him to treat it at a length far out of proportion to other topics of the same rank. In the *Medical Flora* his "monograph" on the genus *Vitis* is an example. In these sixty pages of small type he not only distinguished more "species" of grapes than anybody else, before or since, but, imbued with the sagacity of the Psalmist that wine "maketh glad the heart of man," he rattled on to produce a pretty good manual on viticulture and oenology. Having copies of these pages printed apart as a pamphlet, he realized a modest profit by selling them at twenty-five cents each to those who agreed with the pamphlet's epigraph: "Let every Farmer drink his own Wine."

Rafinesque probably realized little profit on the *Medical Flora* itself, because he complained bitterly about the cost of having the woodcuts prepared; but these, printed in green, usually were singled out for praise whenever the book was mentioned by his contemporaries. Some of the woodcut illustrations now have been added to Flannery's article, and others have been added to Beverly Seaton's article that follows.

Beverly Seaton had begun a study of the curious language of flowers craze that swept over Europe late in the eighteenth century and was exported to America in the next when she came across Rafinesque's "School of Flora" contributions to the popular press. Her paper on his periodical articles and the woodcuts that illustrated them was one of two given at the 1983 Transylvania Bicentennial Celebration worthy of permanent preservation. It was preserved, though without illustrations, in the pages of *Bartonia,* and from that she went on to publish in 1995 the definitive book on the subject of the language of flowers. Thus was opened up to scholarship for the first time a consideration of Rafinesque's writing for entertainment and instruction. Much more could be done in this area about his contributions, especially to newspapers and other periodicals, that were intended to be both amusing and educational.

My own paper on "Fugitive Publications" traces various avenues that have led to the discovery of hitherto unknown publications by Rafinesque and suggests that several may still await discovery.

Introduction

VI

The foundation for the legendary Rafinesque was laid during his own lifetime. When Audubon was writing his *Ornithological Biography*—ostensibly, as the title implies, about birds—his friend John Bachman suggested that the book would sell better if it included more human-interest anecdotes. One response to this advice was Audubon's sketch of "The Eccentric Naturalist," whose central figure, M. de T., has been identified ever since as Rafinesque. Exploiting the ever-captivating stereotype of the bumbling, comic scientist whose field of vision is so constricted that he lives in a world of his own, Audubon did enliven his book with this, one of his finest prose sketches.

In 1989, when Audubon's holograph manuscript came up for auction, the Sotheby's catalog revealed for the first time that Audubon had not written "M. de T." at all! Was it now necessary to scrap a century and a half of received wisdom? The catalog opined that Audubon's publisher must have thought that "using even one of the real subject's initials was too risky" and stated without a trace of dubiety that "in Audubon's manuscript Rafinesque is identified as 'M. de C.'" After the Filson Club History Society acquired the manuscript, I confess that my heart skipped a beat when I had a chance to examine it and found an asterisk immediately following the first mention of M. de C. But, alas, M. de C. was not identified at the foot of the page or anywhere else in the four and a half pages of manuscript, despite the misleading assertion of the Sotheby's catalog.

I would as soon lessen Virginia's innocent faith in Santa Claus as dwell on the discrepancies between Audubon's account of the "Eccentric Naturalist" and Rafinesque's own report of his visit in a letter to his sister: such differences as Audubon's assertion that he entertained the guest for three weeks; Rafinesque's recital of merely the eight days of his visit. (Playing host to an eccentric naturalist for eight days may have seemed like three weeks.) Like "Honest Abe" Lincoln who trudged four miles to return the pennies he had overcharged a customer at Denton Offutt's store, if Rafinesque did not sit for Audubon's portrait he should have, for the demeanor of the "Eccentric Naturalist" fits exactly the image we have of "the gentleman from Constantinople" or "the gentleman of Transylvania"—take your choice.

The final essay is a jeu d'esprit I enjoyed putting together some years ago that has given me more pleasure than it has the administration of Transylvania University, not to mention its students. No, Transylvania has not supplanted Rafinesque Day with Mary Passmore Day as a result of it. The odor of frying hamburgers drifts heavy in a basement hangout still called "The Rafskeller," and local newspapers still find "The Curse of Rafinesque" handy filler at Halloween time. Despite the belief of William Cullen Bryant that "Truth, crushed to earth, shall rise again," some fables are too soul satisfying to be nullified even by flawless facts. This yarn strikes me as a fitting finale to the story of Rafinesque, a man not only a legend in his own time but in ours as well.

VII

In closing I want to thank for permission to reprint their essays those of my contributors who are living, and to acknowledge no less gratitude to those who are not. The living authors have had the opportunity, like that afforded our federal legislators, to "amend and extend the record," and most of the reprinted essays also differ somewhat from their original versions by the necessity of excising what otherwise would have been tedious repetition. Formal acknowledgment is given elsewhere to the publishers of reprinted material, as well as to the sources of illustrations. Those illustrations not so attributed have come from my own collections.

One of the two reviewers of the manuscript of this book, John C. Greene, made several helpful suggestions for which I am grateful. I am happy also to register my gratitude to Scot Danforth of the University of Tennessee Press. Having adroitly copyedited an earlier book of mine, he immediately saw virtue in my proposal for this one, shepherded it through the long process required by academic publishers, and again performed a low-key but highly professional copyediting job.

For the book that has resulted, I must explain—if not apologize for—the inclusion of more of my own writing than of anybody else. As remarked earlier, most of whose who have written about Rafinesque took up a particular aspect of his work as an adjunct to some larger area of concern. By contrast, the life and times of Rafinesque has been a central interest for me for the past fifty years, since I first started reading him in graduate school. During sixteen years that I lived, with my family, in South Asia, carrying out Rafinesque research was difficult in the total lack of appropriate reference materials in Iran, Pakistan, and India. Without ever having been within thousands of miles of the subcontinent, Rafinesque himself had several interests there nevertheless: he read Sir William Jones, founder of the Asiatic Society of Bengal, and he gave the scientific name *Delonix regia* to the flamboyant Gold Mohur tree, which I enjoyed in Hyderabad (Deccan). But as far as I know, the subcontinent has never had any interest in Rafinesque.

Introduction

During those years, however, several friends in the United States—including my two daughters when they returned here for college and my wife during home-leave visits—photocopied and shipped to me research materials as needed, not infrequently by bending the rules covering what can be transmitted through the diplomatic pouch. I shall always be grateful for this assistance. Living at such a distance from American research centers turned out also to have an unforeseen benefit, though. During home leaves at two-year intervals I could stop off in European countries where it seemed possible Rafinesque papers had lodged. Otherwise difficult to finance, this travel resulted in the discovery in eight countries of hitherto unknown Rafinesque letters and other manuscripts, written in four languages. Finally, when further residence abroad proved impractical, Transylvania University obligingly gave me office space and other facilities for five years while I worked on Rafinesquiana full time. This book, then, is a partial accounting—but a representative one—of what has been concluded during the past century and a half about the life and achievements of C. S. Rafinesque.

Part I

The Man

The Life and Work of Rafinesque

Francis W. Pennell (1886–1952)

I HAVE FOUND unpublished material in various libraries—letters at the Historical Society of Pennsylvania, philological manuscripts at the American Philosophical Society, a most illuminating day-book of expenses at the Library Company of Philadelphia—but by far the most valuable discoveries were made at my own institution, the Academy of Natural Sciences of Philadelphia. There in 1849 Dr. S. S. Haldeman, who was Rafinesque's zoological critic and had purchased his manuscripts *en masse*, deposited apparently all but the strictly zoological papers in his possession.[1] I found many botanical manuscripts, also a long one dealing with the Negro races over the earth, etc. There were letters to Rafinesque from various botanists and, most unexpected of all, ones to him in French from his sister and in Italian from his daughter, both semi-mythical characters and the latter hitherto known to us only from her father's will. Finally, by a most happy chance, there has come to the academy this very summer [1940] the correspondence between Rafinesque and Zaccheus Collins, letters telling particularly of the naturalist's experience in Kentucky.

While I have not had time to deal as adequately as I should desire with the 932 publications[2] of Rafinesque or to absorb all this other new matter with its intimate glimpses of his personality, I am most glad for such an opportunity as this offers in order to attempt a new appraisal of the man and his work. It is now over a hundred years ago, September 18, 1840, that he died in Philadelphia, in poverty and nearly alone. The city's daily press, discursive of many things, had only this to say of him: "Died, on the 18th inst. Professor C. S. Rafinesque, for many years a resident of this city and the author of several scientific and literary works."[3] Within two years his main scientific work was duly reviewed in the *American Journal of Science*, where were two articles, one by Dr. Asa Gray[4] and one by Dr. S. S. Haldeman,[5] considering, and mostly condemning, his botanical and zoological papers, respectively. These truly expressed the considered scientific judgment of the time. Slightly over fifty years after the naturalist's death, when only a few old men and women could bear personal testimony of him, there came the first appreciative account of Rafinesque, the memoir of Dr. Richard Ellsworth Call[6] published in Louisville, Kentucky, in 1895. It has been followed by the scholarly bibliography of Prof. T. J. Fitzpatrick,[7] issued in Des Moines, Iowa, in 1911. Excellent as these are, I think there is yet more that needs be said in answer to the following questions: Why was Rafinesque so condemned and neglected in his later years? Why, in spite of this, did he pursue steadily his own divergent road? Why, after a century is he not forgotten but increasingly appreciated?

I

Rafinesque himself traced for us the outline of his life—a career as varied as its subject was versatile. Published at Philadelphia in 1836, only four years before his death, *A Life of Travels and Researches in North America and South Europe*[8] enables us to share his experiences intimately. "I opened the eyes in the fine grecian soil and climate, at the eastern end of Europe, and in sight of Asia, since I was born at Galata, a suburb of Constantinople, inhabited by Christian merchants and traders. My father was G. F. Rafinesque,[9] a French merchant of Marseilles, settled in the levant, being a branch of the firm of Lafleche and Rafinesque of Marseilles. My mother was M. Schmaltz a Grecian born, but of a German family from Saxony." [*A Life of Travels*, 5] The date of birth, not mentioned by Rafinesque, was stated by Haldeman as October 22, 1783. In a year or so the family went to Marseilles in southern France. In 1791 his father made a voyage by the Cape of Good Hope to China, but on the return in 1793 his ship only escaped the English cruisers by putting in to Philadelphia, where the father succumbed to yellow fever. At Marseilles his family became frightened by the excesses of the French Revolution, and escaped to Leghorn (Livorno) in Italy.

> We remained at Leghorn from 1792 to 1796. I was taught by private teachers
> Geography, Geometry, History, Drawing, &c., as well as the English language.
> The Italian language became quite familiar to me by mere practice. I learnt

many other things by myself, as I was greedy for reading any book that I could get; but I liked travels above all. Those of Leguat, Levaillant, Cook, &c., gave me infinite pleasure: I wanted to become a traveller like them and I became such. They increased also my taste for natural history, in which Pluche and Bomare became my guides and manuals. Before twelve years of age I had read the great Universal history, and 1000 volumes of books on many pleasing or interesting subjects.[10] [8]

Then came boyhood trips in the Italian mountains, and his return to Marseilles where at the home of his grandmother Rafinesque he studied to himself "travels and natural Sciences; but I added now natural and moral Philosophy, Chemistry, Medicine, &c." [10]. On his grandmother's death in 1800, it developed that the family fortune was gone; Rafinesque blames Mr. Laflèche, the surviving partner, for never making for his brother and himself any settlement of their father's estate, but Laflèche could plead the firm's losses due to the Revolution; in those troublous times it seems that no guardian had been appointed for the three Rafinesque children (Constantine Samuel, Anthony Augustus, and their sister, Georgette Louisa) while their mother had married again a Mr. Lanthois, another merchant. The boys went to their mother, who was again at Leghorn in Italy, and there for two years Constantine learned "Commerce with Mr. Lanthois, studying Botany and other sciences." One remark shows the boy's tender nature. "I began to hunt, but the first bird I shot was a poor *Parus* [titmouse] whose death appeared a cruelty to me, I have never been able to become an unfeeling hunter, [but] I sent accounts of rare birds to Daudin" [12]. In 1802, when eighteen years of age, Constantine, accompanied by his younger brother, was sent to the United States of America.

The days of preparatory study were now over. The story of Rafinesque's childhood shows how wide and yet how desultory this preparation must have been. He learned modern languages early and well but Latin construction he never mastered fully. His scientific study seems to have been wholly unguided. Before the loss of the family fortune "I was to go to Switzerland, into a College to finish my education; but this project was not fulfilled" [9]. The brilliant boy needed such training, and one wonders if it might not have forestalled much of the adverse criticism of later years, criticism which must have made its recipient keenly unhappy but which caused him to smart under a sense of injustice done him rather than to make desirable changes in his practices. I believe in the independence of the self-made man, but the discipline of sustained and logical study, and the equally important matter of friendly criticism during one's formative years, were in this case lacking.

Francis W. Pennell

It was on the ship *Philadelphia* bound for Philadelphia that Constantine and Anthony Rafinesque left Leghorn in March 1802. Of the Atlantic Ocean Constantine says:

> It afforded me a new study by its fishes and mollusca. I drew and described all those that we caught. It was more difficult to procure Birds, but Turtles could be taken while sleeping on the waves.
>
> We had a favourable passage, without accidents nor storms. In forty days we obtained the first sight of America, the Capes May and Henlopen forming Delaware Bay. These shores are so low, that the trees are seen before the soil, and give a sylvan impression of this continent. In two days we ran up the Bay and River to Philadelphia, where we landed on the 18th April 1802.
>
> Here I was then upon a new Continent, where every thing was new to me. The first plant that I picked up was also a new plant, then called *Draba verna,* and that I called *Dr[aba]. Americana,* altho' the American Botanists would not believe me; but Decandole has even since made with it the new Genus *Erophila!* This is the emblem of many discoveries of mine, of which ignorance has doubted, till science has proved that I was right. [13–14]

Alas, for this illustration which had even the support of the Swiss De Candolle, possibly the most experienced botanist of his time! Subsequent students are now convinced that the American plant of cultivated and waste ground, to which Rafinesque gave a distinctive name, is simply the European *Draba verna,* a weed that has followed civilized man to many lands.

On this first visit to America, Rafinesque remained three years, from 1802 to 1805. He tells us of his pleasant reception in Philadelphia, "particularly [by] the brothers Clifford, owners of the Ship that brought us, the brothers Tarascon[11] formerly of Marseilles, [and] Dr. Benjamin Rush" [14]. He might have studied medicine under Rush, but felt that he should enter the countinghouse of the Cliffords. But the fear of yellow fever, from which his father had died in this very city, scared him from Philadelphia, and led him to commence his series of botanical peregrinations over the northern United States. Year by year the *Life of Travels* tells whither he went, what botanists or other naturalists he met, and recounts some of his discoveries. His brother Anthony seems to have worked more steadily, but Constantine tramped afoot from the New Jersey seacoast inland to the Peter's Mountain of the Pennsylvania Alleghenies and from the Delaware Water Gap in the north to the Eastern Shore of Maryland on the south. From Fitzpatrick's bibliography it appears that Rafinesque's first publications date from the first year of this American visit, being two papers concerning new species of birds from Java, which he had studied in the Peale Museum in Philadelphia; these were sent to a scientific journal in Paris,[12] where they appeared in 1803, when their author was barely twenty years old. Both brothers returned to Italy in 1805.

II

Less need be told of the ensuing ten years, 1806 to 1815, which Rafinesque spent in Palermo, Sicily [actually, he reached there in May 1805]. There this man of many backgrounds counted as an American, and for a while had employment with the American consul. He had various business projects, but I suspect that these were neglected when the urge of the naturalist was strongly upon him. He had evidently taken to Sicily a good herbarium of plants gathered in the United States, for in 1806 and 1808 there appeared in the New York *Medical Respository* six communications concerning American plants. We may fairly complain of the brief and sketchy nature of the descriptions of new species, but they testify to Rafinesque's keen observation of our flora, and most of his new names have been received into all our books. One paper, appropriately for this journal, deals with the medical properties of some North American plants; while another is an admirable sketch of those plants that have become naturalized in America from Europe. Yet another is a "Prospectus of . . . two intended Works on North American Botany," one to include everything new since the time of Linnaeus, the other to deal with "Funguses," both projects so grandiose and impossible of achievement from Sicily, that I wonder why the editor allowed their mention.[13]

Three other works of this Sicilian period concern us. His *Principes Fondamentaux de Somiologie*[14] gave with excellent clarity the logical rules that he thought should control the naming, in Latin, of newly described species of animals and plants; while his *Analyse de la Nature* gave his own synoptic outline of the whole world of nature—mineral, animal, and plant. As to the last, throughout his life Rafinesque was a firm adherent of the French system of plant families, which has since become universally adopted, but in this work he developed his own series of such families. To the standards of these early works he remained faithful throughout life, and much that his contemporaries misunderstood finds ready explanation there. The third work is his venture of a detailed natural history journal, *Specchio delle Scienze (View of Science)*, appearing in the vernacular Italian. It contained over two hundred articles from his pen,[15] on almost every kind of subject, and these included descriptions of new species of all sorts, even animalcules, though the most important were flowering plants and fishes. It was the forerunner of many like ventures by Rafinesque in America.

Francis W. Pennell

One other bit of information from his stay in Sicily, and we are ready to return permanently to the United States. It appears that in 1809 Constantine Rafinesque was married, though not regularly so, since the decrees of the Council of Trent forbade the union, to Josephine Vaccaro.[16] A daughter and a son were born to them, but the latter lived only from 1814 to 1815. We shall hear again of Emilia Rafinesque, born in 1811 and so only four years old on her father's departure for America. Rafinesque sailed from Palermo for New York in July 1815, with all his scientific collections and manuscripts, but without his family. On the second of November the ship was sunk on the Race rocks, between Fisher Island and Long Island, and the passengers escaped in lifeboats to the light house of New London, Connecticut. In his *Life of Travels* Rafinesque only tells us of the loss of all his collections and unpublished scientific labors, but in Sicily the shipwreck had yet a worse repercussion for him. Josephine suddenly married a comedian, "dissipated the property left in her hands," and in the next two years twice refused to send on Emilia to her father. All this we learn from Rafinesque's will, composed eighteen years later. The *Life of Travels* makes no mention even of the marriage, and for years Rafinesque closed the door firmly on that experience. But in his autobiography he says of Sicily that it offers "a fruitful soil, delightful climate, excellent productions, perfidious men, deceitful women" [27].

III

It was as a mature man of thirty-two that Rafinesque once again came to the United States, into which he was to be completely absorbed for the quarter century of work that still lay before him. In spite of the destitute state in which he arrived there seemed every reason why his should have been a brilliant and scientifically successful career. He soon settled in New York, where he was one of the founders of the Lyceum of Natural History. He began to make a "new Herbal," and doubtless found some assuagement for his trouble in a long journey up the Hudson and to the mountains west of Lake George. In 1816 he applied unsuccessfully for the professorship of natural history and botany in the University of Pennsylvania, and also for the chair of chemistry; it is a presage of later years that the former application was through a letter to James Mease. (In the same year he became a member of the Academy of Natural Sciences of Philadelphia.) His scientific papers were mostly contributed to a New York journal called the *American Monthly Magazine and Critical Review*. They were sometimes trenchant reviews of botanical works of the period, among which Stephen Elliott's *Sketch of the Botany of South Carolina and Georgia* (1816–24) stands out as alone receiving genuine praise—

any taxonomist today would signal it out, as did Rafinesque, as the most deserving of these various works. Rafinesque was at that time alone in the country in advocating the coming French system of Natural Families, as against the so-called Sexual System of Linnaeus. His other contributions comprised descriptions of new species of plants and animals, the former mostly flowering plants, the latter including sponges, insects, crustacea, fishes, serpents, and mammals. He ransacked works of travel for animals not yet characterized by science, and thus from the descriptive account of Meriwether Lewis, Rafinesque assigned to the prairie dog its still accepted name of *Cynomys* (dog-mouse). One is amazed by such versatility, but also left wondering how adequately the rapidly working author could have consulted the works of his predecessors. Also one notices that the sketchiness of presentation, evident in his earlier papers, has now become a habit. And it is further evident that whatever seems at all peculiar is promptly proposed as something new.

In 1817 appeared a larger botanical work, Rafinesque's *Florula Ludoviciana*. Just as with Lewis's "barking squirrel" Rafinesque had found material in a work of travel. C. C. Robin, a Frenchman who had traveled between 1802 and 1806 near the Gulf coast from our present Louisiana to western Florida, had made careful descriptions of the plants seen, and these he had appended as a "Flore Louisianaise" to the third and last volume of his travels. These French descriptions, made always from the living plants, were accurate, but their describer had only crude ideas of what his plants were technically; some few were identified, but most were left with mere vernacular names. Rafinesque attempted to complete and correct these identifications, and, after comparing the descriptions with the two available comprehensive floras of the United States, those of Michaux in 1803 and of Pursh in 1814, proposed as new 30 genera and 196 species. (Of course, he should also have consulted many journals and other travels, since little from the modern Louisiana had been seen by either Michaux or Pursh.) Robin's, and therefore Rafinesque's, descriptions were unconventional, and often leave one ignorant of the special distinguishing features that a trained botanist would have noted. But, scanning the work now after the subsequent study of nearly a century and a quarter has told us what were necessarily the plants that Robin encountered, the identity of nearly all of them may be decided. Not so did it seem to Rafinesque's contemporaries, and Asa Gray of Harvard University, the greatest taxonomist our country has produced, has this to say of it in the general review of Rafinesque's botanical work to which I have referred earlier. After dwelling upon Robin's errors in classification as revealed by Rafinesque (and, I am confident, without ever looking into Robin's interesting volume), Gray states:

The Flore Louisiane [*sic*], in the state Robin left it, could do no harm, and whatever information it contained was quite as available as at present. As *improved* [the word used by Rafinesque on his title-page] by a botanist who had never been within a thousand miles of Louisiana, and who at that period could scarcely have seen a dozen Louisianian plants, the only result has been to burthen our botany with a list of nearly two hundred *species semper incognitae.* There can, we think, be but one opinion as to the consideration which is due to these new genera and species: they must be regarded as fictitious, and unworthy of the slightest notice.[17]

Since Gray's time botanists, and zoologists as well, have adopted strict rules as to what counts as adequate publication of new species and genera, and by these rules the new propositions of the *Florula Ludoviciana* were validly presented. As modern workers consider special groups, name after name has been adopted from this flora, although an adequate reworking of it is now long overdue. Gray's advice has been all too well followed.

In the next year, 1818, Rafinesque was to obtain ample collections of his own on which to base his studies. The routing of his "great western tour," as of all his travels long or short, is faithfully given in his *Life of Travels.* "In May I went to Philadelphia and Lancaster, where I left the stage to cross the Alleghanies on foot as every botanist ought" [53–54]. The gist of this year of marvels is succinctly brought out in four letters to Zaccheus Collins,[18] a fellow botanist and the vice-president of the newly established Academy of Natural Sciences of Philadelphia.

The first letter was from Pittsburgh, June 1.

I arrived here the 25th May, after a tedious journey through the mountains, being often detained by the rains; the roads were dreadful in many parts where the turnpike is not yet made. I have the pleasure to acquaint you that the Collinsia of Nuttall . . . is a native of Pennsylvania: I have found it plentifully on Turtle creek 12 miles east from Pittsburg[h], and I have collected 100 specimens of it, in full bloom. I think to have discovered 12 new species, at least they are not in Pursh. Here are their names.[19] . . . I have begun to study the fishes of the Ohio; they are all new species, unless described already by Mr. Lesueur. . . .

The second letter was from the Falls of the Ohio, July 20.

I shall now proceed to enumerate some of my most striking discov[eries] while descending the Ohio and since I am here, where I have found so much to do, that I have dwelt here since the 4th of July. This is a beautiful spot for a Naturalist and a center of new Productions. I have paid particular attention to the Icthyology and Conchology by this River, two subjects entirely new, and have been rewarded by the discovery of about 22 New species of fishes out of 32 existing and 3 new genera with 32 New species of Living Shells. . . . It is very extraordinary that all the shells of this River are New ones, that all the bivalves (24 sp) belong to a single genus, which is also a New one! . . .

Then follows discussion and a list of the shells. "I have described 3 New Quadrupeds," and these are listed. "I hand you a complete Catalogue of the Fishes of the Ohio, with their vulgar names. . . ." These comprise 26 species, of which 23 are new. "The fossils are numberless in the valley of the Ohio. I have collected more than 60 species. . . ." And some of these are told of. "In Botany my discoveries are less than I expected, owing to the peculiar features of the Vegetation of the Western States, which consist in a singular monotony, large tracts of Country being often covered with only a few species, which grow by millions covering the whole soil. I have however, detected the following New Species," and there are listed 12 species. "I believe to have also discovered 3 New genera. . . ."

The third letter was commenced at Henderson, Ky., August 12. "I have since the date of my last Letter increased to about 60 species the ichthyology of the Ohio, all new and undescribed except 5 or 6 and of which several appear to be new genera. Here is some of my most remarkable additions," and there follows a list of 12. "Having paid some attention to the Small Quadrupeds of the Western Country, & particularly to the neglected tribes of Bats and Rats, I have already been able to observe or ascertain the existence of 8 new Species of Bats and 10 New Species of Rats or Mice." Then follow lists of these. One wonders which one of these bats was brought down by Audubon's favorite violin![20] The next paragraph was of Audubon—perhaps the thought of bats had brought him to mind: "I have had the pleasure to see here the collection of paintings of the birds of North America by Mr. Audubon, a splendid Work containing more than 300 species, among which 50 at least are not in Wilson's, many are new species and there are 2 or 3 New genera of Birds; all the birds even the largest are painted on natural size, it might perhaps be purchased for a valuable consideration." Then he continues: "Among the Reptiles, I have described many new Snakes, Lizards, Frogs, &c." There follows a list of these. "I have added some few shells to my former Collection. . . ." "In Botany, my discoveries are equally extensive, I have collected or seen nearly 600 sp of plants within 2 months, among which are about 20 N. Sp. Such are the following," and there comes another list. "I have studied some of the most remarkable Insects of this Country, such as the Ticks, Ants, Gnats, &c—and I have many species of each genus."

Turning over the page, the third letter continues from the Falls of the Ohio, September 12.

> Since writing the above, I have visited the Wabash and Green river, the Prairies of Indiana and Illinois, the barrens of Kentucky, &c; but want of time and the bad roads have prevented me from reaching the Mississip[p]i and Missouri, yet I am perfectly satisfied with my share of discoveries, as they have continued to exceed my most sanguine expectations. I hope to have nearly completed the Ichthyology and Conchology of the Ohio, having detected, figured, described (and often preserved) 64 species of fishes, and 48 species of Shells! all new, except very few.

Then are listed the latest finds of both. "I have increased to 10 New species the Bats of the Western Country," and three are mentioned as recently ascertained. "The Barrens and Prairies afford a peculiar Vegetation & many plants unknown or uncommon elsewhere, among which I have reaped many new species, and I have collected many more which are not yet ascertained to my satisfaction; I will mention some of the new ones, from Kentucky, Indiana & Illinois." Then follows a list of 24 species. "In Geology, my researches have been various. I have made a Map of the Valley of the Ohio from Pittsburg[h] to the Wabash, probably its ancient bed, and very different from its bason. I have ascertained where the old falls were or the highland passage of the Ohio, where the river and valley is reduced to one quarter of a mile, it is below Salt river; these highlands form the Barrens of Kentucky and the Silver hills of Indiana. I have collected one thousand specimens of fossils at the falls and elsewhere. . . ."

The fourth and last letter was from Pittsburgh, November 6, 1818. "It happens as usual whenever I travel, that a variety of interesting subjects have engaged my attention, & that I have been rewarded by many unexpected discoveries, particularly among the fossils." Then, after telling of a few plants and of meeting in Lexington Mr. Bradbury, the botanical traveler up the Missouri in 1811, he continues:

> I spent 3 whole weeks in Lexington, being delighted with Mr. Clifford's Museum, where I observed, described and drew nearly 220 species of fossil Shells, Polyps, &c., all from the Western States, and nearly all new: I had to establish more than 12 New genera among them. Mr. Clifford who is an old acquaintance of mine, has declared himself my friend & mecenate,[21] and has offered me to travel next year at his expence through Tennessee, Alabama,

The Life and Work of Rafinesque

Missouri, &c. in order to collect for his museum. His name is ominous, since you know that Clifford was Linneus's first mecenate. But moreover I have been offered the professorship of Botany, Natural history and Chemistry in the University of Lexington, and my immediate election was only prevented by the difficulty of forming a quorum of the trustees; it will however take place as soon as they can meet, the difficulty being merely whether they will be able to offer a salary. Under those circumstances I would have spent the winter in Lexington, if I had not felt obliged to come back to Philadelphia, to settle my former concerns, fetch my library, and publish my Journey and Observations.

Although this letter continues to list some new plants, a remarkable new genus of freshwater shrimp, many new shells, both living and fossil, and many other fossils, this seems the logical place to cease telling of this journey of 1818. But what a year it had been! Whoever heard of one man in a single season finding so much that was new, and of such wide diversity? We may say that Rafinesque in the field could not have had with him the books[22] to decide what was new or not, and in not a few cases he must indeed have erred on that account. But especially with the fishes and shells he was actually in a new and almost untouched fauna, and my colleagues assure me that his discoveries were largely valid. But, they say, as one might expect of so much undertaken in so little time, that the descriptions were sadly sketchy; some are readily recognizable, but only in recent years have other names replaced ones given by later men who worked more slowly but more thoroughly, and finally many must remain permanently unidentifiable.[23]

Fortunately, from a letter of John D. Clifford to the trustees of Transylvania University on Rafinesque's behalf, we have recorded what terms they likely offered him: "Mr. Rafinesque is willing to accept the appointment of Professor of Botany, Natural History and the Modern Languages without any salary from the Institution except the privilege of boarding in Commons free of Expence—any remuneration he may receive for his Lectures to be at such reasonable rates as his Students and himself may voluntarily agree to." Clifford's note was written February 11, 1819. I suppose that it was late March or early April before the trustees dispatched their invitation to Rafinesque. His acceptance, dated from Philadelphia April 25, is printed here (fig. 1.1).

High were Rafinesque's hopes as he journeyed to Kentucky in the early autumn of 1819. He wrote to Zaccheus Collins from Lexington on September 25: "I arrived safely in the State by the way of Baltimore, Harpers ferry, Cumberland, Brownsville, Pittsburg[h], Maysville, &c.; but my baggage & library is aground in the Ohio. . . ." The *Life of Travels* tells us that he had again crossed the mountains afoot, and that he had got one hundred dollars in Pittsburgh for his map of the Ohio River made the preceding year. He sends Collins his usual list of new plants collected, this time on the journey out and a special series on a trip into the sandstone Knob Hills of eastern Kentucky. "In Zoology I have added about 15 New fishes & 7 new Shells . . . but in Orcytology[24] we have (together with Mr. Clifford) made wonderful additions, we have not less than nearly 80 New fossil Shells & Polyps, which we had not seen before. . . . Our Western [Review and Miscellaneous] Magazine has begun. I write something for it every Month. My first paper was on the lightnings of the U. States."[25] On the first of December he wrote:

> I am very much engaged, since besides writing all my Lectures, I have begun to publish my natural history of the fishes of the Ohio (100 Species) in the Western Review & [Miscellaneous] Magazine. I shall have some printed apart for my friends. I am also going to publish next Month the first Number of my Annals of Nature, or an Annual Account of my discoveries, you shall receive it by mail. And we are describing all the fossils of Kentucky with Mr. Clifford—Therefore you see I am not idle, & shall not disgrace the Western states.

On April 20, the date of the second letter of 1820 to Collins, he has begun to botanize, and reports for plants two new genera and six new species, and a new genus of fish. It is the next letter of June 25 that tells of "the death of my friend and worthy col[l]aborator Mr. Clifford. It has been a very severe loss for me and Science in general. He had just begun to write on the Geology of the Western States in the Western review. Our work on the Western fossils was in forwardness; but far from being concluded. His museum will probably be annexed to our University."

The loss of Clifford became, in retrospect, as dwelt upon in the *Life of Travels* and elsewhere, a calamity of grand proportions for Rafinesque, but at the time I think he took it much more wholesomely. He seems to have hoped that Clifford would have ever stood to him in the role of a Maecenas, and yet he counted him a working companion. Never again did Constantine Rafinesque have another such intimate associate. One wonders how different the future might have been could Clifford's advice have remained with him. But do we know that John Clifford had the judgment, as well as the skill, to advise Rafinesque?

It was inevitable that after so ardent collecting there would have to come a great bout of preparing and publishing papers. And now came difficulties for Rafinesque. Where could these be presented? Instead of having increased opportunities for publishing his discoveries, his papers were being refused by most

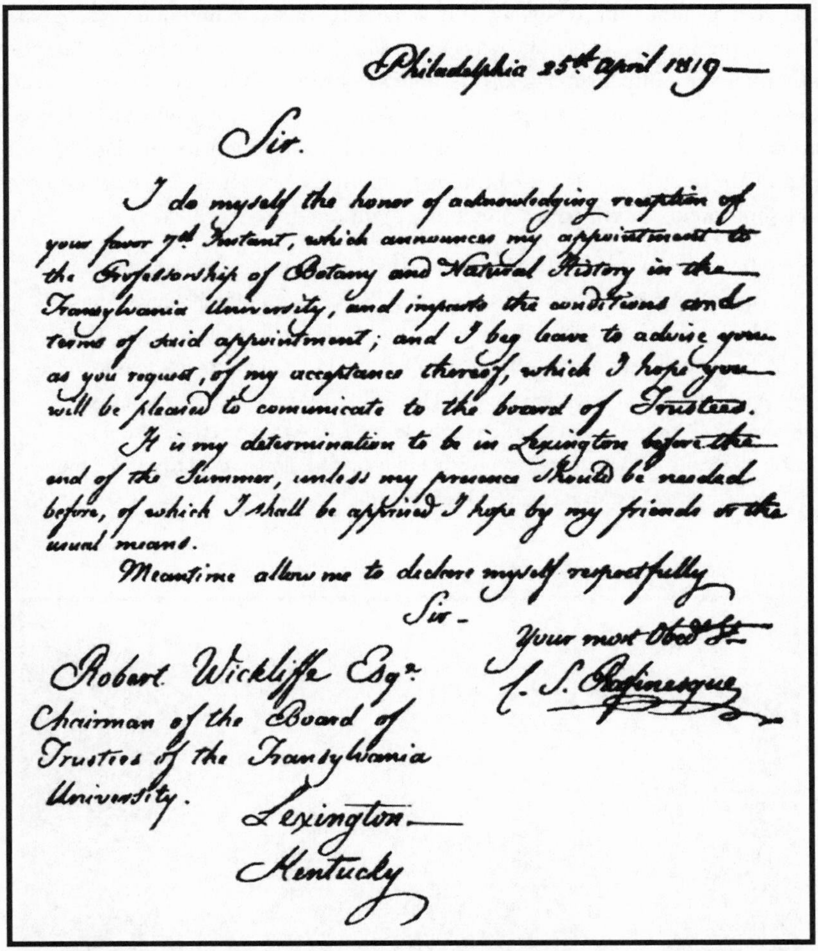

Philadelphia 25th April 1819—

Sir.

I do myself the honor of acknowledging reception of your favor 7th Instant, which announces my appointment to the Professorship of Botany and Natural History in the Transylvania University, and imparts the conditions and terms of said appointment; and I beg leave to advise you as you request, of my acceptance thereof, which I hope you will be pleased to comunicate to the board of Trustees.

It is my determination to be in Lexington before the end of the Summer, unless my presence should be needed before, of which I shall be apprised I hope by my friends or the usual means.

Meantime allow me to declare myself respectfully

Sir—

Your most Obedt St

C. S. Rafinesque

Robert Wickliffe Esqr

Chairman of the Board of Trustees of the Transylvania University.

Lexington—

Kentucky

Fig. 1.1. Rafinesque's acceptance letter. The formal copperplate chirography is not representative. A more typical example of Rafinesque's handwriting is that found in fig. 13.1 or fig. 19.1. Courtesy of Transylvania University Library.

American journals. Instead of developing amply comprehensive and convincing studies, it was his habit to prepare a host of short dissertations. In 1818 and 1819 ten such had appeared in the *American Journal of Science,* but now in 1820 Rafinesque comments to Collins: "I have been surprised to find that Prof. Silliman has not published any of my essays in his late Journal—he has had 12 Memoirs of mine, some for 2 or 3 years! Is not this strange? Why am I used so? Is it through jealousy, neglect, ignorance or wilful intent? It is well that my zeal is above this paltry & sorry usage: But should I not be right to retaliate or complain?" In rebuttal let us hear Professor Silliman's view of the case as presented in a footnote to Gray's paper previously mentioned:

> It was in this year (1819) that I became alarmed by a flood of communications, announcing new discoveries by C. S. Rafinesque, and being warned, both at home and abroad, against his claims, I returned him a large bundle of memoirs, prepared with his beautiful and exact chirography, and in the neatest form of scientific papers. This will account for the early disappearance of his communications from the Journal. The step was painful, but necessary; for, if there had been no other difficulty, he alone would have filled the Journal, had he been permitted to proceed.[26]

Fig. 1.2. Transylvania's principal building, where Rafinesque lectured and had his lodging, in 1820. Courtesy of Transylvania University Library.

The Life and Work of Rafinesque

Continuing again, Rafinesque to Collins: "Meantime I must publish my discoveries at my own expense, or send them to Europe where they meet with respectful notice. Many of my Memoirs have been published in the Journal d'histoire Naturelle, Annales generales des Sciences physiques, Journal of the Royal Institution, &c., and I am requested to send more, which I will do, since I am almost excluded from American publications." So it is that many of Rafinesque's most important botanical discoveries from Kentucky came to be published in journals of Brussels or Paris. But a surprising amount, nearly wholly zoological and antiquarian, appeared in the *Western Review and Miscellaneous Magazine,* the short-lived and rather diverting *Western Minerva,*[27] and the *Kentucky Gazette* newspaper, all published in Lexington.[28]

Of his lectures at Transylvania University Rafinesque says in his *Life of Travels:* "Then began [in the autumn of 1819] my lectures on Natural history to a class of ladies and students, and in the spring of 1820 those on botany." He prepared his addresses carefully, and we are fortunate in the fact that a number of them have survived at the Philadelphia Academy. A manuscript book of them, drawn up in the chirography so admired by Professor Silliman, contains a "Lecture on Knowledge," delivered November 7, 1820; "Lecture on Materia Medica," delivered on the November 21, 1821; "Lecture on the human Mind," delivered in April 1822 (especially interesting); "Lecture on Medical Botany," delivered in November 1822; "Lecture on Phrenology," undated; "Lecture on Natural Sciences & Botany," delivered December 16, 1823. Inserted are a series of lectures on Botany, which seem to be the earlier ones of the course of 20 lectures announced on a prospectus pictured here (fig. 1.3). The lectures are all well prepared, well pondered, and interestingly presented. Dr. Call was still able in 1894 to hear from those who had attended these discourses; among them Gen. George W. Jones remembered vividly Rafinesque's descriptions of ant behavior (which is clearly outlined in one of these lectures), and remarked: "I would now give any reasonable sum to hear him repeat one of his lectures that I listened to in Transylvania University."

As a student at Transylvania from 1821 to 1825, General Jones had seen much of Rafinesque. "He was a man of peculiar habits and very eccentric, but was to me one of the most interesting men I have ever known." Dr. Call has published the reminiscences of four students of that period [62–67], and all agree in portraying the professor's total absorption in his sciences, his profound abstraction (commonly called absent-mindedness), his carelessness as to attire and appearance, and his inability to parry the many practical jokes of his students. He was "absorbed in his books and his bugs, his researches and his writings, a genius

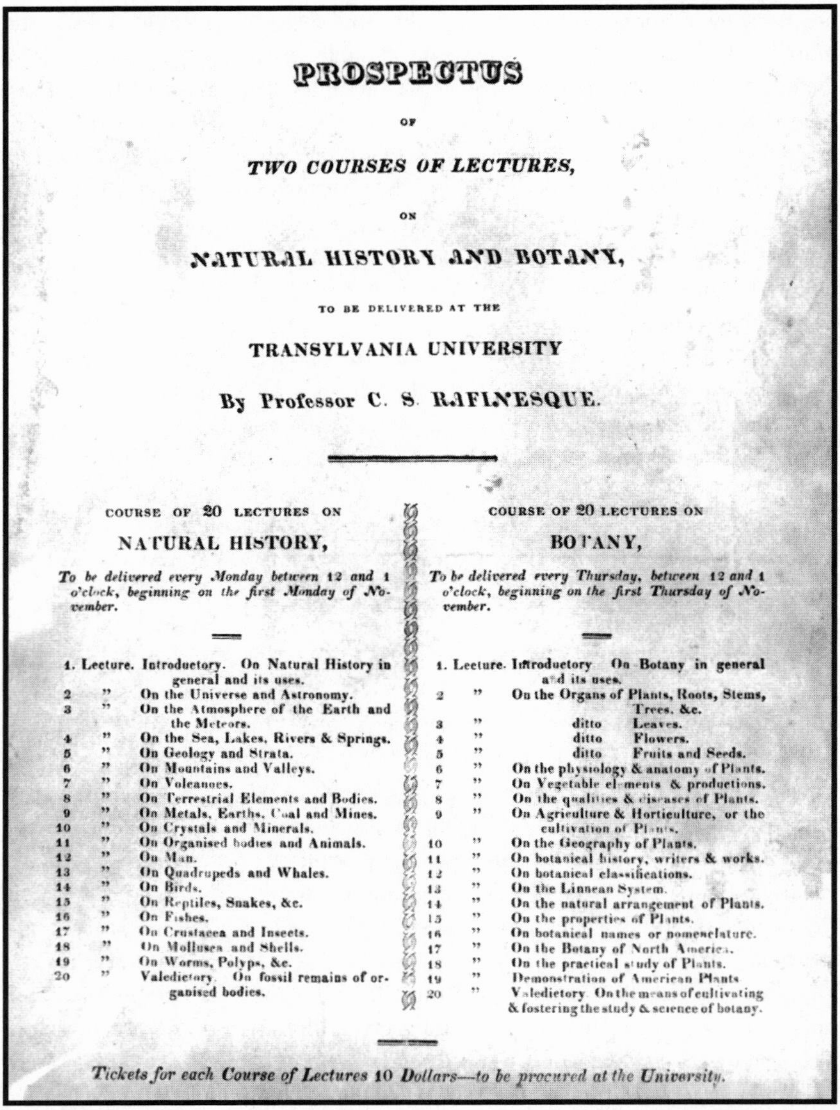

PROSPECTUS

OF

TWO COURSES OF LECTURES,

ON

NATURAL HISTORY AND BOTANY,

TO BE DELIVERED AT THE

TRANSYLVANIA UNIVERSITY

By Professor C. S. RAFINESQUE.

COURSE OF 20 LECTURES ON	COURSE OF 20 LECTURES ON
NATURAL HISTORY,	**BOTANY,**
To be delivered every Monday between 12 and 1 o'clock, beginning on the first Monday of November.	*To be delivered every Thursday, between 12 and 1 o'clock, beginning on the first Thursday of November.*

1. Lecture.	Introductory. On Natural History in general and its uses.	
2	"	On the Universe and Astronomy.
3	"	On the Atmosphere of the Earth and the Meteors.
4	"	On the Sea, Lakes, Rivers & Springs.
5	"	On Geology and Strata.
6	"	On Mountains and Valleys.
7	"	On Volcanoes.
8	"	On Terrestrial Elements and Bodies.
9	"	On Metals, Earths, Coal and Mines.
10	"	On Crystals and Minerals.
11	"	On Organised bodies and Animals.
12	"	On Man.
13	"	On Quadrupeds and Whales.
14	"	On Birds.
15	"	On Reptiles, Snakes, &c.
16	"	On Fishes.
17	"	On Crustacea and Insects.
18	"	On Molluscas and Shells.
19	"	On Worms, Polyps, &c.
20	"	Valedictory. On fossil remains of organised bodies.

1. Lecture.	Introductory On Botany in general and its uses.	
2	"	On the Organs of Plants, Roots, Stems, Trees. &c.
3	"	ditto Leaves.
4	"	ditto Flowers.
5	"	ditto Fruits and Seeds.
6	"	On the physiology & anatomy of Plants.
7	"	On Vegetable elements & productions.
8	"	On the qualities & diseases of Plants.
9	"	On Agriculture & Horticulture, or the cultivation of Plants.
10	"	On the Geography of Plants.
11	"	On botanical history, writers & works.
12	"	On botanical classifications.
13	"	On the Linnean System.
14	"	On the natural arrangement of Plants.
15	"	On the properties of Plants.
16	"	On botanical names or nomenclature.
17	"	On the Botany of North America.
18	"	On the practical study of Plants.
19	"	Demonstration of American Plants
20	"	Valedictory. On the means of cultivating & fostering the study & science of botany.

Tickets for each Course of Lectures 10 Dollars—to be procured at the University.

Fig. 1.3. Like professors of Transylvania's medical faculty, Rafinesque had the privilege of selling tickets to his lectures. "Ladies" were admitted free. Courtesy of Transylvania University Library.

with many peculiarities and not much dignity." Three of them testify to the pleasantness and intimacy of his relations with President Holley and his family, one remarking that he was "always a favorite" in their house, and a lady saying further: "Mrs. Holley, the wife of the President, took a motherly supervision over this lone, friendless, little creature, and saw that he ate his dinner, that the mud of his various expeditions was removed from his garments, that his hair was combed and his face was washed, as often any or all of these particulars would be forgotten by the oblivious scientist." The same lady gives this description of him: "In appearance Professor Rafinesque was small and slender, with delicate and refined hands and small feet. His features were good and his eyes handsome and dark, or apparently so from the long, dark eyelashes. His hair, which he wore long, was dark and silky."

Though rarely seen to laugh, he could unbend enough to write descriptive and even amorous verses in French, Italian, and English for the *Western Minerva*, while his skill in drawing is shown in the sketches of Lexington folk which have so happily been preserved in the library of Transylvania University.[29]

In spite of his oddities, Professor Rafinesque seems to have got on fairly easily with his fellow teachers. Various records of his intercourse with the faculty and trustees have survived. There are bills for cost of candles bought independently, for board outside of commons, etc. On finding under the date of May 1, 1822, a memorandum with a formal array of items running up to "Tenthly" addressed to the Board of Trustees, I wondered if there were serious difficulties, but apparently his request for space for a museum, etc., was more or less met, although certainly not "Seventhly," which was that he might be appointed also "Professor of Public Economy in the Law School," then in contemplation. By 1825 he was both librarian and secretary to the faculty, for which positions he received some compensation;[30] this is evident from the fact that on asking leave of absence in 1825 he agreed that this salary should go to his substitute.

On July 10, 1822, Transylvania honored its professor of Natural Science and Modern Language with the degree of Master of Arts. In the institution's catalogue of 1824 Rafinesque's name is followed by the mysterious sign "P.D.," the which is properly expanded to "Ph.D." on the signature of a letter to Collins on April 10 of that year. The Transylvania Catalogue attributes this to "Imp. Acad. Bonn." and sure enough it appears on a certificate given to Rafinesque on November 28, 1820, by the learned society at Bonn, Germany, known by the appalling name of the "Cesarea Leopoldina-Carolina Academia Naturae Curiosorum." The purport of this document, which is at the American Philosophical Society, was to confer on Rafinesque the appellation of "Catesbaeus," signifying that for discoveries in natural history he was a very Catesby. (Mark Catesby, you will recall, was an English naturalist who visited Carolina early in the eighteenth century.) The "Philosophiae Doctor," after Rafinesque's name and before the statement of his position at Transylvania, looks to me like a bit of misinformation on the part of the learned society. At any rate it seems to have supplied Rafinesque with a second degree, that appears after his name on the title-pages of many of his subsequent works.[31]

In 1822 and 1823 there were but a single letter each to Collins. That of 1822 has further reports of new species of plants, and especially a new tree which in 1823 has been ascertained to be a new genus which Rafinesque proposes to call *Cladrastis*. This was one of his first papers in yet another journal, as it appeared in 1824 in volume 1 of the *Cincinnati Literary Gazette*. We may be glad that his generic name stands unchallenged for the Yellow Wood.[32] In 1823 from May to July he went as far as the Tennessee River near its mouth and "detected 7 new fishes, many other new animals, 2 New Shrubs of a New Genus, . . . about 15 new plants, and collected nearly 2000 specimens." He adds

> My zeal is unabated, altho' I often have moments and weeks of dire despondency, my labours are not appreciated, and I have been treated with great injustice. Metaphysics and idle vapid talk is the fashion in our University. We have lately established a Kentucky Institute, but do not perform much as yet, I am the most zealous member & have already read 4 or 5 Memoirs. I am thwarted in every thing, I have failed in establishing a Museum and Botanical garden—I have some serious thought of leaving Lexington, as soon as I can find a suitable situation elsewhere. This is no place for me as yet, I can't publish my works nor Discoveries, and hardly make a living while idle rhethos wallow in sumptuous luxury.[33]

The report of a Committee appointed to consider some "Communications of Professor Rafinesque"—probably those of December 1823, concerning which another brief memorandum has survived—contains among several items this: "Resolved that the Board highly appreciates the liberality of Professor Rafinesque in offering to present the University with his Collection of Minerals, Plants & Animals, and are willing to receive them—and appoint him Keeper of the Public Museum, provided a satisfactory arrangement can be entered into." I suppose it was the matter of cases and room that unfortunately prevented this gift from being fulfilled. The entire space needed, if the collections were to be adequately disposed, would have been considerable. There has survived at the Philadelphia

The Life and Work of Rafinesque

Academy a summary of the Herbarium alone, dated January 1, 1824, which is divided into five main divisions: 1. North American Plants, 5,000 species & 23,050 specimens; 2. South American Plants, 350 species & 450 specimens; 3. European Plants, 4,375 species & 11,170 specimens; 4. African plants, 2,150 species & 2,300 specimens; 5. Asiatic plants, 870 species & 970 specimens; making in all 12,745 species & 37,740 specimens. "Value at 5 Cts. per Specimen $2387 in specie." Each part of the world is variously subdivided, and the whole shows that Rafinesque's herbarium was much more cosmopolitan than has been supposed. In North America his collections were mainly from New England to Virginia and Missouri, although there were plants from the southern as well as the northern states. Kentucky heads the list, with 1,850 species and 8,000 specimens.

The Botanic Garden was a chief desire and for it Rafinesque sought help from the state legislature at Frankfort, which, to quote from the *Life of Travels,* p. 72,

the Senate granted me, but the House refused. My friends however succeeded in procuring the incorporation of a company for the purpose. Here I took the measles then prevailing, and was very sick on my return to Lexington; but I recovered in spite of the Physicians, by taking none of their poisons, antimony and opium, while many died in their hands. To divert me from the garden I was appointed Librarian of the University, and keeper of the Museum, where Clifford's Cabinet was then deposited. I accepted, but without giving up the Botanic Garden, that I proposed to found. I never owned an acre of ground, this garden would have been my delight: I had traced the plan of it, with a retreat among the flowers, a Green house, Museum and Library; but I had to forsake it at last, and make again my garden of the woods and mountains.

The organization of a stock company finally did get the Garden on 10 acres of ground within Lexington; in March 1825, Rafinesque began to plant it, but that summer he was away from Kentucky, and he left the state finally in the autumn of the next year.

The letter of 1823 to Collins tells of a wholly new interest, that for some years threatened to crowd out natural history. Rafinesque thus summarizes it:

Having within 2 years deeply studyed the Ancient history Antiquities & languages of America, I wish to know if the Philos[ophical] Soc[iety] proceed in their plan, what have they done since the publ[ication] of Heckenwelder [*sic*]? I wish they would publish Vocabularies of all the

American languages.— My labours in Amer[ican] Archeology are extensive, they consist principally of—

    1. Descriptions & Maps of about 100 Monuments in Ken-
tucky & ab[ou]t 50 elsewhere (never published or fig[ured].)

    2. Enumeration of all the Sites of Ancient Monuments in
North-America, with a short notice of each, amount[in]g to
ab[ou]t 500 sites

    3. The Ancient history and Chronology of North America
reduced to a system, with a short history of each Indian
Nation.

    4. The history of the invasion of N[orth] Am[erica] by the
Spaniards between 1539 & 1543, discovery of Tennessee, Ky,
Missouri, Arkansas

    5. Comparative Numerals of all the lang[uages] of N[orth]
Am[erica] & short Compared Vocabularies, to prove the affiliation of Nations &c.

    Could any thing be done with these labours? they were
intended for the Amer. Antiq. Soc. but they have also treated
me unjustly.

The first paper[34] in this new field of interest likewise appeared in 1824 in volume 1 of the *Cincinnati Literary Gazette,* and was followed during the year by the
"Ancient Annals of Kentucky" in Marshall's *History of Kentucky.* The only permanently valuable part of this work is a tabular enumeration of prehistoric sites, both in
Kentucky and elsewhere, which Rafinesque had spent much time and effort locating.

Yet another new idea was to seize Rafinesque, and this was one by which he
suddenly dreamed of passing quickly from his modest financial estate to affluence.
I suspect that it was the organizing in 1824 of the stock company to develop the
Botanic Garden that suggested something much more worth while to Rafinesque's
fertile mind. Why could not divisible certificates be issued for bank stocks and
deposits, and these made to circulate freely like money? I do not know if he waited
for experienced financial approval of what he called his "Divitial Invention," but
Mrs. Norton (the librarian of Transylvania) brought to my attention a letter from
Daniel Raymond of Baltimore, author of a two-volume work of political econ-
omy, which Rafinesque reproduced in the *Kentucky Reporter* for December 26, 1825.
This letter seems to me to lay a finger upon the weakness as well as the merits of
Rafinesque's idea.

I have examined your Divitial Invention by which you propose to invest
Certificates of Stocks and deposits with the qualities of a circulating medium;

The Life and Work of Rafinesque

I have no doubt but what your Invention may be applied to Stocks of various kinds, with advantage to the Stockholders and benefit to the public, although the perpetual fluctuations of the value of Stock might operate to prevent the great mass of the people from knowing their value. Your invention nevertheless may be very useful in facilitating the transfer and remittance of Stocks, and then give them some of the qualities of a Currency.

I wonder how correctly Rafinesque can be held to have anticipated the modern system of coupons, since the latter are attached to bonds of set value rather than to stocks.[35]

About this new interest he wrote to Collins on June 16, 1825:

I have at last turned my attention to something practical and extensively useful, and have succeeded to achieve 4 Discoveries or Inventions of the utmost importance and Magnitude: Each of them is sufficient to change my State for the better; but I am going to apply myself to one after another in Succession and shall begin with the most valuable or profitable which I call the Divitial Invention, being a new Principle of wealth, which gives rise to a new Art, the Divitial Art, and a new banking System, calculated to cause a revolution for the better in Money Matters.

I have applied & obtained a Patent for the same and I am going to Washington City to carry my Specification, which is so important that I would not trust it out of my hands.— And from thence I mean to visit Baltimore, Philadelphia, Newyork, Albany & Boston, in order to spread my Discovery & put it into practice every where.— I have also applied for Patents in England, France & other Countries & mean to put it in operation at once in all those Countries.

When we meet, I shall have a great deal to communicate & I anticipate much pleasure in our interview. I hope that you will not find my plans too gigantic, and will befriend me as well as able as usual. Meantime I remain respectfully

Your friend
C. S. Rafinesque, Ph.D.

Rafinesque was aglow with his great idea, and was evidently writing widely to possible supporters or to those who might extend the project. I have chanced upon a most remarkable letter to Joel Poinsett,[36] the U.S. ambassador to Mexico, in which both of Rafinesque's new interests figure. The first half is an appeal for vocabularies from native tribes of Mexico, and remarking: "The importance of these fragments to history are beyond belief. I have been able to ascertain by the Aztec lang[uage] that the Mexican Nations had origin in Tulan or Turan in Asia, while the Poconchians came from the East. . . . The vocab[ularies] or fragments you will send me will enable me to trace the origins of all the Mexican Nations." The second half of the letter tells of the Divitial Invention which should aid Mexico in three ways: (1) to raise a Mexican loan in the United States, which Rafinesque is willing to undertake; (2) to form a company to unite the Pacific and Atlantic Oceans, of course by canal; and (3) to establish a great bank in Mexico. He appoints Poinsett his agent to obtain a patent from the Mexican government. Then follows this:

> As a further inducement for the Mexican Government you may acquaint them that I have made a dreadful Discovery in the Art of Defensive War. Or invented a *New Kind of Artillery*, a single discharge of which will destroy *One thousand Men in Arms*, one mile off, or sink a large Ship of War. This awful Invention will be communicated Secretly to all such governments who will grant me a Patent or Privilege for my *Divitial Invention*. I hereby authorize you to offer the knowledge & use of it to the Mexican government if they grant me the privileges asked above.

And Rafinesque concludes:

> I recommend again these important and almost national matters to your good Care. It will be sufficient to add that you may realize $100,000 yourself by helping me to establish my Invention in Mexico. Please to write me speedily, and send me a Duplicate of your Letters, under cover of the Secr[etar]y of State H[bl] Henry Clay my personal friend and fellow townsman.
>
> C. S. Rafinesque, Ph.D.
Professor in Transylvania University
Central Divitial Office
Lexington, Kentucky
5th May, 1825

As full of hopes as when he came to Kentucky, Rafinesque set out at the end of June for Washington. But he was not in too much haste to cross the Alleghenies again on foot, this time from Wheeling to Cumberland. In Washington he was detained until September 10, first to get the signature of Henry Clay, the Secretary of State, who was out of town until near the end of August, and then of Attorney General William Wirt, who also was away, both of which were necessary for the patent of the Divitial Invention. Several letters about it came to Zaccheus Collins. "I can talk of little else now." "It is of so extraordinary a Nature as to stagger belief." "I hope you have understood my Improvement; if not you will soon when you hear me. By find[in]g the means of investing Stocks with several new properties—

1 Divisibility ad libitum, 2 Transmutability into each other, 3 Transferability at all times & places &c.—I have increased the value of an article already valuable—& I have in my power to coin for 14 years a new Money better than Gold & Silver, because bearing Interest!" He looked to the formation of new companies and new banks on his plan. "My profits are to be realized upon a Small Commission of 1 to 5 per Cent on the Amount of divitial operations, according to their latitude & utility." "If a new Divitial Bank can be established in every State, either by Charter, or patent right, the Amount of Stocks for them, might be 50 millions, which at 1 per Cent only for patent right would be half a million! Then there is no end to this improvement. It was suggested to me by the pecuniary difficulties of Kentucky,[37] and will be applied I hope to restore its sound Currency."

From Baltimore, on September 12, he wrote hopefully. "I have taken the proper Steps to have my Invention tryed in the District [of Columbia]. A Meeting of my friends was held & 3 of them appointed Agents protem. I have applied to the Corporations of the District—tendered them the use of it, & those of Washingt. & Georgetown have appointed Committees to report on the same. The trial will be made & even a new Bank is talked of on this plan. . . . I have had a very satisfactory Interview with Mr. [Richard] Rush Sec^{ry} of the Treasury." How easily optimism can set aside adverse criticism is shown by the next paragraph: "I have now explained my Invention to 50 intelligent persons, and all think well of it; but the Banks are afraid that my Stocknotes will be so good as to interfere with their Notes!"

The next letter to Collins was written in Philadelphia October 10, 1825, and was significantly marked "Confidential."

> I am compelled to return to Kentucky, where I am wanted, without conclud-
> ing any definite arrangement concern[in]g my Inventions (except appointing
> Agents) & give up my journey to Newyork. . . . My long delay and unlucky
> detention in Washington City has been fatal to my immediate projects, &
> above all has caused me unexpected expences, so that I find myself with only
> $20 left in my pocket, out of $200 I had provided for this journey. I have 600
> miles to go and my two boxes of books & clothing not being arrived, I am
> compelled to ask you the loan of such a Sum as your friendship can spare
> me, to help me on & provide Winter apparel, which I promise you to return
> as soon as I can, & pledge you my Library, my patents and my emoluments
> in the Univ^y for the same. . . .

Collins responded, for there has survived this: "Received Thirty five Dollars from Zaccheus Collins which I promise to repay to him as soon as I can. Philad^a—11th Oct. 1825 C. S. Rafinesque."

Francis W. Pennell

Back in Kentucky, Rafinesque writes on November 22 from Frankfort where he has that day "given a public lecture by appointment to the Legisl[ature]," and is further "called to a public Conference with the Committee of Ways & Means." By January 12, 1826, although "Our Legislature have done nothing but quarreling and have of course laid aside Internal improvements, I have at last issued my Circulars & established a Divitial Institution here." (The letter is written upon stationery headed "Patent Divitial Invention," and giving considerable gratis information about it.) He continues: "I am delivering now my last course on Botany in this University. I mean to travel again next summer to establish divitial Institutions. I shall probably visit Mr. Owen's settlement [at New Harmony, Indiana] and then see you in Philad^a afterwards. I have been invited to join the Co-operative Society of Valley forge. Pray what do you think of it? Is it likely to become a useful & enlightened Community? if so it might be a good resting place to spread my Labours, Inventions & Works." "I have serious thoughts of removing from Kent^y which is not a suitable place for my talents; but I am hardly decided where to settle: It will depend upon Circumstances & where the largest Div. office can be established—I incline for Philadelphia. Your advice will be acceptable. My valuable Collections must move with me. I have the finest herbarium in the United States, upwards of 25000 specimens.— An inducement would be the possibility of publishing my Numerous Manuscripts to advantage." He adds this, perhaps by way of reconciling the present to his former self: "Money is every thing in this selfish World & Money must be made somehow—The Social System of Owen even must have Money to begin with, altho' they hope to do without it afterwards, a desirable object but scarcely attainable for a while. When the Divitial Inv. is well understood Money will be as plenty as labour & property is now of which it will be the real representative." On March 1 there is another letter from Lexington, continuing the same thoughts, from which I quote: "Please to enquire about the situation of the Friendly Society of Valley forge & how they proceed. Has Mr. Haines[38] joined it?— Mr. Maclure has advised me to join them, & they have offered to pay the expences of my removal (not personal) or rather my Collections & Library which will load One Waggon at least. I have 25 Boxes & Trunks packing up of Plants, Shells, Minerals, Books & Manuscripts, Draw^{gs} &c. Could I safely trust or deposit these valuable Objects & fruits of my labours since my Shipwreck, into the hands of this Society? I will if Mr. Haines is there." And "When my Collections reach Philada you will be astonished how many fine New plants &c I have got."

In his *Life of Travels* Rafinesque attributes his departure from Transylvania to the way he had been treated by President Holley, but no hint of this appears in the frank correspondence that we have been perusing. Perhaps science was unappreciated, but if his room had been appropriated and his belongings piled in a heap together, there was certainly the possibility that the President had gotten wind of some of the rather indiscreet promotion of the Divitial Invention that had been going on.[39] I wonder if Rafinesque was in good state for his teaching in the session of 1825–26? He says that after his break with your President: "I took lodgings in town and carried there all my effects: thus leaving the College with curses on it and Holley; who were both reached by them soon after, since he died next year at sea of the Yellow fever, caught at New Orleans, having been driven from Lexington by public opinion: and the College has been burnt in 1828 with all its contents"[40] [78]. (But Clifford's and Rafinesque's collections were saved by previous removal.) Rather vigorous and vindictive, but I suspect with the dramatic heightened as he brooded over the matter in after years! However that may be, I can scarcely doubt that it was actually not so much the smart of ill treatment as the lure of the Divitial Invention that took Rafinesque from Lexington.

Rafinesque's next letter to Collins was from Germantown, which is today counted a part of Philadelphia. It is dated August 14, 1826. He had arrived but in what condition let the letter tell.

> I called on you 3 times when I visited Philad[a] last, without meeting you. I had to ask your friendly Advice on the strange Circumstances in which I am placed. I have given 2 Lectures in Philad[a] on my Invention and we are going to organize a Benevolent Mutual Institution among the 300 families that wish to Co-operate, a Committee of 15 is named to frame a Constitution[41] &c. You know I came here invited by the friendly Association, who were to pay for my removal. But this Assoc. having deviated widely from its genuine purpose of benevolence &c., is going to be dissolved. Part of my Library, Apparatus & baggage is deposited with them: the remainder is with Mesr Gratz to whom the expences were not paid. I thus find myself encumbered with 40 Boxes of books, plants, minerals &c., heavy charges to pay, no place yet to put them, and what is worse at the end of my Cash. I left claims in Kent[y] for $250 which I could not recover without litigation, I have now here another claim on the Fr. Assoc. which also should require coercion & I am disinclined to use it.

Then he wonders about becoming an itinerant lecturer: "I think that I could collect handsome audiences in the states of N. York and New England for short Courses of 12 Lectures for $2 per ticket, on the new & popular subject of the Ancient & Modern history of America upon which I am prepared to *lecture extempore*, perhaps a Course could also be given in Philad[a]." A paragraph is about his collections, where to deposit them in safety; he values them at $5,000; perhaps he can sell some duplicates, and he offers Collins "the Box of 500 fossils & 500 plants, at your own price the fossils alone are worth $50 at least." Then of prospects:

> I think that $100 would set me up and enable [me] to Lecture in 20 Towns in each of which $50 at least could be collected. This would enable me to repay this Sum, to look out for a Settlement, to procure Subscriptions for my hist^y of America Anct & Modern and to spread my Divitial Invention from which something will be made at last. Thus I can repay any favour conferred on me. Please to help me if you can. I have not funds to advertise, else I could find some Employment. By sending me protem $15 (or even $5 at least) which would increase my small debt to you $50 (or only $40) you would oblige me much. I cannot come to town because I have no money to spare & do not wish to run in debt at board^g houses.

He ends by expressing his readiness to leave all his collections for pledge if needful.

We have not heard from Collins in all this intercourse, although he had kept copies of a few of his letters. He did so of his reply, dated "Philad^a, August 16, 1826," and it is worth giving in full.

> Dear Sir, I received last evening your letter of the 14th ins^t. The circumstances you mention are indeed discouraging. I send enclosed the small sum you ask and wish it were convenient for me to make it greater. I fully agree with you that an itinerant lecturer except under special circumstances should be the very last resource, if one at all. And as to pecuniary advantage derived from that source do not calculate on it. Keep in view how oft your sanguine projects have misled you. You even now seem to look forward with expectations to a contemplated Philadelphia association (Owen or Owen like) altho you are suffering by what you deem the mismanagement of a similar one. With such fond hopes upon these new associations you cleared out from a place where you were well known and respected. New Harmony put you off where your knowledge would probably have most availed you and it. Indulge me in saying thus much. Your letter entitles me to expect you will. With respect & best wishes Yours, Z. Collins.

It was two weeks later that a receipt was sent worded as follows: "Received fifteen Dollars from Zaccheus Collins Esq. as a friendly loan Philad^a, 30th August, 1826. C. S. Rafinesque." That was all, and there ends the Rafinesque–Collins correspondence. As Rafinesque was now also in Philadelphia, there was slight need

for written intercourse, but a painful sequel (to be told in its place), makes me suspect that there was little further contact.[42] Perhaps Rafinesque's proud spirit could
not bear such frank criticism.

IV

Now commences the last chapter of Rafinesque's life, the fourteen years in
Philadelphia from 1826 to his death in 1840. We shall miss the revealing intimacy
of the Collins correspondence, but we have up to 1835 the narrative of the *Life
of Travels*. That recounts long journeys every year, as it had indeed for all
Rafinesque's years in America. I take it that these trips were ever his greatest joy.
Even in 1826, when coming for the last time back from Kentucky, while he had
sent his baggage directly to Philadelphia, he had gone himself across Ohio to San-
dusky, by steamboat on to Buffalo, seen Niagara Falls, and proceeded to Lockport
from which he took passage on the Erie Canal, then nearly completed. By happy
chance he met Prof. Amos Eaton of the Rensselaer Academy at Troy, a remarkable teacher and a man of singularly wide interests, and together they had a
delightful journey across the state.[43] Coming down the Hudson he stopped at West
Point to see Dr. John Torrey, a foremost American botanist and an unusually
kindly man, with whom Rafinesque was to maintain a friendly though occasional
correspondence[44] till death. (It was from such a trip that he came to the trying
conditions that awaited him in Philadelphia.)

We must hasten over the ensuing years. The first winter, that of 1826–27, he
gave "a course of natural history on the earth and mankind to a large class in the
Franklin Institute." He says in the *Life of Travels* [83–84] that, "I became afterwards
during 1827 Prof. of Geography and Drawing in the High School of the same
Institution." This he gave up apparently within a year. So far as I have discovered,
that was his last teaching engagement. But he was always ready with the pen, and
now began a series of illustrated accounts of individual plants published as "The
School of Flora"[45] in two Philadelphia magazines, the *Casket* and the *Saturday
Evening Post*. These ran from 1827 to 1832, showing over 140 plants in all; it must
have brought in something to the author. In 1828–29 he developed a most unexpected source of income, about which we shall let the *Life of Travels* tell [87–88].

> Having cured myself completely in 1828 of my chronic complaint, which
> was the fatal Phthisis, caused by my disappointments, fatigues, and the
> unsteady climate; which my knowledge in medical botany enabled me to sub-
> due and effect a radical cure: I entered into arrangements for establishing a

Chemical manufacture of vegetable remedies against the different kinds of Consumption. This succeeded well. I introduced also a new branch of medical knowledge and art. I became a Pulmist, who attended only to diseases of the lungs, as a Dentist attends only to the teeth. Being thus the first Pulmist, and perhaps the only one here or elsewhere. This new Profession changed my business for a while; yet enabling me to travel again in search of plants or to spread my practice, and to put my collections in better order, publishing many pamphlets, &.

In 1829 I gave a public proof of my art, in printing a small book called the *Pulmist or the art to cure the Consumption,* and many hundreds of individuals, whom I have cured or relieved are another striking proof of the beneficial results of my new practice.

Whatever Rafinesque could save from any source went into two things, trips afield or scientific publication, the latter now almost wholly financed on his own account. His trips took him through the Middle Atlantic states from northern Virginia (now northeastern West Virginia) to the upper Hudson River and once to Boston, Massachusetts. His largest publication was his two-volume *Medical Flora,* 1828 and 1830, the most widely distributed of all his books. Into its pages, although with restraint, were introduced many of the plants that he believed new. Doubtless it contains many that he had found locally used in Kentucky, and which are not included in later works of the sort. On the frontier men depend upon home remedies, the knowledge of which is forgotten as these become replaced by store preparations.

What Pulmel, this remedy for consumption, did for Rafinesque's finances is clearly shown both by the appearance of the *Medical Flora* and in the release of his collections. At the time of his arrival in Philadelphia from Kentucky in 1826 he had been unable to meet the costs of transportation of these, and, being wholly unsuccessful in obtaining purchasers or any of them, he had been forced to place them "for a while in store and under a mortgage." It was not until after Pulmel had prospered for several years, and the relatively expensive *Medical Flora* had been paid for,[46] that Rafinesque finally released the last of his possessions. Of this he wrote to Dr. Torrey on January 2, 1832: "I have at last withdrawn from Stores all my herbarium, collections, books, &c., at great expence." But this long storage, some of it for over five years, had been damaging: "I am now overhauling the whole of my immense herbarium and I have not yet gone through half of it.— I find upon an average one tenth of the plants lost or spoiled but not many of the new and rare ones. This loss will amount at least to 4000 specimens: but I shall have 36,000 left large and small." (He has no space for the large ones and wishes help in selling them.) That this storage was so detrimental must in part explain the poor reputation that the herbarium later bore.

Our picture of Rafinesque becomes more intimate again for the years from 1830 to 1834. To begin on a material plane, at the Library Company of Philadelphia is the [manuscript] "Day-Book of C. S. Rafinesque 1832 to 1834." It commences with an "Introduction to my Day Book or Statement of my Properties &c on 1st January 1832." This consists of 22 items, the largest of which are: "Value of 200 Wood Cuts of flowers & Plants . . . $2400." "Value of my Herbarium 36000 Specimens of Plants at $5 the 100 . . . 1800." "Amount of Pulmel & medicines out in the hands of Agents in Pittsburg[h], Cincinnati, Louisville, Baltimore, Washington, Norfolk, Leesburg, Wilmington, Lancaster, York, Trenton, Easton, Newyork, Hudson, Albany, Utica, Boston, Newhaven & other places as per schedule in Pulmel Book abt $1520." "Value of my Manuscripts as per Schedule chiefly Tellus, hist. of America, Journals, Tracts of many kinds, &c. ab^t $1000." Among the smaller, but still substantial, items is one concerning the Zaccheus Collins estate, to which we shall revert. Other items place a valuation of $360 on his maps and engravings; $240 on original maps, drawings, etc.; $570 on copies of his works for sale; and on his collections, $500 for shells, $400 for fossils other than shells, $200 for minerals, and $100 for other objects of natural history. His assets so reckoned[47] total $9,262, to which he adds the remark: "From $5000 to 20000 more may be the value of my Secret of Pulmel, Patent Divitial Invention & other patents."[48] "Besides these available properties, I have yet many old claims & bad debts to collect," and as instances he gives a list of eight, heading which is "1. On Transylvania Univ^y— balance due me." And below is one that confirms the genuineness of Rafinesque's account of his early years: "Besides the many uncertain & obsolete claims on Id. of France, Marseilles &c for paternal & my uncle's Estate. . . ."

At the beginning of each year Rafinesque summarized his money on hand: on January 1, 1832, cash in hand $65, in United States Bank $120, and in Saving Fund Society $75, giving $260 in all; on January 1, 1834, cash in hand $26.50, in United States Bank $110, in Savings Bank of Philadelphia $5.50, and in Savings Bank of Baltimore $24, giving $166 in all. It is not easy to analyze the records, since for 1832 and 1833 all financial matters, both income and outgo, as well as material sent widely for sale, are entered in lineal sequence. In 1834 the record is kept separately as "Cash received" and "Cash paid," but still with the uncertainty of materials advanced or the entry of unclassifiable items. Apparently however over the period from January 1, 1832, to August 31, 1834, Rafinesque's income, almost wholly derived from two sources, was: in 1832 from sale of Pulmel $206.15

and from that of publications $157.72, making $363.87; in 1833 from Pulmel $234.51 and from publications $19.41, making $253.92; and for the first eight months of 1834 from Pulmel $96.30, from publications $8.00, and from plant specimens $26.00, making for two thirds of the year only $130.30. I suspect that some entries must either have failed of record, or have been actual payments for material sent out for sale, as the expenditures for these periods total: for 1832 the sum of $549.31, for 1833 the sum of $300.71, and for eight months of 1834 the sum of $171.26, all figures in advance of the income recorded for the same periods. Of course Rafinesque's reserve of money was depleted in two years by nearly $100, but these discrepancies amount to $273.19. Further, remembering Zaccheus Collins, one may suspect unpaid obligations, but so far as the matter is capable of being checked Rafinesque regularly paid his bills promptly. Expenses for rent, food, clothing, and postage must have been promptly met. He kept a careful record of those for travel, this item costing $92.80 in 1832, $81.00 in 1833, and $5.50 in 1834. All printing bills, where one might suspect delay, were expeditiously met. These, inconceivably low as they seem to us, were duly curtailed with diminishing income; thus in 1832 he spent for paper to be used for his publications $50.00, for printing $128.00, for wood cuts $7.00, for folding & stitching $2.72, and for binding $3.00, making $190.72 in all; in 1833 for paper $14.50 and for printing $25.31, making $39.81; and in 1834 in paper $3.75 and for printing $14.11, making $17.86. One gets the impression that Rafinesque had learned care as well as economy of living.

But what of the publications which came as the result of real scrimping? (His food amounted in 1832 to $98.15, in 1833 to only $74.60.) The publications of this period were: in 1832 the *American Florist*, both a first and second series, a small thing called the *Atlantic Alphabet*,[49] and the *Atlantic Journal*, of which four numbers were printed in 1832 and two more in 1833. The last, much the most important, was a subscription journal covering an indefinitely wide field of interest. New species and other groups appeared in it, but very sketchily presented.

Now we come to an incident that I wish we could purge from the career of Rafinesque. I have said that one of his assets concerned the estate of Zaccheus Collins. That gentleman had died June 12, 1831. Not long after this Rafinesque presented an extensive and itemized bill for scientific specimens of various kinds and for publications; these entries are from 1818 to 1831, although there are only a few items, and those mostly papers and books, after 1826. The whole totals $402. This is followed by the information: "At various times Mr. Z. Collins paid me small sums of money on account or anticipation of these collections of Natural Objects made for him, for all of which I gave him receipts; but he never settled with me for any one of them nor ever had a receipt in full from me—To the best of my recollection the whole Am^t of money rec'd from him was

The Life and Work of Rafinesque

	$56
And 60 plants (whole number rec'd) at 10 Cents	6
Deducting these from the whole bill the	$62
Balance due me is	$340—

Gen. Daniel Parker, executor for the Collins estate, refused to honor the bill, and Rafinesque brought the matter to the District Court of Philadelphia, to which Parker was summoned by a writ of September 10, while the case was actually taken up as No. 176 of the term of December 1831. Three arbitrators were appointed on January 20, 1832, and they held meeting after meeting, Parker producing the correspondence we know and Rafinesque his business records. According to the notes of the defending side, which is all I have, Rafinesque's attorney early advised him to drop the case. From the papers with us now it would seem hopeless, and the contentions of the defense that Rafinesque had never presented any bill in Zaccheus Collins's lifetime, nor suggested remuneration for the specimens he had sent unasked, that the supposed payments made by Collins were really loans and were clearly so receipted by Rafinesque, would seem unanswerable. I fail to see what could have been produced to cause the arbitrators in early June of 1835 to bring in a decision awarding $173 to the plaintiff.[50] In a codicil to his will under date of June 16 Rafinesque says: "I further add and solemnly declare that the late award of $173 made in my favor by the Arbitrators in my claim on Collins' Estate is less than is justly due me, & if the Administrator appeals this claim must be pursued to the utmost and papers found to prove $306 and beyond." (Whatever its merits, Rafinesque had thoroughly persuaded himself of the justice of his claim!) The case did go on, though now with the new development that Rafinesque wanted the valuable herbarium of his late friend. In the first part of his *New Flora of North America*, apparently issued in December 1836, he remarks about "being in hope of obtaining the Herbarium of my late friend, Z. Collins." I suspect that General Parker, who we may presume was no botanist, welcomed this solution; the last note from him, written December 23 from Washington, to John W. Ashmead, admits as much: "[I am] requesting your continued attention to the vexatious suit of Doctor Rafinesque. Dr. Harlan writes me that Raff. has seen Dr. Pickering and talked of taking the Herbarium on apprisement &c. Dr. Harlan also says that Dr. Griffith and a Mr. Durand offer $150 for the Plants. If Raff. will not take the plants and give a full discharge from his suit had we not better sell the plants and defend the suit at the bats end!" Rafinesque took the Collins Herbarium—he could hardly have asked for more—and in the second number of his *New Flora* [12], appearing late in 1837, he says in speaking of Herbaria, "That of Collins was very valuable, and is now added to mine."[51]

Francis W. Pennell

V

Another story, never told before, is that of Rafinesque's search for his daughter. The letters concerning her are dated from 1830 to 1834. You remember that in 1815, when the child was four years old, she remained with her mother in Sicily when her father came to this country; then that, after news came of Rafinesque's shipwreck, the mother promptly married a comedian. Also please recall that Constantine Rafinesque had a sister, Georgette, and that their mother had been remarried to a French merchant named Pierre Lanthois. Now Mr. Lanthois, Sr., evidently had already a son Paul, who duly married Georgette. Our story opens with the four of the Lanthois name living together at Bordeaux, France.

Fig. 1.4. Rafinesque's sister as a young girl, drawn from her brother's memory of her when he was living in Kentucky. Courtesy of Transylvania University Library.

Constantine wrote on July 28, 1830, to his sister, evidently asking that she and her husband make inquiries for his daughter through French consuls in Italy. Her reply of September 28, in French of course, says that they have made such through the consuls at Naples and Palermo, but so far in vain. She adds that one is frightened at having to seek Emilia among traveling comedians, a class that are considered everywhere, and especially in the two Sicilies, as wholly corrupt; and it is feared that the truth about her will never be known, for Georgette is told that those into whose hands Emilia is likely to have fallen would never even report her dead, if they could extract money by having her alive. But Rafinesque succeeds by direct inquiries, apparently also to the French Consuls, and our next letter, from Tunis, in Italian, and written over a year later on December 11, 1831, is from Emilia Rafinesque herself. Her father had written her on July 15. Emilia's letter commences: "My dear and loved Father I can not possibly describe to you the inexpressible pleasure which I have experienced in seeing for the first time your venerated handwriting, how many tears I have shed reading and rereading the tender lines dictated by your paternal solicitude. My dearest prayers have therefore been heard. A father, whom they had always taught me to love, whom an absence of 15 years has not been able to remove from my heart, calls me into his arms." She tells him of her past life: how with her mother she went to Naples in 1825 (when she was fourteen) where she became second actress in the Teatro Fenix; then to Rome with both her mother and stepfather to become first actress in another theatre, and how there she had yielded to the importunity of a young English nobleman, Sir Henry Winston ("I was not far from loving him"); how he had deserted her before the birth of their daughter, Enrichetta ("I have paid with many tears for my error"); how later she had left her child with a friend in Naples, and with her mother had gone to Malta, where she had been two years in the role of first or second actress; then how chance had taken her to Tunis where she is now engaged. Her mother, whom she loves so much, tells her that it is her duty to go to her father, but how can she do so when that mother has a boy of six years old and her stepfather is too old to support the family? "Good-bye, my dear father; your Emilia repeats every day your name. She will try with her conduct, with her tenderness, to cancel the past, to beautify the ending of her career. I beg you to send your paternal benediction, and with tears I kiss humbly your hands and call myself your tender daughter Emilia Raffinesque."

Written only three days later, on December 14, 1831, is a letter from Georgette, to whom her brother had written some four times during the year. It seems that he had missed receiving a message from Mr. Lanthois informing him of the death of his mother on September 8. Georgette tells how on the arrival of Constantine's letter of August 5, their mother had said: "Tomorrow thou wilt read it to me, it is long, to-day I have a headache, it suffices me to know that Constantine is getting along well and that he is content." That evening she was stricken with apoplexy, and was unconscious for her last two days. Then the letter attends to business, by telling of the arrival of a box with some bottles of Pulmel, which had been so poorly stoppered that much was lost in transit; this happily prevented the Lanthois' getting into trouble with the customs, as little inquiry was made, whereas prepared medicines were prohibited. She also acknowledges receiving two copies of the *Medical Flora,* six of the *Pulmist,* and some other papers. Mr. Lanthois is going to read the *Pulmist* and translate it into French. Can not the Pulmel be made in France? In reply to inquiries she gives details of French postal rates and of the communal society of Saint-Simonism.

In April 1832 Mr. Lanthois writes at length about the Pulmel, to which Georgette appends a statement that they still have no word about Emilia, although they have again taken up the matter with the consuls in Italy; that her husband's business after two years is still in difficulties; and that there is cholera in Paris and fear of it in Bordeaux.

In the meantime, on March 27, Emilia's father had written making it imperative for the daughter to choose between her mother and himself. This called forth a vigorous reply (June 5) from Emilia. It is a cruel alternative.

> You impose upon me to choose between you and my mother, between being a lady or an actress. In addition to the love which I have for you there would be the illusion of happiness to make me choose you, as if felicity consisted in comforts and riches and not in contentedness of soul alone. . . . How would I ever be able to leave my mother in poverty and live myself among comforts? How would I ever be able to leave her who had done so much to give me an education worthy of you. . . . If you have need of my work to sustain you and if she were in your place I would not waver a single moment in the choice, there would be neither obstacles nor troubles that I would not face to join you.

Still, it is a pleading letter, for she yet longs to go. "Oh, father, if Heaven will grant me the favor of living near you, all my efforts will be to give thanks to a tender father for the love he has shown to me; and in what way can I show thanks if not to love him, serve him and to soften the last moments of his life? Tears obscure my sight. Oh, father, love always your daughter and do not forget her ever." Then, in reply to another point: "Although I have not the pleasure of knowing my aunt who lives in Bordeaux, I will not fail to write to her and do my duty toward her. You tell me that I should go to her house, but you are not sure that she will receive me and my daughter Enrichetta. How then, could I decide to go to her?" She becomes suddenly docile: "Advise me how I ought to conduct myself for the future. I beg that everything should be said clearly and definitely. Every command of yours shall be law for me. Remember everything that is necessary for my sailing." The letter ends with fervent expressions of her love, while a postscript reminds him that her mother has two children whom their aged father can not maintain and suggests that some allowance be sent on year by year. It is evident that Emilia supposes her father to be financially prosperous.

Turning now to Georgette, the next letter, the last from her, was written from Bordeaux on December 6, 1832. After speaking of how much more happily her brother had fared in his search than had she and her husband, there is this candid paragraph:

> Reflections present themselves in a crowd on thinking of Emilie. I am arrested by thy last letter; since thou hast allowed her to decide her own fate in choosing between her mother and thee, thou hast done everything that thou oughtest to do, now her decision will teach thee what she is, I fear that she does not seek to deceive thee. Her trip to Tunis does not please me, and this agrees with certain hearsay which came to me about her conduct and which was not favorable to her. Let us see if she will write me as she says she wishes to do: I have delayed to write thee hoping every day to receive a letter from her which might be interesting for thee, but nothing arrives and I doubt whether she be worthy of thee.

Then, changing the theme she makes a remark that would be very apt today: "Thy hesitancy to leave the United States would seem to me very reasonable, thou art in a favored country (aside from cholera), thou doest business there, and our old Europe is on a volcano." She adds: "I should very much like to see thee again, but I can not advise thee to come now."

The letter goes on to tell of her meeting Mr. Frederic Cailliaud, who had spent four years in Egypt where he had discovered the ruins of the ancient Meroë: "when I saw him he was seeking fluvial shells to adorn the museum of Nantes; we talked natural science; I told him that thou hadst just received word that thy memoir on the origin of the negroes had won a prize and that a gold medal had been decreed thee; that it was Mr. Jomard of the institute who took thy part. It seemed that Mr. Jomard is the protector of Mr. Cailliaud, that made us acquainted, I gave him thy address, as well as several of thy works. He was immediately interested in thee. . . ." And so she has arranged for her brother a scientific correspondence with "a good acquaintance" and one who "has the cross of the legion of honor." This is interesting now because at the Academy of Natural Sciences in Philadelphia we have the manuscript of Rafinesque's "Mémoires sur l'Origine des Nations Nègres," signed "Philadelphie, 1831," while there are letters from Jomard from 1832 to 1834 on behalf of the Société de Géographie of Paris, which had

Fig. 1.5. Gold medal awarded to Rafinesque in 1832 by the Société de Géographie for his memoir on the Asiatic Negroes. Courtesy of the College of Physicians of Philadelphia.

conferred the award.[52] In Rafinesque's will the medal was left to his nephew Jules, but was never delivered to him; it is pictured here.

Returning to Georgette's letter we immediately make the acquaintance of Constantine Rafinesque's nephew and niece, the children of his brother Anthony Augustus, who lived long at Havre but who had died in 1826.

> I am very happy that thou hast received the letters of Jules and Laura, these poor children are delighted to have received a letter from thee. Jules burns with the desire to go to America when he shall be old enough, to go and see his uncle he says. Everyone praises them both, as to physique they say that Jules will be handsome and that Laura will be a very pretty woman, it would seem that she will look a good deal like our good mother. It seems that their studies go well and that Laura does very well in drawing and music. Their mother's relatives are very wellbred, well born; perhaps a little scheming as a result of their uncertain position; the grandmother has a pension from the state, the grandfather has a place in one of the ministries; they move in good society and both of them live in Paris.

Their uncle had evidently asked about these relatives and what influence they would have on the children, to which his sister replied somewhat ambiguously:

"I think that it will be very feeble, they content themselves with loving them very much."

Here we have to take leave of sister Georgette. Her letters, except for knowing little punctuation except commas, show her a well-educated and truly interesting person. Could she have accompanied her brother to America I suspect indeed that she could have advised him well. His *Life of Travels* was written for her in the ensuing year, 1833, and so was first drawn up in French.[53] Constantine had expected to make her his chief heir but she died before him. I presume that this was early in 1835, as on June 15 of that year he added a codicil to his will about the matter.[54]

But we must return to Emilia, whose future is still undecided. On September 15, 1832, her father had again written her, and she had received the letter at the French consulate November 16. She replied December 4, again from Tunis.

You say that I prefer to remain an actress and not a lady, and I say that you deceive yourself, dear father if you believe this. I have always prayed God to take me out of a career so contrary to my sentiments, but at the same time in it I have gained my bread honestly. . . . The reason I have stayed in Tunis is to assure myself the position for a year. I could have gone to Naples as you say, but how to do it if the lack of money would not permit me to take such a voyage? And then who knows if I was going to find a position very soon? You know well that in Italy there are many companies and so often they remain on tour a long time; therefore you can well see that I could not leave the certain for the uncertain and hence I have had to stay here despite everything. . . . My dear father, if you desire my happiness, if you desire that I leave this profession, which I scarcely exercise, first because I hate it, second because not even my physique can stand it, it depends on you alone. You tell me that I should give an annuity to my mother, and with what means? On what funds ? I live by my own hard work, father, and am scarcely able to maintain myself; [living] with her, with what I earn. It is two years since I have sent anything to my daughter and I am embarrassed by this. . . . In Italy the greatest pay that a Second Lady can have does not amount to 50 pesetas per month. At present I receive 45 and have also to think of dressing myself properly in the theatre at my own expense. My engagement terminates October 1833. . . . Father, I have informed you of all and have put my affairs under your eyes. You alone can help me and make me happy. I feel that you have reproved me unjustly saying that your first letter had not moved me, but that which most grieved me was your telling me that my words are sometimes false and dramatic. Truly, I do not know what to answer you about this, fearing to be at fault. I say to you only that if there comes the day, *as I hope*, when you will know fully your daughter, you will say to yourself *how much I have deceived myself.* You say to me in what can I help you, and I answer in everything if you want to. I depends entirely upon you. Decide my fate. If you wish my daughter I will send her to you; if you want me with her I will come; if you want me to come with my mother then I would be fully happy and I am sure you will find her very different from what you imagine her. I swear to you that they have informed you wrongly about her. No one can complain of her conduct because she has been always honorable and an example to others, and if there has been some question of a dispute with anyone it has been only to defend her House and her family. . . .

Finally: "if you think I am asking for money at the inspiration of some of my relations, you are much mistaken; I have nothing to do with them. . . . I stay with my mother only. I think of my own affairs and no one counsels me upon my resolutions."

A shorter letter dated May 22, 1833, next came from Tunis telling of the death of Emilia's stepfather, and that she is leaving tomorrow for Livorno and Naples. It seems that the manager had abandoned their company, and also a war had broken out. "Think, father, that the decision regarding my fate is in your hands. Now my mother has been left a widow with three children by the deceased. One, a boy of 13 years, is in the Conservatory of Music in Naples . . . a boy of 8 years, and the girl . . . of 16 years, who has come into my family since the death of her father." Emilia is the support of all this family. The voyage to Italy will cost, and she will have to take a contract as soon as she arrives. In closing are the usual sentiments of her fate being entirely in her father's hands, and the request for his benediction on his obedient daughter.

We have one more letter from Emilia to her father. It was from Bari, Italy, January 28, 1834, where she has an engagement until the middle of the next month. Beginning "My most affectionate Father," it is filled with concern at not having heard from him since his letter of the fifteenth of September 1832, nearly a year and a half ago. She reminds him of his understanding pity toward her, and assures him that her "present life does not merit any reproach. . . . Patience. I am content to suffer everything so as not to incur the accusation of being unappreciative and ungrateful toward one of my own blood." But she reminds him of his first letter to her, underscoring his message that "*a loving father would not ever have forgotten her. . . .* Saddened beyond measure because of your silence of more than a year, I decided to write you this letter, begging you, for all that is the most sacred in Heaven and for the most dear you have on the earth, to let me have news from you as soon as possible." Then of her own condition she says, "In my house complete desolation rules." Her stepfather has died, but "his children form an almost unsupportable load upon me, and without my filial devotion and the sensibility of my heart, they would be the prey of the most horrible misery." But she turns to something brighter:

> I also have a little daughter of about 6 years, beautiful as the angels, and I hope she will be good, and I will neglect no means to succeed in such a worthy object. Tender father, if you can make me happy do not hesitate an instant; take me from this barbarous position. I feel I can not bear up under the strain of my work, which is not adapted to my physical condition, and indeed for it my days will be shortened. Tell me sincerely your thoughts, give me a decisive answer and do not make me more tremulous about my fate. You must pronounce either my happy lot or my unhappiness. In the meanwhile I will pray Heaven that you may have a long life, and that it may not be disturbed by the least evil. If my prayers are heard I hope some day to be near you. Then I will have nothing else to desire. I am sending this letter to your sister, my aunt, in Bordò, through whose kindness I hope it will be sent to you.

It is evident that Emilia is now in touch with Georgette. "I live in Naples," and she gives a third-story address. "Finally, I kiss humbly your hands, and sign myself, your affectionate daughter, Emilia Rafinesque."

There was a tragic pathos to this letter that may well have moved her father to real action. As we have seen from his financial record, his own resources had been rapidly diminishing from 1832 to 1834, so that he could hardly bring over Emilia and her child. But he did arrange another way, and there has survived a contract drawn up in Italian in April 1834 between him and the Impresario of the Italian Opera in New York by which, subject to her approval, the Sig^{ra} Emilia Winston was engaged in the capacity of a supplementary to sing, gesticulate, and act in all kinds of opera in all the cities of the United States north of Virginia. The pay was to be at least $100, or at most $150, every month for ten months, from the first day of September 1834 to the last day of June 1835. "All the expense of the voyage from Livorno to New York by the said virtuosa and her daughter Enrichetta of about 5 years and perhaps a servant will be paid by the Impresario. All the expense of travel in the United States is put to the charge of the Impresa." There are various further charges, but the way would seem fully opened for Emilia to come.

One can not know definitely what further passed between father and daughter. We only know that she did not come. Perhaps the opportunity was too late and she felt afraid that she could not fulfill the contract. Perhaps her mother's family in some way interfered. This would seem to have been her father's view of the matter, for a codicil to his will dated June 15, 1835, alludes to Emilia as "bereft from me and in the power of rapacious relatives"; he made her then but a joint heir with his nephew and niece, with the stipulation that she should receive only interest on her third of his estate, the principal reverting at her death to her cousin Jules. Emilia's child, his little granddaughter, was not mentioned then, though two years earlier, in the body of his will, Jules had been requested to "allow something" to her.

VI

In the *Life of Travels* [130–31] we have the story of the eventual success of one of Rafinesque's most cherished business schemes.

> In 1835 my former plan of a Six per cent Savings-Bank was resumed, matured and accomplished. . . . I invited in May the whole public to join me in founding a beneficial Institution, which I called *Divitial Institution* (meaning Institution of wealth) and *Six per Cent Savings-Bank,* dividing a first Stock of $50,000, in 5000 shares of $10, whereon from 7 to 10 per cent could be made per annum; while depositors were to be allowed 6 per cent, and to become stockholders after awhile.
>
> This plan succeeded at last. About 50 subscribers took nearly all the shares, and on the 1st of June we began to organize and manage the Institution; which has given quarterly dividends ever since of 2 or 2½ per cent, being 9 per cent for the first year.

This was written in 1836. I know not how long the bank prospered nor what income Rafinesque drew from it, but certainly it accounts for his last and greatest period of scientific publication.[55] He doubtless had to pay the whole cost of publishing the comprehensive volumes about to be mentioned.

To 1836 belong *The American Nations,* no. 1 of 260 and no. 2 of 292 pages; *The World, or Instability, a Poem,* of 248 pages, with a second edition dated the same year; *A Life of Travels and Researches in North America and South Europe,* of 148 pages; *New Flora and Botany of North America,* part I of 100 pages; a total of 1,048 pages for the year.[56] To 1837 belong the *New Flora,* part II of 96 pages; *Flora Telluriana,* parts I of 104, II of 112, and III of 100 pages; *Safe Banking, including the Principles of Wealth,* of 138 pages, printed by the Divitial Institution;—a total of 550 pages, privately printed. To 1838 belong the *New Flora,* parts III of 96, and IV of 112 pages; *Flora Telluriana,* part IV of 136 pages; *Genius and Spirit of the Hebrew Bible,* of 264 pages; *Alsographia Americana,* of 76 pages; *Celestial Wonders and Philosophy,* of 136 pages; *Sylva Telluriana,* of 184 pages; *The Ancient Monuments of North and South America,* 2nd ed. of 28 pages—a total of 1,032 pages. To 1839 belongs only the *American Manual of the Mulberry Trees* of 96 pages. To 1840 belong the *Autikon Botanikon,* of 200 pages; *The Good Book, and Amenities of Nature,* of 84 pages; and *The Pleasures and Duties of Wealth,* of 32 pages, printed for the Eleutherium of Knowledge, a total of 316 pages, privately printed. These figures afford their testimony that funds were running amazingly well for Rafinesque until at least the end of 1838. Even the final year, 1840, shows a good output. I suppose that the pagination of these five years together outbalances all of his earlier work.

As to fields of interest, of course banking is present, with 2 out of the 14 titles; ancient or philological history with 3 titles; philosophical considerations, 2 titles; travel and miscellaneous information about nature, 1 each; zoology none; while botany now has the largest share, 5 titles. While there are journalistic papers that I have not listed, these are very few, for Rafinesque has now turned almost entirely to the production of larger works. He was bringing out what had been maturing for years.

What worth is there in all this work? I am little able to judge about the economic and philosophic portion of it. Presumably *The American Nations* was on the lines of the *Ancient Annals of Kentucky,* but likely much more perfected, since Rafinesque came to view the first as an imperfect, premature performance. I feel more at home in considering the botanical works, and as these are doubtless the most enduring of the series let us turn to them.

Dr. Asa Gray, Rafinesque's unsparing critic, has this to say concerning these later studies:

> A gradual deterioration will be observed in Rafinesque's botanical writings from 1819 to about 1830, when the passion for establishing new genera and species, appears to have become a complete *monomania.* This is the most charitable opinion we can entertain, and is confirmed by the opinions of those who knew him best.[57] Hitherto we have been particular in the enumeration of his scattered productions, in order to facilitate the labors of those who may be disposed to search through bushels of chaff for the grain or two of wheat they perchance contain. What consideration they may deserve, let succeeding botanists determine; but we cannot hesitate to say that none whatever is due to his subsequent works.[58]

Gray's verdict was emphatic! And yet, were it not for the views put forth in these works of his last five years, Rafinesque would be to me a much less interesting, less worthwhile figure.

Francis W. Pennell

The earlier fretfulness and despondency at finding his opinions neglected have now largely changed to a calm assurance in his own position. He looks widely over the botanical field and with mature insight builds on the standards set by the papers published years earlier in Sicily. He dares to differ from the conservative course, dares in fact to be wholly alone among American botanists, because he is convinced that he is right. Take this matter of establishing new species and genera, treated by his critic as if a mental aberration. It springs logically from the definitions of these categories that he has well framed in the *Flora Telluriana*.

> *Individuals* alone have a separate physical existence, all the other clusters [an apt, but unusual, word for groups of related beings] are useful botanical groups of ideal abstractions, based on physical characters, by successive proportions of affinities. *Species* are the collections of individuals perfectly alike in all their parts. *Varieties* are slight casual deviations. Breeds or *Proles* are permanent Varieties. Therefore Species are natural altho' variable. *Genera* are the collective groups of Species, that agree in the Characters of the fructification. No Species belongs to a Genus unless it agrees with all the others therein included. . . . Therefore proper Genera are also natural.[59]

Before these definitions are reached, the *Flora Telluriana* has a remarkable introductory discussion. Speaking of plant relationships, Rafinesque writes that

> If 40 years of botanical observation, with herborizations in similar spots of North America at a distance of 32 years, may entitle me to state my impression on this abstruse subject, and add my testimony thereto, I must declare my conviction that
>
> 1. Vegetation produces only individuals! whose permanence is limited by their life.[60]
>
> 2. Plants vary gradually, in features, aspect, size, color, &c. by a natural spontaneous deviation from seedlings. This may happen quicker in annuals, less quick in perennials, slower still in trees, except when the tendency has already become active. These deviations may gradually form distinct varieties, next Breeds, at last becoming separate Species, when they assume a striking difference, and peculiar specific characters of a more permanent nature. The disparities in the descriptions and figures of old and modern botanists amply verify this.[61]
>
> 3. Even perennials may vary slightly in annual shoots from the same root, and trees in different branches or annual

growth. When a tendency to deviation by monstrosity, hybridity or variety is taken by an individual, the seeds produced will unfold them when growing, particularly if removed from the native place into gardens and new soils.

4. Pelorian Genera, or Generic Deviations in flowers and seeds happen slower or more seldom; being often unnoticed, or the produced seed is not always fertile. When it is, the offspring may become the type of a New or distinct Genus. Many such perish before they reproduce the deviation by fertile seeds; but a few survive and are the types of akin Genera.[62]

5. The periods of these deviations are doubtful, much fluctuating and various in length or existence. But we may assume as an average 30 to 100 years for the deviating or splitting range of specific deviation, and 500 to 1000 years for the Generic deviation; altho' their real permanence is much longer. Specific and generic Lives have not yet been calculated.[63]

6. Therefore many of our actual or newly described Genera and Species, may be of recent origins and all may have once sprung at the last rinovation or cataclysm of this Globe, from a lesser number of original types, perhaps found in the fossil plants of our Earth, which are far from being all known as yet, and whose seeds were preserved in mountains, earth, mud or water till the catastrophe was over.[64]

7. It is even possible to ascertain the relative ages and affinities of actual species and Genera, sometimes their very parents or connections in the Genus or the tribe. Those we call hybrids are not always such, they may arise from other deviations; but artificial hybrids are evidently such. All these deviations are still less permanent.[65]

8. As a general rule the real Genera (not the false ones of blending Botanists) of single or few species are the newest in order of time, and the most prolific the oldest in the Series. . . . Species prolific in individuals and varieties are always the oldest, and rare Species probably the newest of all, unless they are fragments of extinct groups.[66]

Considering the propositions as a whole we can scarcely fail to be impressed by their anticipation of later thought. Rafinesque continues:

Such exposition of my principles, and explanation of motives were perhaps needful, when I am going to increase the generic groups perhaps beyond any thing ever done of this kind. Linneus had only 1444 Genera. . . . Jussieu in 1789 had nearly 2000. . . . Ever since 1815 I had ascertained and classified nearly 3000, whereof 500 were my own. It is this labor, indicated in my Analysis of Nature, that I now propose to enlarge, rectify and publish: whereby as many as Linneus ever had will be added or revised, and about 1000 will be totally new, even now, as late as 1836, or not yet generally adopted.

Altho' this attempt may astonish or perplex some timid Botanists, my labor will be duly appreciated ere long, and my unceasing efforts to improve the science meet with a kind reception from the new improving school. The axiom that a multiplication of names enlarges our ideas, holds true in all cases and sciences, since they are based on facts or mental entities.[67]

The last theorem I certainly cannot hold, although there is a logic behind it. But—contrary to Asa Gray—Rafinesque's course here outlined, with the principles on which it was based, do not seem to me the plan and justifications of a monomaniac. There is vision as well as daring here. But I would call in question Rafinesque's fundamental ground for establishing genera as being too dogmatic; he needed to realize that there could often be great diversity in floral features among plants that were unquestionably nearly akin.

Was the fulfillment of this program as well achieved in the *Flora Telluriana* as it had been conceived? Here I must vigorously say no. Genera innumerable were proposed with but a few phrases to tell of them, and these plants were from every part of the earth. Their proposer had but limited collections at hand, and he had set himself to a task that needed to be a life-work of someone laboring in one of the largest herbaria in the world. So many of Rafinesque's proposals are vague, and so long have they lain neglected, that to the present time little use has been made of this great undertaking. But by our rules of nomenclature there must be many genera proposed in it that ought to supersede those adopted from later works.

The remaining botanical works are more of a class, descriptions of species and genera, haphazard or by limited groups. The *New Flora and Botany of North America*, the chief of these, confines itself to plants of our continent, and embodies much seen on the field trips of its author. It reflects Rafinesque's sanguine belief that our species are innumerable, and that whatever looked at all different would assuredly prove so. I think that it contains much of his best work. It has more consistent studies of special groups, although it does not at all merit the name of a flora.[68] We must remember that Rafinesque was beyond question the best field botanist of his time. In the introduction to this work is a section on botanical geography (wherein he incidentally lists his favorite collecting localities), and there is also so charming an account of one's experiences afield that even Gray could not resist quoting it. I am sure I cannot.

Let the practical Botanist who wishes like myself to be a pioneer of science, and to increase the knowledge of plants, be fully prepared to meet dangers of all sorts in the wild groves and mountains of America. The mere fatigue of a pedestrian journey is nothing compared to the gloom of solitary forests, when not a human being is met for many miles, and if met he may be mistrusted; when the food and collections must be carried in your pocket or knapsack from day to day; when the fare is not only scanty but sometimes worse; when you must live on corn bread and salt pork, be burnt and steamed by a hot sun at noon, or drenched by rain, even with an umbrella in hand, as I always had.

Musquitoes and flies will often annoy you or suck your blood if you stop or leave a hurried step. Gnats dance before the eyes and often fall in unless you shut them; insects creep on you and into your ears. Ants crawl on you whenever you rest on the ground, wasps will assail you like furies if you touch their nests. But ticks the worst of all are unavoidable whenever you go among bushes, and stick to you in crowds, filling your skin with pimples and sores. Spiders, gallineps, horse-flies and other obnoxious insects will often beset you, or sorely hurt you. Hateful snakes are met, and if poisonous are very dangerous, some do not warn you off like the Rattle-snakes.

You meet rough or muddy roads to vex you, and blind paths to perplex you, rocks, mountains, and steep ascents. You may often lose your way, and must always have a compass with you as I had. You may be lamed in climbing rocks for plants or break your limbs by a fall. You must cross and wade through brooks, creeks, rivers, and swamps. In deep fords or in swift streams you may lose your footing and be drowned. You may be overtaken by a storm, the trees fall around you, the thunder roars and strikes before you. The winds may annoy you, the fire of heaven or of men sets fire to the grass or forest, and you may be surrounded by it, unless you fly for your life.

You may travel over a[n] unhealthy region or in a sickly season, you may fall sick on the road and become helpless, unless you be very careful, abstemious and temperate. . . .

Yet although I have felt all those miseries, I have escaped some to which others are liable. I have never been compelled to sleep at night on the ground, but have always found a shelter. I have never been actually starved, nor assailed by snakes or wild beasts, nor robbed, nor drowned, nor suddenly unwell. Temperance and the disuse of tobacco have partly availed me, and always kept me in health.

In fact I never was healthier and happier than when I encountered those dangers, while a sedentary life has often made me unhappy or unwell. I like the free range of the woods and glades, I hate the sight of fences like the Indians! The free constant exercise and pleasurable excitement is always conductive to health and pleasure.

The pleasures of a botanical exploration fully compensate for these miseries and dangers, else no one would be a travelling Botanist nor spend his time and money in vain. Many fair days and fair roads are met with, a clear sky or a bracing breeze inspires delight and ease, you breathe the pure air of the country, every rill and brook offers a draught of limpid fluid. What delight to meet with a spring after a thirsty walk, or a bowl of cool milk out of the dairy! What sound sleep at night after a long day's walk, what soothing naps at noon under a shaded tree near a purling brook!

Every step taken into the fields, groves, and hills, appears to afford new enjoyments. Landscapes and Plants jointly meet in your sight. Here is an old acquaintance seen again; there a novelty, a rare plant, perhaps a new one! greets your view: you hasten to pluck it, examine it, admire, and put it in your book. Then you walk on thinking what it might be, or may be made by you hereafter. You feel an exultation, you are a conqueror, you have made a conquest over Nature, you are going to add a new object, or a page to science. This peaceful conquest has cost no tears, but fills your mind with a proud sensation of not being useless on earth, of having detected another link of the creative power of God.

Such are the delightful feelings of a real botanist, who travels not for lucre nor paltry pay. Those who do, often think only of how much the root or the seed or the specimen will fetch at home or in their garden.

When you ramble by turns in the shady groves, grassy glades, rocky hills or steep mountains, you meet new charms peculiar to each; even the gloomy forest affords a shady walk. Every rock, nook, rill has peculiar plants inviting your attention. When nothing new nor rare appears, you commune with your mind and your God in lofty thoughts or dreams of happiness. Every pure Botanist is a good man, a happy man, and a religious man! He lives with God in his wide temple not made by hands.[69]

VII

After 1835, the last year concerning which the *Life of Travels* informs us, we have no further record [from this source] of how much Rafinesque was able to go afield. On October 7, 1836, he writes to Dr. Torrey, "my good friend," telling him about an excursion of three days "among the Pine barrens of Mullica hill," evidently beyond the town of that name in southern New Jersey, from which he "brought 500 specimens of Autumnal plants, some very rare & even new to me."[70] These are discussed with all his usual zeal, and a number of new species from this trip find entry in the *New Flora*. There must have been later field trips, but I suspect at more remote intervals, during the remaining years. The vast amount published makes it unlikely that Rafinesque ever again strayed far from Philadelphia.

He worked indomitably on, surrounded by his specimens and books. How early the serious internal trouble that was to prove fatal manifested itself we do not know. To judge from the record of his published output, he was still working vigorously during the early half of 1840. Of the *Autikon Botanikon* three parts were published of the projected five, and none was dated earlier than this year. The manuscript of the last pages is in the find at Philadelphia, but there is no copy beyond that point—a matter easy of detection, since all the species are in one running sequence of numbers. The latest dated paper[71] I have noticed was of "May 1840," and is entitled: "Copy of the Botanical letter to Dr. Lindley printed in the Amenities of Nature N 2"; the journal was Rafinesque's own, and this second number was never issued. In these two manuscripts the hand is firm and steady, as of old. The very fact that he should have started as new in 1840 the *Autikon Botanikon* and *The Good Book, and Amenities of Nature*, both as serials, is evidence enough that he did not anticipate the end so soon. Also, knowing his care in financial matters, I am sure that he thought himself possessed of sufficient funds to carry these projects through.

It seems likely therefore that his complaint, cancer of the stomach and liver, did not become evident, or at least dominant, many months before his death. It is a sad picture to think of him in his garret,[72] awaiting, mostly alone, the inevitable end. It seems doubly sad when we remember his daughter and the ardor of her words: "If Heaven will grant me the favor of living near you, all my efforts will be to give thanks to a tender father for the love he has shown to me; and in what way can I show thanks if not to love him, serve him and to soften the last moments of his life?" He likely had a few friends who stopped to see him. Presumably the druggist and botanist Elias Durand was one, for he told long afterwards that he had bought Rafinesque's herbarium, and Thomas Meehan, who had known Durand, adds the word "charitably." I suppose that this sickness was using up all

Francis W. Pennell

the small working margin of money on which Rafinesque habitually operated; but he must have thought the situation desperate when he parted with his most prized possession. Whatever money he got for this, or perhaps for books sold from his library, could not have been much, and was all but used up before the end. Apparently, Dr. James Mease, the physician through whom twenty-four years earlier he had made his first application for a professorship, was the most faithful in visiting him, but it seems unlikely that his busy life left him much time to attend to the sufferer. I think it is said that one morning he came and found his patient dead.[73]

It was thirty-seven years later that a story was told in a Philadelphia newspaper,[74] by someone who had heard it from Mr. Bringhurst, an undertaker. It was that, when Dr. Mease had brought him to the house (to a garret on Race Street, between Third and Fourth), they found that Rafinesque's body had been removed and locked in another room, supposedly for sale to a medical school; "in the presence of Dr. Mease and Mr. Bringhurst . . . the door was forced open, and the body was let down by ropes into the back yard and conveyed to its last place of rest."[75] Although H. H.'s introductory paragraph is unusually inaccurate (containing some fifteen errors about Rafinesque's career), this which is based on direct information seems likely true, at least in essentials. I have checked it so far as to ascertain that Mr. Bringhurst was the undertaker; in the records of the Bringhurst firm,[76] which is still operating in Philadelphia, is a bill, dated September 20, 1840, to Dr. James Mease for a walnut coffin for C. S. Rafinesque, $7.00; for winding sheet and shroud, $3.00; and for hearse and carriage, $6.00: a total of $16.00. But, as to the story, H. H. informs us that Mr. Bringhurst told it to him "some sixteen years ago," and one fears how well the details may have stuck in such a mind as H. H.'s. However that may be, Rafinesque was buried in the Ronaldson Cemetery, and in the part of it "set apart for strangers dying away from home and friends."[77] H. H. told of seeing there a "small painted head-board," but, characteristically I fear, he reported that the initials on it were "S. C. R."

The elaborate will, drawn up in 1833 and with codicils in 1835, was registered on November 16, and on the twenty-eighth James Mease alone took oath as executor. Two others had also been appointed. Constantine Rafinesque had first designated his sister, Georgette Louisa Lanthois, his sole heir, but after her death he had made his nephew and niece Jules and Laura Rafinesque, with his own daughter, Emilia, equally joint heirs, but the last to receive only interest during her lifetime and the principal of her share to revert to Jules. His patents and secrets were to be sold. His collections were to be offered for sale first to the Museum of Natural History of Paris, with obligation to publish all his manuscripts; then to other institutions and societies in Europe, as he thought that his collections could be sold better abroad than in this country. Private sale was enjoined. The gold medal from the Geographical Society of Paris was left to Jules, "at the condition to keep it forever in the family of Rafinesque as a honorable record of a reward of merit."

It is hard to charge such a faithful friend as Dr. Mease with breach of trust; rather I suppose that he considered the will fantastic and inconceivable of execution. We think that at least such heirlooms as the gold medal should have been delivered as intended.[78] But it may have seemed one of the few things on which something could be realized among what was else in his eyes but old books and junk. The herbarium, so far as it had been wanted, had already gone to Durand.[79] The other possessions were sent to public auction. It is to Thomas Meehan's zeal that we now owe our knowledge of both the will and the sale, as he was the instigator in bringing this information out of its resting place in the public archives.[80] He tells us that, in addition to the plants which Durand had bought, "the balance of the herbarium went with the waste paper noted in the executors' return given below, from which, however, some specimens were rescued by Mr. Isaac Burk, still living, which he presented as part of his herbarium to the University of Pennsylvania"; evidently this information is directly from Burk, and so can hardly fail to be authentic. The executor's return, which was termed the "The Inventory," should be seen in full.

In the inventory and account of sale and expenses appears the following: Cash taken from one of two rouleaux of half-eagles found on the person of the deceased at the time of his death and $1 from his purse, $6. To cash received of Joseph S. Clay from one of the said rouleaux in the evening for funeral expenses, and 62 cents, $25.62. To cash from Lewis Bebe for waste paper, $20.25; for eight copies of Medical Flora, $3.16; for several rough boxes and demijohns, $2.55; for black bottles to Mr. Burk, Lombard street, $10.25; for lot of Medical Flora to Dr. Sweet, $19.75; for gold medal, valued at the Mint, to C. A. Poulson, $16.55; for manuscript books, $5; for books and clothes, sold by C. C. Mackey, $22.29. Total, $131.42. The chief expenses were, to N. W. Robertson, for arranging the library, cleaning the books and finding purchasers for the bottles and some books, $8; for removing the minerals and shells to the auction and making a catalogue of them, $7; to P. Maguire, for assisting in removing library to sales, $4; to John Gilfillan, eight loads of books, minerals and shells to Lord's auction and thence to Freeman's, $6; for services of James Dallins in ascertaining at the different banks whether any money was to the credit of the deceased, $5;

Register's fee and Orphan's Court fee, $5.25; commissions of executors, 10 per cent, $12.14. Other expenses when added footed up an amount that left the estate indebted to the executor to the amount of $13.43.[81]

It seems well-nigh inconceivable that there should have been so little financial return. The specimens are said to have been poorly preserved, often not labeled, badly mixed, etc., but I seek in vain for any money brought in from their sale, even though a catalog had been made of the minerals and shells. Can it be that the books, sold with the clothes by Mr. Mackey for $22.29, comprised all that remained of the library that Rafinesque had valued in 1832 at $1,250? How could so much have disappeared for food or other necessities? Even now books and specimens together filled eight dray-loads on their way to auction, and it cost some $25 to put them in order and get them there. Comparing returns for these two items with Rafinesque's estimate of their value eight years earlier, the difference is amazing.

In spite of all that he had published, many of Rafinesque's studies were left in manuscript form at the time of his death. Whatever there was of this nature seems to have been included in the lot of "manuscript books" sold at the auction of 1840 for $5.00. Dr. Haldeman tells us that in 1842 some unpublished manuscripts and drawings were in the possession of Mr. Poulson, and some in his own; so I presume that one or the other was the actual purchaser of this bargain item of the sale. According to an accompanying slip, the academy's series of these manuscripts was presented by Dr. Haldeman on November 13, 1849, but one wonders if these may not have been augmented by Poulson's series also. Evidently Dr. Haldeman retained the zoological manuscripts, in which lay his own special interest.

Rafinesque's manuscripts are now too widely scattered for us to attempt to follow them. Philadelphia, Washington, Lexington, and Lincoln each have their share, and I am sure that a search would uncover many more.[82] I have mentioned the Academy of Natural Sciences of Philadelphia as possessing a manuscript of what appears to be the prize paper on the origin of the Negroes. From the botanical ones with us I have been interested in getting a little insight into Rafinesque's method of recording the results of his studies. Apparently, whenever he detected something new, he entered his brief description of it in a manuscript book, where they follow a sequence that is regardless of their actual classification. We have three such books, with such entries as these: "Neocloris," begun in New York in 1817 and continued to 1825; "Neophyton," in three parts, all dated 1830; and a "Mantissa," in three parts, dated 1832 to— . Apparently, in preparing text for publication, he copied these descriptions into the works that he later had printed. It is always interesting to happen upon an author's sourcebook.

The Life and Work of Rafinesque

Time gives queer changes to what it considers valuable. Durand has told us[83] that he bought Rafinesque's herbarium, not for his own plants, "I knew that his specimens were miserable, but he had come into the possession of Mr. Collins' herbarium, had corresponded with D[rs] Riddell & Carpenter of Louisiana, and from those sources I expected a good harvest of botanical matter. My hopes were not frustrated. Mr. Collins' collections were to me a most valuable acquisition. . . . In his herbarium I found the largest collection of southern plants perhaps extant at the time. . . ." Durand's strictures on Rafinesque's own specimens were doubtless correct. Long pedestrian enterprises, and especially Rafinesque's temperament for hasty work, were not conducive to the preparation of neat or well-documented material.[84] Durand, like a loyal Frenchman, gave his herbarium to the Museum d'Histoire Naturelle in Paris, where it has been kept apart, according to his bequest, as the Herbier Durand. Now for the change of values with time! Some five years ago Mrs. Agnes Chase, a leading American student grasses, as five years earlier I had, consulted the Herbier Durand solely to see what plants of Rafinesque might be in it. She found no grasses saved; I was more fortunate, but eight out of sixty species in [the Scrophulariacae,] the family I am studying, is a meager survival.

The disappearance of Rafinesque's specimens, which I suppose has happened in all the fields to which he contributed, is the most serious impediment to solving the identities of the organisms that he described. They largely went as waste either at the time of sale or as such conservatives as Durand sorted over what they had acquired from him. Any museum would now gladly welcome a find of Rafinesque specimens, for the light that it could shed upon his descriptions. But, in the main, such help will never be forthcoming; we shall have to plod along, making with effort the best identifications that we can, when certainty might easily have been within our grasp.

Some specimens of Rafinesque survive though. Probably most of them are in Europe, but we have a few in our country. Of plants some are in the New York Botanical Garden, for that institution has inherited the herbarium of Dr. Torrey, who was very retentive of whatever came to him. The largest series I know is at the Academy of Natural Sciences of Philadelphia.[85] These are mostly the plants of a "Medical Herbarium," sold, as Rafinesque's day-book tells us, on April 14, 1834, to Mr. Hembel for $26.00; they are accompanied by little slips bearing in Rafinesque's handwriting merely the name of the plant with "M fl" below, while someone has later added: "Pres. by Wm. Hembel Esq." There are also a few other specimens scattered through our herbarium, which either arrived with the collections of Dr. Short or were given by Rafinesque directly. But these survivals altogether are very few, and scarcely ever do they come from the last years of his life for which his work so much needs whatever authentication actual specimens could give.[86]

Of Rafinesque's zoological collections, at least some shells are preserved at the academy in Philadelphia. These are ones that in 1842 were in the possession of Mr. Poulson.[87] I wonder if anywhere there remains more than the scantiest material of other groups of animals. So far as I am aware the collections of no other naturalist, in America at least, have been dissipated so willfully as have these of Rafinesque.

VIII

I have shown you, almost too frankly, the man Rafinesque in the light of intimate records hitherto unknown. There was both frailty and strength in him. He retained to the end of his days a child's freshness of outlook, a love of discovery for its own sake, and, whether an animal, a plant, or an idea, the fact that it was new made it appeal mightily. In this appraisal I am better satisfied to let my further remarks be wholly of his scientific work, for I believe in it lies his most enduring contributions to knowledge. On coming to Kentucky he simply reveled in seeing and describing a new world of living things. He rushed from one group of plants or animals to another in an ecstasy of discovery. He could scarcely spare the precious time for thorough work, and the tedium of minute dissections always irked him. Hence it is that so many of his presentations remain unsatisfactory. Then, and all his life long, his work was too hastily done for his volumes to become masterpieces. As he matured, his views widened, and he came to see some things that were beyond the vision of his contemporaries. As he thought life through, he ceased to be so worried at lack of sympathy or at open criticism, for he felt confident that the time would come when men would think as he was thinking. In much they do, and a later generation welcomes as prophecies some of his utterances.

I can hardly close better than with Rafinesque's statement in the *Flora Telluriana* [I: 99–100] of what he conceived to be the universal laws controlling life.

> These great laws that rule living bodies and vegetation, are, *Symmetry,* that gives the bodily forms to Genera, casting the mould of typical frames— *Perpetuity* that by reproduction perpetuates these original primitive forms— *Diversity* that bids and compels all living bodies to assume gradually a variety of slight changes when reproduced, and never evolves two individuals perfectly alike, nor two leaves quite similar in all points on the very same tree. Lastly *Instability* that does not allow any forms nor frames to be perpetual nor ever the same, giving to plants as to animals birth, growth, decay, and death! in succession, within a term of a few hours, a day, a month, a year, or 1,000 years.

The Life and Work of Rafinesque

Notes

1. [Most of these have been transferred since for safekeeping to the American Philosophical Society, which already had a notable collection of Rafinesque manuscripts. There is no way of knowing just what Haldeman purchased, but there are good reasons to believe that at least as many of Rafinesque's papers were destroyed as were purchased by Haldeman and others. In 1847, seven years after the naturalist's death, George Remsen, a collector, reported that the Philadelphia bookseller Leary "got possession of the major part of Rafinesque papers and books" and had "sold for old papers" the manuscripts "as being of no use" (Draper MSS., 32cc18, State Historical Society of Wisconsin). **Ed.**]

2. [This sum results from construing the last index number in Fitzpatrick's bibliography as a total, without examining critically the entries that lead up to it. Fitzpatrick gave a serial number to the title of each magazine Rafinesque edited, then another serial number to the title of every article contained in it. A careful look at the titles Fitzpatrick has listed reveals that many are reprints, some are translations of titles previously listed, and a few are by persons other than Rafinesque, who never claimed for himself the authorship of more than 220 "Works, Pamphlets, Essays and Tracts." **Ed.**]

3. [*Philadelphia Inquirer*, Sept. 19, 1840. **Ed.**]

4. Asa Gray, "Notice of the Botanical Writings of the Late C. S. Rafinesque," *American Journal of Science* 40 (Mar. 1841): 221–41.

5. S. S. Haldeman, "Notice of the Zoological Writings of the Late C. S. Rafinesque," *American Journal of Science* 42 (Mar. 1842): 280–91.

6. Richard Ellsworth Call, *The Life and Writings of Rafinesque*, Filson Club Publication No. 10 (Louisville, 1895).

7. T. J. Fitzpatrick, *Rafinesque: A Sketch of His Life with Bibliography* (Des Moines, 1911). [Both the Call and the Fitzpatrick volumes were reprinted in facsimile in 1978 by Arno Press in its Biologists and Their World Series, along with Rafinesque's own *A Life of Travels*, under the title *Rafinesque: Autobiography and Lives*, ed. Keir B. Sterling. **Ed.**]

8. For references to Rafinesque's own works the reader to referred to Fitzpatrick's bibliography [1911, revised and enlarged 1982, supplemented 2001. **Ed.**]

9. [For the full and correct spelling of the names of both parents, see Reynaud's article included here. As Reynaud also shows, the mother was born in Constantinople, when Greece was part of the Ottoman Empire. Her son, however, did not want to be identified with that Muslim state. **Ed.**]

10. [*Voyage et avantures de François Leguat, & de ses compagnons en deux îles desertes des Indes Orientales* (London, 1720) is the first of the books that gave pleasure to the youthful Rafinesque. François Le Vaillant published in Paris in 1790 his *Voyage dans l'intérieur de l'Afrique* and also wrote on ornithology. Although young Rafinesque was being tutored in English, he might also have read about the adventures of James Cook in the collection compiled by John Hawkesworth that had been translated as *Relation des voyages entrepris par*

Francis W. Pennell

ordre de Sa Majesté britannique . . . pour faire des découvertes dans l'hémisphere méridional . . . exécutés par le commodore Byron, le capitaine Carteret, le capitaine Wallis & le capitaine Cook . . . rédigée d'après les journaux tenus par les différens commandans & les papiers de M. Banks (Lausanne, 1774). His taste for natural history was whetted by a book especially designed for boys by Noël Antoine Pluche, *Le spectacle de la nature; ou, Entretiens sur les particularités de l'histoire naturelle qui ont paru les plus propres à rendre les jeunes-gens curieux, & à leur former l'esprit* (Paris, 1700), while his guide was Valmont de Bomare's *Dictionnaire raisonné universel d'histoire naturelle* (Paris, 1768). This last may well be what Rafinesque meant by "the great Universal history," but Arthur J. Cain speculates (pp. 135–36) that Rafinesque may have had in mind the French translation of Benjamin Martin's *Philosophical Grammar,* and, in a section of his book not reprinted here, Cain calls attention to the influence at the time of such comprehensive surveys of knowledge as Diderot and d'Alembert's thirty-five-volume *Encyclopédie ou dictionnaire raisonné des sciences, des arts et des metiers* (Paris, 1751– 80). More about authors who influenced Rafinesque appears on pp. 125–28, 134–37. **Ed.**]

11.   Both the Tarascons, as well as one of the Cliffords, he was to meet again years later in Kentucky.

12.   [The "papers," a paragraph each, appeared in the *Bulletin des Sciences, par la Société Philomatique,* published in Paris in the month of "Veuemiaire, an 11 de la République" and "Brumaire, an 11 de la République" respectively. The first describes a woodpecker and a quail, the second a swallow—all three specimens from Java. From a bibliographical standpoint, their authorship should be assigned to the French zoologist François Marie Daudin, whose initials do sign each entry, though the title of each states that the description is "par C. Rafinesque." It is likely that Daudin, to whom Rafinesque had been writing since boyhood, culled these snippets from a letter. **Ed.**]

13.   [In all, nine of Rafinesque's contributions appeared in the *Medical Repository* during the period 1804–11. **Ed.**]

14.   "Somiology," a term newly coined by Rafinesque, was an admirable one by which to express in a single word knowledge of the organization of both plants and animals. It has not been adopted, however. [As so often happens in respect to Rafinesque, others disagree. In a portion of his book not included here, Arthur J. Cain said of *somiology* that "It is characteristic of Rafinesque that this new term should be incorrectly constructed, of dubious application, and superfluous." Superfluous because Lamarck, whom Rafinesque read with approval, had used the word *biologie* as early as 1802. **Ed.**]

15.   [Once again Fitzpatrick's serial numbers have been misleading. He gave the title of the magazine itself serial number 26, and this is followed by serial number 27 for the first article in it. Four of the remaining articles are clearly labeled letters from others to editor Rafinesque, so that the last serial number, 229, by no means represents the total of the editor's own contributions. It is true, however, that Rafinesque was the magazine's chief contributor. **Ed.**]

16. [Because she was a Roman Catholic and he was a Protestant. **Ed.**]

17. [Gray, "Notice of the Botanical Writings," 233. **Ed.**]

18. The whole Rafinesque-Collins correspondence, which was presented to the Academy by members of the Rawle family, shows well Rafinesque's shifts of interest from 1817 to 1826. It opens with his comments on having a paper rejected for publication in the academy's journal due to lack of supporting specimens.

19. Rafinesque was amazingly fluent in apt Latin names, and always had an appropriate one available. About this he says in his *Flora Telluriana* of 1836: "I am never at a loss for names as Linneus was when he named *quisqualis;* I could readily supply 20000, *all good.*" All his novelties, now being reported, were given technical names. [But see the Cain essay included here for an opposing view of the "appropriateness" of Rafinesque's scientific names. **Ed.**]

20. [In his footnote, not included here, Pennell cited Audubon's "Eccentric Naturalist" sketch and quoted bits from it about the naked visitor knocking bats out of the air with his host's violin. Pennell was convinced that "'M. de T.' could be no other than Rafinesque," but allowed that the depiction was "to some degree a caricature." Audubon's entire essay may be found in this book as chapter 19. **Ed.**]

21. This word, presumably coined by Rafinesque, would of course mean "patron," being derived from the name of the patron of Virgil and Horace. [Without going back to Latin authors, Rafinesque merely had anglicized the word *mécénat* of his mother tongue. **Ed.**]

22. A letter to Dr. Charles W. Short (from Louisville, July 17, 1818, and published in the *Filson Club History Quarterly* 12 [Oct. 1938]: 214–15) speaks of "having with me Pursh's Flora & Persoon's sinopsis," adding "I am able to determine any plant that I see."

23. In the library of the Smithsonian Institution are three field-books from Rafinesque's tour of 1818. These contain descriptions in French, mostly accompanied by drawings, of many kinds of animals and of some fungi.

24. [Rafinesque's coinage for the science of organic fossils. **Ed.**]

25. ["On the Different Lightnings Observed in the Western States, by C. S. Rafinesque, Professor of Botany and Natural History in Transylvania University," *Western Review and Miscellaneous Magazine* 1 (Aug. 1819): 60–62. It was Asa Gray, who concluded his obituary article cited before with the slander often repeated since about "a paper which Rafinesque many years ago sent to the editor of a well known scientific journal, describing and characterizing, in natural history style, *twelve new species of thunder and lightning!*" This fabrication has been chortled over almost annually ever since by writers who, like Gray, never bothered to look at the article. The *Western Review* was neither well known nor a scientific journal, and all Rafinesque had done there was describe the appearance of lightning discharges under such terms as "spark," "darting," "ball," "sheet," etc., and call attention to the unremarkable fact that sometimes lightning appears to jump from cloud to ground, or from cloud to

cloud, and even from ground to cloud. Being much in the field, he was well experienced with the spectacular thunder storms that strike the Ohio Valley. As the bibliographer T. J. Fitzpatrick primly observed: "They only discredit themselves who charge that Rafinesque deliberately described twelve new species of thunder and lightning." **Ed.**]

26. [There was more to Rafinesque's banishment from the pages of Silliman's *Journal* than could meet Pennell's eye in 1940. See "The Fall from Grace of that 'Base Wretch' Rafinesque," included here as chapter 11. **Ed.**]

27. In a letter to Dr. John Torrey of February 1, 1821, now at the Historical Society of Pennsylvania, Rafinesque says: "The 1st. Number of the Western Minerva is printed, but the publication is suspended, altho' we have 2 or 300 subscribers, because a literary Inquisition, censorship or cabal (call it what you please) has threatened to declare war against it as soon as it should appear. Some proof sheets have circulated and raised their wrath, because they were hoaxed in it &c. I shall tell you more about it anon; it will be a curious anecdote & the number a great literary curiosity, if it should never be published; however we are endeavoring to compromise and suppress some parts." According to the *Life of Travels* only three copies were saved. Another story, as given by Call (*Life and Writings of Rafinesque*, 169) is that the printer [Thomas Smith] destroyed the remainder because he was not immediately paid. That Rafinesque felt a grievance in the matter is shown by his inventory of January 1, 1832, where among "old claims & bad debts" appears one "on Th^s Smith of Lexington for broken contract on subject of Western Minerva." [One of the three saved copies lodged at the Academy of Natural Sciences of Philadelphia, where Fitzpatrick examined it and listed its contents as Rafinesque "publications," even though these never became available to the public until that copy of the proof sheets was issued in photo-offset by Peter Smith in 1949. **Ed.**]

28. [Pennell was unaware that Rafinesque used all three Lexington newspapers, especially to encourage an interest in natural science, and Pennell overlooked the fact that while in Kentucky Rafinesque also published several articles in the *Cincinnati Literary Gazette*. **Ed.**]

29. Reproduced by Harry B. Weiss in *Rafinesque's Kentucky Friends* (Highland Park, N. J., 1936). [Three are included here: Rafinesque's sister, fig. 1.4; his mother, fig. 2.1; and a woman named Juliet, fig. 17.6. **Ed.**]

30. [Three hundred dollars a year. **Ed.**]

31. [I am unable to explain why Pennell found so "appalling" the Latin form of the name of the oldest learned society on earth, an institution said to be the only surviving remnant of the Holy Roman Empire. Anyway, the Leopoldina at that time did have the right to award the honorary Dr.phil. degree, which it usually reserved for distinguished foreign scientists. Unknown until 1987 was a news item appearing in Lexington's *Western Monitor* (Dec. 30, 1823), which stated that Rafinesque had received *two* certificates: one (which still exists at the American Philosophical Society) from the Leopoldina, and the other

(presently unknown) described in the newspaper as "a Diploma of Doctor of Philosophy from the University of Bonn on the Rhine in Germany." So there really is no reason to doubt that the naturalist was "Dr. Rafinesque," even if his degree had not been earned by conventional means. As for the mysterious P.D. after his name in the catalogue, since no other member of the faculty was distinguished by the degree of doctor of philosophy it was not known at first how to abbreviate one. **Ed.**]

32. [Missing from the Fitzpatrick bibliography when Pennell wrote, an even earlier description had been published by Rafinesque in the Lexington newspaper, the *Kentucky Gazette*, Nov. 7, 1822. **Ed.**]

33. A letter from Rafinesque to Charles Wilkins Short, dated February 1, 1822, mentions the request of Mr. James Campbell that Rafinesque would be head of the Western College of Kentucky, lately erected at Hopkinsville. Dr. Short, then resident of that town, was one of the trustees. Evidently nothing came of the idea.

34. [What Pennell had in mind was "Clio No. I," the first of five such articles by Rafinesque in the *Cincinnati Literary Gazette* in 1824; but he was misled by the arrangement of Fitzpatrick's bibliographical entries. Actually, Rafinesque's published articles on the "antiquities" had begun to appear in the *Western Review and Miscellaneous Magazine* as early as 1819 and in the *Kentucky Reporter* as early as 1820. **Ed.**]

35. [The answer is: Not at all. It was the conceit of R. E. Call that the Divitial Invention was "the basis of the claim of Rafinesque that he was the inventor of the coupon system now so common on bonds and similar instruments" (*Life and Writings of Rafinesque*, 41). Rafinesque made no such claim, but Fitzpatrick (*Rafinesque*, 32) inflated Call's misconception by stating that Rafinesque "always claimed to be the inventor of this now widely used system of divisible commercial paper," which Fitzpatrick averred is called the "Coupon System." Both Call and Fitzpatrick were so impecunious it is doubtful that either ever had seen a bearer bond having coupons to be clipped to collect interest. **Ed.**]

36. Letter now at the Historical Society of Pennsylvania.

37. [Brought on by the Panic of 1819, when the unsecured paper money issued by Kentucky banks was discounted to as much as half its face value. Rafinesque's scheme was intended to support paper currency with an underlying tangible value. **Ed.**]

38. Reuben Haines was a philanthropist, and a prominent member of the Academy of Natural Sciences of Philadelphia. [Rafinesque's earlier expression of an interest to "visit Mr. Owen's settlement" in Indiana misled Pennell to believe that William Maclure, who was helping to bankroll the New Harmony experiment, had advised Rafinesque to settle there. Accordingly, I have excised Pennell's bracketed insertion which identifies New Harmony as the place suggested. Rather, it was the Friendly Society of Valley Forge, modeled on the English Friendly Societies that had been established on Owenite principles, where Maclure though Rafinesque might find happiness. However, the Owenite settlement at Valley Forge,

Pa., was in the process of dissolving even before Rafinesque reached there. As a result, it could not carry out its promise to pay the freight on the shipment of his belongings (which had swollen from 25 to 40 boxes). Neither could Rafinesque, so the boxes were put in storage. By the time he saved up enough money to retrieve them some years later, his dried plant specimens had mostly been ruined by heat, dust, and insects. **Ed.**]

39. [Although Holley tried (unsuccessfully) to discharge Rafinesque more than once, as a good a reason as any for his turning the naturalist's rooms to other purposes was that Rafinesque grossly overstayed the leave he requested and had been granted. It appears that President Holley thought he had decamped for good. **Ed.**]

40. [The fire actually occurred in 1829. For more about the Curse of Rafinesque, see "Who's Buried in Rafinesque's Tomb?" in this book, pp. 376–79. **Ed.**]

41. Among Rafinesque's papers is a draft in his handwriting, of such a document for the "Benevolent Mutual Institution of Philadelphia." Although undated, the name makes me confident that it is of the society of 1826. There are 22 sections. Accompanying it, but in another hand, is a list of members classified by occupations, with a summary by Rafinesque giving the numbers in each pursuit, a total of 122 persons.

42. In his *Life of Travels* Rafinesque does allude to Collins's advice that he should take his collections "to Europe where they might sell better or be published" [84], likely a suggestion made after 1826.

43. [This episode is fully narrated in Samuel Rezneck, "A Traveling School of Science on the Erie Canal in 1826," *New York History* 40 (1959): 255–69. **Ed.**]

44. [Pennell, whose manuscript sources were mostly those he found in Philadelphia and Lexington, was unaware that the extant Rafinesque–Torrey correspondence bulks larger than that of Rafinesque with Collins. Most of it is at the New York Botanical Garden Library, and, as it true of all of Rafinesque's letters, his side is better represented than that of his correspondent. Even if he did save his correspondents' letters— as is probable—it is likely that after his death they were "sold for old papers" and were pulped. Other possessions of his that did not sell at public auction went to the Philadelphia dump. **Ed.**]

45. [See Beverly Seaton's "Rafinesque's Sentimental Botany: 'The School of Flora,'" in this book, pp. 323–39. **Ed.**]

46. [Though both volumes were issued by commercial publishers, they caused Rafinesque considerable expense for the 100 woodcuts which illustrated them. **Ed.**]

47. [Anyone who goes to the trouble of adding up the items will see that Rafinesque's assets listed here total only $9,090. It appears that the total Pennell gives is taken directly from that stated in the manuscript day-book. Rafinesque seldom added accurately. **Ed.**]

48. In his *Life of Travels* [86] Rafinesque says: "I ought to have made another fortune by my inventions, which comprised so many useful things, such as a Steam Plough, an aquatic

Rail road, the Divitial Invention, &c.; but I could not meet with cooperators. Meantime my useful Divitial Invention was stolen or modified in Baltimore by establishing new Savings Banks partly on my plan, without consulting me nor asking my leave."

49. [This *Atlantic Alphabet* is the chart reproduced in fig. 15.5 of George Stuart's article included here. It was intended to illustrate Rafinesque's "Second Letter to Mr. Champollion on the Graphic Systems of America . . . ," *Atlantic Journal and Friend of Knowledge* 1 (Summer 1832): 40–44, and perhaps because he considered it more decorative than substantive, Elmer D. Merrill shifted it from the article to the position of frontispiece for the volume when he reprinted that journal in 1946. Rafinesque thought so highly of his *Atlantic Alphabet* chart that he had one thousand copies struck off as separates, which he valued in his day-book at a penny a piece. **Ed.**]

50. Amount as shown in the transcript of Rafinesque's will appended to Call's *Life and Writings of Rafinesque*. It is not stated in the Parker notes. [The administrator's motive for settling is revealed in the notes. Since the defendant was assessed court costs every time the arbitrators met, Parker could see the estate being whittled away by the interminable delays and postponements and may have urged some reward to the plaintiff in order to bring the case to a conclusion. **Ed.**]

51. [For an even fuller account of this episode, see Ronald L. Stuckey, "The First Public Auction of an American Herbarium including an Account of the Fate of the Baldwin, Collins, and Rafinesque Herbaria," *Taxon* 20 (Aug. 1971): 443–59. **Ed.**]

52. [The manuscripts have been transferred since to the American Philosophical Society. **Ed.**]

53. [Until the publication of the original French version of the autobiography in 1987, it was commonly believed that the *Life of Travels* was a kind of long letter, written to apprise his sister of the events of Constantine's life. This misconception arose because Rafinesque said in a "Preamble" to the English version that "These outlines . . . are chiefly a translation from a sketch of my travels written in French in 1833 and sent to France, to my only sister. . . ." He had sent the French manuscript to Georgette for her to transmit it to the Société de Géographie in Paris, hoping the society would publish it; but she died in childbirth before she could carry out her bother's wishes. Three years later, he translated it himself, added some additional material, and published it in Philadelphia at his own expense. **Ed.**]

54. [Such presumptions—in this instance a false one—about crucial dates of Rafinesque family members need no longer be made. See the next chapter, the English translation of Georges Reynaud's genealogy. **Ed.**]

55. [Pennell was only partly correct in this assertion. No doubt profits from the bank did help pay for the printing of some of Rafinesque's books and pamphlets, but others were subsidized by Charles Wetherill, a wealthy Philadelphia paint manufacturer. **Ed.**]

56. All parts of the *New Flora* and the *Flora Telluriana* were dated 1836, but the actual years of issue have been ascertained by Dr. J. H. Barnhart; see *Torreya* 7 (Sept. 1907): 177–82. Presumably the second part of *The American Nations* likewise appeared later than 1836.

57. [For copious documentation of the opinions of those contemporaries "who knew him best," see Ronald L. Stuckey's article included here as chapter 8. **Ed.**]

58. [Asa Gray, "Notice of the Botanical Writings of the Late C. S. Rafinesque," *American Journal of Science* 40 (Jan.–Mar. 1841): 237. **Ed.**]

59. [*Flora Telluriana*, I: 38–39. **Ed.**]

60. An evident truth that all would endorse then or now.

61. This is a positive statement of evolution, the germ of which might have come from the French school of Lamarck; but now said at a time when safe conservatism stuck by the fixity of species as separate creations. The relative rate of evolution in annuals, perennials, and trees, is a point advanced just eighty years later by Professor E. W. Sinnott ("Comparative Rapidity of Evolution in Various Plant Types," *American Naturalist* 50 [1916]: 466–78), a process to which Dr. Agnes Arber ("The Tree Habit of Angiosperms: Its Origin and Its Meaning," *The New Phytologist* 27 [1928]: 69–84) has applied the term, the "Evolutionary Lag" of woody vegetation. We now interpret the discrepancies between old and modern figures of the same plant by the comparative crudity of the earlier work; Rafinesque could not anticipate the vastness of geologic time as held since the day of Lyell, and so tried to fit evolution into a relatively small time-scale.

62. Rafinesque, by maintaining a qualitative difference between species and genera, the latter only showing differences in the fructification, made it essential that they should be produced by a different type of evolution. Today we would not recognize any such fundamental distinction. The thought that sterility should more likely occur where there was change in flower type seems plausible enough, although we now know it to be largely associated with hybridity.

63. As stated under 2, Rafinesque could not anticipate the vastness of geologic time as since universally held, and so he made his time-scale altogether too short. He was nearly unique in having formulated it at all.

64. Correct in the surmise that some genera and species are relatively modern compared to others.

65. Correct, and a remarkable statement for Rafinesque's time.

66. An excellent statement of the central thesis of J. C. Willis (*Age and Area: A Study in Geographical Distribution and Origin of Species* [Cambridge, 1922]), and made eighty-five years before that recently much discussed book. I do not think it holds universally, however, as if we suppose the newer species to be the better adapted to the life-conditions of the present, why should they not become the most widespread? So far as I have been able to check this matter from structural evidence, I think the latter condition the more general in nature.

The Life and Work of Rafinesque

67. [*Flora Telluriana*, I: 12–15. **Ed.**]

68. [Pennell means "flora" in the technical sense of a comprehensive inventory of the plants of a particular region, one often restricted to a particular segment of the plant kingdom, which provides a means of identifying the plants it describes. Rafinesque, however, made no pretense that his book fulfilled any of these criteria for all of North America, as the title implies. Rather, he said (I, 4), it is "a kind of mantissa [i.e., supplement] and appendix to all my predecessors, avoiding thereby the trouble in accumulating all the scattered materials and fragments already published. . . ." **Ed.**]

69. [*New Flora of North America*, I: 12–15. **Ed.**]

70. [This letter is at the Historical Society of Pennsylvania. If Pennell had consulted the Rafinesque–Torrey correspondence now at the New York Botanical Garden Library, he would have found the naturalist's travels for the last four years of his life quite well documented in fifteen extant letters. His assumption that Rafinesque did not stray far from Philadelphia, however, is correct. **Ed.**]

71. [The last document now known is a letter (at the New York Botanical Garden Library) to Elias Durand, dated August 5, 1840, seven weeks before Rafinesque's death. In it Rafinesque says "I have been unwell for 2 months of a dyspeptic complaint." **Ed.**]

72. [Not knowing the New York Botanical Garden's Rafinesque–Torrey correspondence, Pennell was unaware that in March 1837 Rafinesque moved to 172 Vine Street; and there, until his death, he had the entire rented house to himself. It is unlikely, therefore, that he chose the attic as the site for his deathbed. **Ed.**]

73. [Most of what follows about the death of Rafinesque and the disposition of his effects is a romantic myth that has become so firmly entrenched in popular lore that naked truth has no capacity to supplant it, so it has been left stand here. However, a later essay on "The Last Days of Rafinesque," chapter 3, narrates what actually did happen, so far as it can be reconstructed now. **Ed.**]

74. *Philadelphia Ledger* Supplement, Mar. 5, 1877; reprinted in the *American Naturalist* 11 (Sept. 1877): 574–75. The article, signed by "H. H.," concludes with the sentence: "Two years since I forwarded his last work to the National Library of France, and received acknowledgment through the French Consul, Philadelphia."

75. [About a year after the publication of this essay, Pennell learned of the report on the autopsy performed on the body of Rafinesque and published in a Philadelphia medical journal. This astonishing discovery showed that two physicians—not one—attended the naturalist during his last illness, and that neither of them was Dr. James Mease. The report described the patient's condition throughout his illness, the treatments given, and the exact time of death. It also established that far from being lowered from a window by ropes, the body was moved to another room for the autopsy. Pennell announced the discovery in "The Last Sickness of Rafinesque," *Bartonia* No. 25 (1949): 67–68. **Ed.**]

76. R. R. Bringhurst & Co., 2000 Walnut St., Philadelphia. [These records are now at the Historical Society of Pennsylvania. **Ed.**] Rafinesque's will called for cremation, but likely this request was not seen until after the burial.

77. [Whether concocted by H. H. or someone else, this phrase, "set apart for strangers dying away from home and friends," also has become part of the cock-and-bull story of Rafinesque's demise. The philanthropic founder of Ronaldson's Cemetery set aside *all* of it for "strangers"—i.e., persons not qualifying for burial in the cemeteries churches reserved for their members—and indeed the disproportionate number of touring actors and other theatrical people buried there is accounted for by the fact that they were away from home when they died. **Ed.**]

78. [I do too. In October 1983, when Transylvania University sponsored its bicentennial celebration of the birth of C. S. Rafinesque and it was learned that the naturalist's great-great-grandnephew and present head of the Rafinesque family, Dr. Jean Rafinesque, would grace the occasion by his presence, I suggested to the College of Physicians of Philadelphia, in whose Mütter Museum the medal had come to lodge, that it would be an admirable beau geste to present the medal to le docteur and thus, after 143 years, carry out at least one of the wishes of the man we had gathered to honor. The college respectfully declined, but it put the medal on loan to Transylvania for the duration of the celebration so that participants could see it. **Ed.**]

79. [But, as Pennell was to learn later and report on in *Bartonia*, the herbarium was acquired by Durand three years after Rafinesque's death. See "The Last Days of Rafinesque" included here. **Ed.**]

80. *Philadelphia Public Ledger,* Feb. 19, 1891.

81. [*Ibid.* But see, subsequently, "The Last Days of Rafinesque," for how the estate probably was pillaged before this accounting took place. **Ed.**]

82. [Since Pennell wrote, searching has indeed uncovered more Rafinesque manuscripts, principally letters. They have been found in Connecticut at New Haven; Delaware at Greenville; Indiana at New Harmony; Kentucky at Louisville; Maryland at Baltimore; Massachusetts at Boston, Cambridge, Salem, and Worcester; New Jersey at New Brunswick; New York at Albany, New York City, and Syracuse; North Carolina at Chapel Hill and Durham; Ohio at Athens, Cincinnati, and Marietta; Pennsylvania at Bryn Mawr, Haverford, and Pittsburgh; Rhode Island at Newport; and Virginia at Williamsburg. In Europe Rafinesque manuscripts—again, principally letters—have been found in Berlin, Geneva, Le Havre, Leiden, London, Paris, Pisa, and Sienna. **Ed.**]

83. [Agnes Chase, "The Durand Herbarium," *Bartonia* No. 17 (1936): 40–45. **Ed.**] This article by Mrs. Chase reproduces Durand's own account of the development of his herbarium.

The Life and Work of Rafinesque

84. In this connection, although I think that the passage has allusion chiefly to zoological and fossil material, may be quoted one of Rafinesque's manuscript sheets, at the Philadelphia Academy, entitled "Scientific Exploration of North America and the South of Europe, during 32 years from 1800 to 1832." "I have since [my shipwreck of 1815] partly repaired this loss; but disgusted with the expence, trouble and fears of new losses, I have often neglected to collect in my late travels those Objects which are too difficult to carry and preserve. How could I carry Bottles and Liquors to preserve Animals when I was exploring Mountains on foot and alone? Or how could I carry fossils weighing 20 lb. or more. My collections are nevertheless copious and rich in rare or new Objects of a portable Nature." On his dried plants Dr. Haldeman said in 1842: "Judging from the appearance of the specimens, his method of preserving plants was more simple than any recommended in books, as it consisted in placing the newly gathered examples between papers (without pressure) where they were left without being disturbed, until required." (Adequate technique requires removing the moisture of the plant by repeated changes of absorbent paper, while keeping the specimen under pressure so that it does not shrivel.) Even without the damage of storage Rafinesque's specimens were evidently none too good, as witness this short note from Dr. Charles Pickering to William Rawle about the quality of a bundle of Rafinesque's plants, presumably the ones sent directly to Zaccheus Collins from Kentucky in 1826: "It contains between two and three hundred specimens, mostly of little or no value. There are however a few tolerable specimens of good plants. It is difficult to fix a moneyed valuation upon such objects, but were I called on, I certainly should not estimate the whole collection at a higher price than five dollars." Rafinesque's bill against the Collins's estate charged the plants at 10 cents each, making these from $20 to $30. [Durand's letter about the miserable state of the Rafinesque plant specimens he bought is quoted more extensively later, by Croizat, pp. 180–81. **Ed.**]

85. [Two hundred seventy-five specimens were uncovered and discussed by Ronald L. Stuckey, "C. S. Rafinesque's North American Vascular Plants at the Academy of Natural Sciences of Philadelphia," *Brittonia* 23 (1971): 191–208. **Ed.**]

86. I have recently found more interesting specimens of Rafinesque in the herbarium of the State Teachers College at West Chester, Pa. These are part of a large series of plants that Durand gave Dr. William Darlington. Mostly they are from the Old World, and some bear Rafinesque's special "Autikon Botanikon" labels. [They number 303; the collection is described in William R. Overlease and Diane Rofini, "Rafinesque's Specimens in the Darlington Herbarium of West Chester University," *Bartonia* No. 53 (1987): 24–33. Apparently overlooked by botanists was the report by D. P. Penhallow long ago that the herbarium of Laval University contains "plusieurs plantes sont étiquetées de la main même de Nuttall et de Rafinesque," Sec. IV, *Proceedings and Transactions of the Royal Society of Canada* 2nd ser., 3 (1897): 3–56. **Ed.**]

87. As stated by S. S. Haldeman in *American Journal of Science* 42 (1842): 287. Dr. H. A. Pilsbry tells me that Mr. Poulson's specimens, including these, are in the conchological department of the academy. [These shells were first discussed in E. G. Vanetta, "Rafinesque's Types of Unio," *Proceedings of the Academy of Natural Sciences of Philadelphia* 67 (1915): 549–59. A comprehensive study of Rafinesque's and those of others is in Richard I. Johnson and H. Burrington Baker, "The Types of Unionacea (Mollusca: Bivalva) in the Academy of Natural Sciences of Philadelphia," *Proceedings of the Academy of Natural Sciences of Philadelphia* 125 (Dec. 27, 1973): 145–86. Finally, Richard I. Johnson also published "The Types of Unionidae (Mollusca: Bivalva) Described by C. S. Rafinesque in the Museum d'Histoire Naturelle, Paris," *Journal de Conchyliologie* 110 (1973): 35–37. **Ed.**]

Genealogical Note on the Rafinesque Family

GEORGES REYNAUD

OF THE FOUR GENERATIONS of Rafinesques we find during a period of 110 years dating back from 1783 to 1673, the last two were from Marseilles and the first two from Ganges. Although the older, Ganges (Department of Hérault), 24 km north of Montpellier, is uncertain as the place of residence, for it is not impossible that the cradle of this Protestant family could have been Saint-Hippolyte-du-Fort, still on the edge of the Cévennes mountain range, but 13 km farther to the east (Department of Gard). Indeed, the marriage contract of Pierre Rafinesque, great-grandfather of the naturalist, was signed before a notary of this village, where one notices, in addition, the traces of numerous Rafinesques. In particular, at the end of the sixteenth century, at the time of the wars of religion, there lived in Saint-Hippolyte a notorious Protestant, the "bailliff" Paul (or Pol) Raffinesque, Lord of Blaquière, the name of a place located in the district of Croix-Haute. But the seigniory passed afterward to his son-in-law Pierre de Nogarède, Squire of Valleraugue, which leads us to suppose that he did not have a male descendent. At the same time, another Raffinesque—named Louis and also carrying the title of Sieur of Blaquière (perhaps a brother of the former)—was owner of the farm of Malet. At the beginning of the seventeenth century, one finds in Saint-Hippolyte a Capt. Jean Rafinesque and a Capt. Louis Rafinesque, as well as six other Rafinesques (two Etiennes, two Jacqueses, a Jean, a Pierre) married and having children. On the other hand, in Ganges at the same time, there seems to have been only one Rafinesque family: that of the forebears of the naturalist. Its genealogy is assembled in the following:

Georges Reynaud

First generation

I. Antoine Rafinesque, shopkeeper in Ganges in 1673, husband of Isabeau Fabrègues, from whom came:

Second generation

II. Pierre Rafinesque, tradesman of Ganges, who contracted marriage on February 27, 1673, with Marie Magdeleine Poussielgue, of Sauve (22 km east of Ganges), daughter of the deceased Abraham Poussielgue and of Marie de Calvas. The wife received a dowry of 1,200 livres, the husband half of all the goods of her father, who kept the usufruct of it (Maître Lacroix, notary in St-Hippolyte). This division lets us suppose that she had a brother, not found. October 11, 1685, just a week before the revocation of the Edict of Nantes, Pierre Rafinesque and his wife "freely [sic] renounce [their Protestant faith] and profess the catholic, apostolic, and Roman religion at the hands of Msgr. the bishop of Montpellier in the church of Ganges," in company with 280 of their co-religionists (Registres de catholicité de Ganges). Marie-Magdeleine Poussielgue died before 1709 whereas Pierre Rafinesque was still alive in 1719, when he resided at Sauve. This couple had seven children, probably all born in Ganges, who follow.

Third generation

III.1. Pierre Rafinesque was born around 1680, became a merchant in Marseilles in 1714, and was groom in his first nuptials at Nimes with Isabeau Bruguière, daughter of David Bruguière and Françoise Calvas. The wife received 5,000 livres in dowry; the husband, who had only one younger brother, Simon (III.7), received half of all the goods of his father (contract of April 21, 1719, private agreement, transcribed by Maître Laurensy in Marseilles on March 30, 1720). Mizac Bousseirolle, brother-in-law of the husband, represented the father of the latter, who resided then at Sauve. Isabeau Bruguière died prematurely, without issue. On June 7, 1734, Pierre Rafinesque remarried; his new wife was Marie Reboul, daughter of François Reboul and of Elisabeth Lacoste (Marseilles, parish St-Martin; contract of November 2, 1734, before Maître Urtis). Jeanne Elisabeth Reboul, the sister of Marie, married Paul Goverts, a trader originally from Hamburg. After living on the Rue du Petit-Mazeau, the couple Rafinesque-Reboul settled on the Rue Coutellerie, still near the town hall; then, after 1750, on the Rue de Rome in the new district. Pierre Rafinesque directed, with his cousin Michel Calvas, a commercial firm that traded with Spain, Malta—where

Genealogical Note

his cousin Abraham Poussielgue resided (power of attorney, February 25, 1716, Maître Lombardon)—the levant, and even China. On November 25, 1745, with his nephew Pierre Bousseirolle, he stood guarantee for the commercial firm in Salonika of the Fougasse Brothers, which then took the name of "Rafinesque et Fougasse." Its prosperity made it possible to buy ships or shares of ships (in 1749, the *Virgin of Grace* renamed the *Minerva*, the *Flora* . . .) as well as a country house with land in the Plombières suburbs, having vines, trees, buildings, and a well. After making his will (on August 5, 1747, Maître Hazard), Pierre Rafinesque died on March 1, 1752. The following year, October 6, 1753, his wife followed him into the tomb (inventory of the goods of Pierre Rafinesque, October 8–22, 1753).

Between 1736 and 1745, Pierre Rafinesque and Marie Reboul had six children, who will follow (fourth generation, A).

III.2. Marie Magdeleine Rafinesque was born about 1680; she became the godmother of her brother Jean on November 7, 1688, in Ganges; married Pierre Afourtit, lawyer of Sauve, prior to 1714.

III.3. Dauphine Rafinesque was born on January 14, 1684 (Protestant Civil Register of Ganges), married Mizac Bousseirolle in Ganges on July 26, 1708. He was the son of another Mizac and Jeanne Deshons. In Ganges seven children were born to them, who will follow (fourth generation, B).

III.4. Jeanne Rafinesque was born on August 3, 1686 and baptized August 8 (Catholic Civil Register of Ganges).

III.5. Jean Rafinesque was born on November 3, 1688, and baptized November 7 (Catholic Civil Register of Ganges; his godfather was Pierre Rafinesque, his elder brother, his godmother was Marie Magdeleine Rafinesque, his sister). Deceased and buried on August 15, 1689.

III.6. Marie-Grâce Rafinesque was born on February 4, 1692, and baptized February 10 (Catholic Civil Register of Ganges); same godfather, Pierre; her godmother was Jeanne Caizergues, "daughter of the gardener of M. de Ganges."

III.7. Simon Rafinesque—the grandfather of the naturalist—was born on September 16, 1694, baptized September 18 (Catholic Civil Register of Ganges; his godfather was Simon Lemosin, his godmother Jeanne Louyse de Ganges).

Settled in Marseilles 1712; he married there, at the church St-Ferréol, on October 23, 1736. His wife was Anne Murat, who was born March 9, 1719 (St-Martin), and was the daughter of the deceased François Murat (died May 31, 1719, St-Martin) and Louise Couliette. The marriage contract was signed on November 27, 1736, "in the dwelling of Lady Couliette, located on the Rue de Rome." The wife received 7,000 livres in dowry, including 2,000 in kind ("clothes, linens, trousseau, furniture") and 5,000 in cash money (Maître Sibon). The Murats and Couliettes also were well-known Protestant merchants. The paternal grandfather of Anne Murat, François Murat, was a native of Montpellier, who had come to Marseilles in 1678, and there took on the office of pastor in 1684; he recanted his Protestant faith at St-Martin on November 2, 1685, after the revocation of the Edict of Nantes. His daughter Marguerite—the aunt of Anne Murat—was married on October 12, 1717, at St-Martin. Her husband was Isaac Tarteiron, originally from Ganges, like the Rafinesques, and was founder of a line of rich Marseilles merchants who moved to Bordeaux. The maternal grandfather of Anne Murat, David Couliette, a native of Rouen (or Rohan) had a son, in addition to Louise (the mother of Anne Murat). The son was also named David and served as a clandestine pastor in 1724 and "missed his creditors in 1731" (i.e., defaulted). The elder Couliette had a second daughter, Anne, who married Abraham Dolier in 1712; from this union were born Louis David Dolier, who married a Mlle Seimandy, and Anne Louise Dolier, who married Georges Audibert, another famous merchant. Simon Rafinesque was assigned on February 25, 1716, by his elder brother Pierre (III.1) to represent the firm Rafinesque and Calvas in Alicant, Cartagena, Almeria and "on all the coast of Spain and other places" (Maître Lombardon). Two years later, on April 7, 1718, he filed a request to reside at Salonika, undoubtedly to pursue commerce. He was then described as having "a good height, a face long and dark, black hair, wearing a wig." Simon Rafinesque died on February 6, 1770, and was buried the same day in the cemetery for "foreigners of the alleged reformed religion," or Protestants, who were once again tolerated. His wife, Anne Murat, survived him another twenty-nine years. She was the grandmother Constantin Samuel Rafinesque speaks about in his autobiography, with whom he lived for twelve years (at Marseilles and at Pisa) and who "died at very advanced age." Indeed, she died at age eighty on April 9, 1799, in the sumptuous mansion of George Audibert (himself having died January 21, 1798) at No. 24 Rue Noailles. Between 1738 and 1750, Simon Rafinesque and Anne Murat had 5 children, who follow (fourth generation, C).

Genealogical Note

Fourth Generation

A) Children of Pierre Rafinesque and Marie Reboul:

IV.1.1. Pierre Rafinesque was born on December 11, 1736, baptized December 12 in Accoules, and died on May 9, 1738, at the Rue du Petit-Mazeau.

IV.1.2. Jeanne Rafinesque was born on May 6, 1738, baptized the same day in Accoules, but died on July 31, 1739.

IV.1.3. Pierre Rafinesque was born on October 15, 1739, and baptized the same day in Accoules. His godfather was his cousin Pierre Bousseirolle. He probably died between August 5, 1747 (date where he is named in the will of his father) and October 8, 1753 (date of the inventory of the goods of the latter, where he does not appear any longer as an heir).

IV.1.4. François Rafinesque was born on July 26, 1741, baptized the same day at St-Martin, and was deceased on July 2, 1750, at the Rue de la Coutellerie.

IV.1.5. Magdeleine Rafinesque was born on January 8, 1744, and baptized January 9 at St-Martin. She was orphaned when nine years old and was the sole legatee of her parents; her guardian was Maître J.-B. Mounier, procurator in the magistrate's court of Marseilles. She married on June 10, 1756, (at the age of twelve and a half!) at St-Martin, Lord Henri-Joachim de Bastide, writer of the Navy, son of the deceased Jean Joachim de Bastide, counselor to the King, lieutenant general (criminal) of the Seneschal's court of Marseilles, and Lady Jeanne Thérèse Maurin. Henri-Joachim was the brother of Jean-François de Bastide (Marseilles, 1724–Milan, 1798), author of many novels, tales, and comedies and the grandnephew of the Abbé Simon Joseph Pellegrin (Marseilles, 1663–Paris, 1745), a poet and prolific but shallow playwright. Henri-Joachim de Bastide died without posterity in 1782, bequeathing to his wife two houses at N.-D.-du-Mont and two country houses at Capelette in the southern suburbs. One of the latter was sold on January 10, 1788, to Jean-Charles Rolland (Maître Porte-Tassy), the other to François Joseph Stanislas Deloutte, September 27, 1797 (Maître Cousinéry). But Magdeleine Rafinesque retained the two houses of N.-D.-du-Mont, where she died on April 13, 1808. The houses were then bequeathed to her niece Julie de Bastide, wife of Pigale de Marilly. After her marriage, the links appear to have been severed between Magdeleine Rafinesque and her paternal family.

IV.1.6. Simon Rafinesque was born on December 19, 1745, baptized the 20th at St-Martin, his godfather being his uncle Simon Rafinesque (III.7), his godmother being his cousin Louise Rafinesque (IV.7.1). He was deceased on June 26, 1746, at the Rue de la Coutellerie.

B) Children of Dauphine Rafinesque and Mizac Bousseirolle:

IV.3.1. Jeanne Bousseirolle was born on May 6, 1709, in Ganges.

IV.3.2. Pierre Bousseirolle was born on June 25, 1710, as inscribed in the will of his uncle Pierre Rafinesque (III.1) in 1747.

IV.3.3. Elisabeth Bousseirolle was born on October 25, 1711, as inscribed in the same will.

IV.3.4. Simon Bousseirolle was born on December 21, 1712, and was the godson of his uncle Simon Rafinesque (III.7).

IV.3.5. Magdeleine Dauphine Bousseirolle was born on March 8, 1714, and was the goddaughter of her aunt Marie-Magdeleine Rafinesque (III.2) married to Afourtit.

IV.3.6. Suzanne Bousseirolle was born on February 18, 1715, and was the goddaughter of her aunt Jeanne Rafinesque (III.4).

IV.3.7. Anne-Grâcie Bousseirolle was born on June 6, 1716, and was the goddaughter of her aunt Marie-Grâce Rafinesque (III.6), as inscribed in the will of her uncle Pierre Rafinesque (III.1) in 1747.

C) Children of Simon Rafinesque and Anne Murat:

IV.7.1. Louise Rafinesque was born on January 2, 1738, and baptized January 3 at St-Ferréol. Her godfather was her cousin Pierre Bousseirolle (IV.3.2), her godmother was Louise Couliette, her grandmother. Died unmarried in Marseilles on August 27, 1783, she was buried the same day in the Protestant cemetery.

IV.7.2. Louis Simon Rafinesque was born about 1739. He was a merchant associated from 1779 with Simon Laflèche in the firm "Laflèche et Rafinesque," specializing in the production and export of wines. In return the firm imported various goods of the Orient, in particular Persian carpets the arrival of which were announced in the *Journal de Provence*. Although it was a limited partnership without large means, the firm managed to fit out the old ship the *Saint-Charles*, bought for 51,000 livres from Dominique Audibert and renamed the *Argonaute* for a trip to China in 1791. The younger brother of Louis Simon embarked on this ship, as a supercargo, which will know many adventures (cf. François George Anne Rafinesque, IV.7.5).

Simon Laflèche, having become first alderman at the beginning of 1789, after March 23 would in fact exercise municipal authority, since the mayor and city's assessor, the Marquis de Gaillard, had to withdraw from the position under popular pressure. But on the night of the following August 19, after a day of rioting near the harbor in Tourette, the mob went to Laflèche's house, Rue Noailles, and plundered it from bottom to top. This event caused his departure for Genoa. Louis Simon Rafinesque remained in Marseilles, where he became part of the citizen militia as a lieutenant to Capt. Jacques Seimandy (55th company), and then became the legal representative of Laflèche. On March 29, 1791, he bought at auction a house from the district "formerly belonging to the convent of the Feuillants" and then was occupied fitting out the *Argonaute* (Maître Bonsignour). On November 12, 1793, he declared having an income from the "capital of 120,000 livres in the company 'Laflèche et Rafinesque.'" Since he was unmarried but had responsibility for his mother (Anne Murat, who lived with him then on the Cours des Capucines), he reckoned to provide as a patriotic contribution the sum of 5,590 livres. A few days later, "although seldom professing an opinion and never having appeared in the districts of the Revolution," Louis Simon Rafinesque ("sensual, prudent and taciturn," according to Lautard), was denounced as an associate of the outlaw alderman Laflèche; "arrested in his house and thrown into a dungeon of the law courts, he falls ill there from exhaustion and grief and is sent to the hospital, where he dies tied down on a mean pallet." Although his death certificate was silent on these circumstances, it confirmed his death at the Hôtel-Dieu on January 19, 1794, at 4 o'clock in the afternoon; that is to say, exactly seventeen days after the execution on the scaffold of Laurent François Tarteiron and Jacques Seimandy, the cousins of his mother.

IV.7.3. Pierre Rafinesque was born on February 6, 1746 and baptized the same day at St-Ferréol; his godfather was Pierre Rafinesque, his uncle; his godmother was Marie Reboul, his aunt. Deceased at the Rue de Rome on April 27, 1750; buried April 28 in the church St-Ferréol.

IV.7.4. Anne Louise Rafinesque was born on April 14, 1748, and baptized the same day at St-Ferréol. Married to Antoine Desmaretz, an English merchant; the couple was living in Leghorn in 1789, the date of the visit of Constantin Samuel Rafinesque (nephew of Anne Louise) accompanied by his parents.

IV.7.5. François George Anne Rafinesque—the father of the naturalist—was born on April 14, 1750, and baptized the same day at St-Ferréol; his godfather was Hugues Moulin and his godmother Marie Ramel. On May 3, 1770, having gained at age twenty the estate of his father, who has just died, he requested authorization to reside in Constantinople as an assistant of the firm Amic, Laflèche et Cie, which afterward became Laflèche et Rafinesque. At the Dutch church in Pera (a suburb of Constantinople) he married on November 29, 1781, Magdeleine Schmaltz, who was born in Constantinople in 1767. She was the daughter of the deceased Benjamin Schmaltz and Marianne Spitzer (transcription of October 11, 1784, in the Protestant Civil Register of Marseilles). The couple settled in Marseilles after the birth of their first child—Constantin Samuel—who told in his autobiography the rest of the history: the return of the father to Constantinople for two years, his own stay with his mother in a country house in the outskirts of Marseilles, the birth of his brother, and the trip to Leghorn in 1789. In February 1791, François George Anne Rafinesque was present at the baptism of his daughter Georgette Louise. In June, he attended as a witness the baptism of Constantin Hayes, son of an English trader of Smyrna visiting in Marseilles. On August 28, accompanied by the navigator Capt. Pierre Blancard (1741–1826), famous for having introduced the chrysanthemum into Europe in 1789, he left this city aboard the frigate *Argonaute* under the command of Capt. Victor Chabert. The ship rounded the Cape of Good Hope on November 29, reached Mauritius on January 2, 1792, and remained there for five months. Then it skirted the Malabar Coast in mid-June, passed through the Sunda Strait by Christmas 1792, and reached China. On the return, off Cape Verde on April 17, 1793, having learned that France was at war with the principal European powers and that many French vessels had been captured, Blancard and Chabert decided to set their course for Philadelphia. Less than a month later, on May 14, the *Argonaute* lay at anchor in the port of Pennsylvania, where it was unloaded and sold, along with all the cargo. François George Anne Rafinesque contracted the yellow fever and died in Philadelphia late in the spring of 1793.

His widow, Magdeleine Schmaltz, a refugee in Leghorn since the late summer of 1792, remarried a few years later (before 1800), and her new husband was the merchant Pierre Lanthois. They went back to France in 1805 and settled in

Bordeaux, where they were located in 1827. Magdeleine Schmaltz died in that city, at No. 42 Rue de la Trésorerie, on September 8, 1831.

From her union with François George Rafinesque, she had three children, who follow.

Fig. 2.1. Rafinesque's mother sketched from memory by her son as she appeared when he saw her for the last time in 1805. There are no likenesses of his father, his brother, or his daughter. Courtesy of Transylvania University Library.

Georges Reynaud

Fifth Generation

V.7.5.1. Constantin Samuel Rafinesque, the naturalist who is the subject of this work, was born on October 22, 1783, in Galata, a suburb of Constantinople, and was baptized the following November 19, "presented for baptism by Jean Arland on behalf of M. Samuel Benoît Boruman, Dutch trader of this town, and by the widowed Mme. Marianne Schmaltz" (transcription of October 11, 1784, in Marseilles). Unable to marry because he was a Protestant and she a Roman Catholic, he lived with Josephine Vaccaro in Palermo, Sicily, in 1809. Believing him lost in a shipwreck, she married the actor Giovanni Pizzarrone in 1815 or a little afterwards.

Having first taken an oath to the Constitution of the United States on January 26, 1816, Constantin Samuel Rafinesque chose permanent American nationality on October 5, 1832, in Philadelphia. Less than a year later, May 1, 1833, he wrote his will in which he first of all entrusted "his soul to the Creator, the ruler of millions of worlds." He wished to be cremated rather than buried (so as not "to contaminate the earth by decay") and that his ashes be "deposited in a[n] urn" and be preserved with his collections. He forgave all his enemies and those who had stolen his property and asked pardon in return of all those who might have something to reproach him with. He bequeathed his goods (scientific collections, books, patents, secrets, and claims) to his sister Georgette Louise and his daughter, Emilie Louise, on the condition that the latter separate from her "unworthy" mother. The will indicated that all his collections—except for his own drawings and manuscripts—must be sold in America or in Europe (the Natural History Museum of Paris having priority) to finance the printing of the manuscripts, drawings, sketches, and maps unpublished at the time of his death. This publication was entrusted to the care of his nephew Jules Rafinesque or, failing this, to Professors Torrey of New York and Green of Philadelphia. The will further directed that if the sale of these collections returned more than 10,000 dollars (or 50,000 francs), the excess should be placed in a bank to the benefit of the first school for female orphans that will be founded in the United States on the model of the school for male orphans of Stephen Girard.

He bequeathed to his nephew Jules the gold medal that was awarded to him by the Société de Géographie and to his niece Laure Rafinesque a necklace of the value of 100 francs as a wedding present. Finally, if his daughter Emilie should die before him, her share of inheritance should go to his nephew Jules whom he asked "to grant something to Henrietta Whinston, daughter of Emilie and of Sir Henry Whinston." Having learned of the death of his sister Georgette Louise at Bordeaux in 1834, Constantin Samuel wrote on June 15, 1835, a codicil to substitute

Genealogical Note

his nephew and his niece for her. He died on September 18, 1840, and was buried the following day in Ronaldson's Cemetery on Bainbridge Street.

Josephine Vaccaro gave two children to the naturalist:

VI.7.5.1.1. Emilie Louise Rafinesque was born in 1811 in Palermo and remained there with her mother after 1815. Raised by her stepfather, G. Pizzarrone, she became second actress with the Phoenix Theater of Naples in 1825. During a tour in Rome, she became enamored of an Englishman— Henry Whinston—who left her before the birth of their daughter Henrietta (Enrichetta) in 1828; she traveled then with her troupe to Malta, then Tunis—where she took up a correspondence with her father—and Leghorn and Bari. In 1834, Constantin Samuel having negotiated for her a contract with the Italian Opera of New York, she was on the point of joining him but finally gave up this project.

VI.7.5.1.2. Charles Linné Rafinesque was born in 1814 and deceased in 1815 in Palermo.

V.7.5.2. Antoine Simon Auguste Rafinesque, brother of the naturalist, was born in Marseilles on April 24, 1785, and baptized the following May 3 (Protestant Civil Register); his godfather was Antoine Desmaretz, an English merchant of Leghorn, his godmother was Anne Murat. He accompanied his elder brother on his first voyage to the United States (1802–5) then to Sicily, where he married on November 18, 1815, in Palermo, Elisabeth Charlotte Amable Merelle de Joigny, born in Paris in 1796, daughter of Jean-Baptiste, pensioner of the State, and his wife, Amable Guillemant (transcription in Marseilles on December 30, 1826).

Returning to France after the Restoration, he became a stockbroker in Le Havre; he died, in that city, at the Rue de l'Hôpital, June 27, 1826. The two children he had in his marriage to Mlle de Joigny are the heirs designated by Constantin Samuel, to wit:

VI.7.5.2.1. Jules Félix Hermann Rafinesque was born in Naples, October 25, 1816, deceased in Paris, June 21, 1886, and his lineage is still continued today.

VI.7.5.2.2. Laure Amable Rafinesque, wife of a man named Démirgian, she was born in Paris in 1818 and died near Jerusalem on April 24, 1867; her lineage also continues today.

V.7.5.3. Georgette Louise Rafinesque, sister of the naturalist, was born in Marseilles on February 28, 1791, baptized March 1; her godmother was Louise Audibert (born Dolier), her godfather Georges Audibert. On February 25, 1827, she married in Bordeaux Paul Lanthois, a stockbroker, born September 27, 1791, in Castelnau (Tarn), son of Paul Lanthois, former merchant, and of the late Jeanne Mialhe. During the last years of Georgette's life, she kept up a regular correspondence with her brother Constantin Samuel, who sent her books to be translated into French (one was indeed by a certain Dr. Lanthois, undoubtedly a relative) and even a sample of his "Pulmel," the anti-tuberculosis drug that he manufactured in Philadelphia. Without leaving live children, she died in childbirth at 21 Rue Poitevine in Bordeaux, August 28, 1834.

References

Acts of Immigration and Naturalization. National Archives of the United States.

Arbre généalogique de la famille du Dr. Jean Rafinesque.

Artefeuil, 1757–86. *Histoire héroïque de la noblesse de Provence*, vol. 2: 545, art. Bastide. Avignon: Girard.

Auction Sales (series A). Wills (series W). Archives of the City of Philadelphia.

B 7, B 16, J 119, J 122. Archives de la Chambre de Commerce de Marseille.

Beaugeard, F. *Journal de Provence*, n° du 14 décembre 1784. Aix: Collection du Musée P. Arbaud.

Blancard, P. 1910. Journal de bord. *In* Les voyages de P. Blancard, navigateur marseillais, importateur du chrysanthème, ed. J. Fournier. *Bull. Soc. Géogr. Marseille* 34: 217–24.

Bourrilly, V.-L. 1956. *Les protestants de Provence aux XVIIe et XVIIIe siècles*. Aix. Pp. 68, 69, 71, 180.

Carrière, C. 1973. *Négociants marseillais au XVIIIe siècle*, vol. 2. Bibl. de l'Institut Historique de Provence. Pp. 721, 731, 798, 901, 903, 933, 936.

Coullaut, P. 1961. *Si Dieu ne bâtit ma maison, histoire de l'Eglise réformée de Marseille*. Marseille. Pp. 48, 67–69.

État civil (XIXe siècle). Archives municipales du Havre.

État civil ancien catholique et protestant (XVIIIe siècle; série GG). État civil moderne (XIXe siècle, série E). États de sections (21 G 25 à 33). Contribution patriotique (2 G 26 à 33). Archives municipales de Marseille.

État civil de Bordeaux (XIXe siècle). Archives de la Gironde.

État civil protestant de Saint-Hippolyte-du-Fort (XVIe et XVIIe siècles). Minutes notariales de St-Hippolyte et de Sauve. Archives du Gard.

État civil protestant et catholique de Ganges (XVIIe et XVIIIe siècles). Minutes notariales de Ganges (XVIIe et XVIIIe siècles). Archives de l'Hérault.

Fassy, R. 1955. Un constructeur de navires ciotaden: Antoine Fassy. *Provence Historique* 6: 142.

Genealogical Note

Ferran, J.-P. 1951. *La haute bourgeoisie protestante marseillaise à la veille de la Révolution*, D.E.S., Fac. Lettres, Aix.

Fournier, J., et Latune, C. 1920. *Un grand négociant négociant d'autrefois, Dominique Audibert*. Marseille. Cf. notamment la *Note historique et généalogique sur la famille Audibert* rédigée par Me Latune. Pp. 27–33.

Lautard, L. 1844. *Esquisses historiques. Marseille depuis 1789 jusqu'en 1815 par un vieux Marseillais*, vol. 1. Marseille. Pp. 46n., 54–56, 152, 368–69.

Malzac, L. 1922. Le deuxième cimetière protestant de Marseille. *Provincia* 2: 126.

Minutes notariales de Marseille (série E). Inventaires (série II B). Archives des Bouches-du-Rhône.

Pennell, F. W. 1942. The Life and Work of Rafinesque. *Transylvania College Bulletin* 15: 10–70.

Peyriat, A. 1951. *Histoire de l'église réformée de Saint-Hippolyte-du-Fort*, Thèse de Doctorat-ès-Lettres, Aix (1951).

Rafinesque, C. S. 1836 [1944]. *A Life of Travels and Researches in North America and South Europe*. Philadelphia. Reprinted *in* Chronica Botanica 8: 291–360.

Register of Ronaldson's Cemetery. Genealogical Society of Pennsylvania, Philadelphia.

Sicard, G. 1975. *Le chrysanthème*, «Massalia» du 3 nov. 1945, repris in *Marseille*, 3e série, 103: 79–80.

Vialla, S. 1910. *Marseille révolutionnaire. L'armée-nation (1789–1793)*. Chapelot : Paris. Pp. 7, 26–27, 477.

The Last Days of Rafinesque

CHARLES BOEWE

I

AS PENNELL HAS NOTED of Rafinesque's activities, the spate of publishing beginning in 1836 and continuing almost up to the point of his death implies that Rafinesque's financial position had taken a turn for the better, since all of these titles were issued by job printers at the expense of the author. Another indication of new-found affluence comes from Rafinesque's ability and willingness to pay cash for plant specimens needed as vouchers for the descriptions he was writing during this period for his botanical publications.[1] Perhaps the best evidence of increased prosperity derives from the way he upgraded his residence between 1833 and 1837. In 1833 he was living "a few doors above Arch Street, East side, next door to a large new Presby[t] Church, there is *Mrs. Howell* on the . . . door."[2] He told John Torrey he could be found there, but added, "I do not keep house nor even board at my lodgings." Only five months later, however, he informed Torrey that he had moved to what sounds like more spacious quarters. "I have now an office & rooms in North 10th street—N 110 near Race Street."[3] And three years after that he moved to 172 Vine Street ("first door above 5th"), where he continued to live for the remaining three years of his life. Here he said he had "a whole house to myself," and, aside from whatever profits might have remained from the sale of Pulmel or be coming in presently from the Divitial Bank, he hinted at another source of wealth when he remarked that "I have at last found a Wise and Wealthy man, who will help me to found a *New Institution of Knowledge* on a liberal

The Last Days of Rafinesque

plan, to be called the *Eleutherium of Knowledge,* to which is to be annexed the *Institute of learning of N. Am[erica]*—& a *Mutual Library*—We are looking out for a location."[4]

The new patron was a retired manufacturing chemist, Charles Wetherill (1798–1838), though Rafinesque chose not to name him in his letter to Torrey. With Wetherill's backing, even the Panic of 1837—the second great American depression Rafinesque had experienced—did little to slow his public and private activities. Indeed, Rafinesque had so many irons in the fire in 1837 that this is one of the thinnest years for his extant letters.

II

Not much is known about Charles Wetherill. He came from an old Philadelphia family; his grandfather, Samuel Wetherill (1736–1816)—already the fourth generation of Wetherills in America—taught himself to be a chemist of sorts in order to make dyes needed for the cloth he manufactured to clothe Washington's soldiers at Valley Forge. Read out of the Society of Friends because he supported military action to oppose British tyranny, he helped to found the Free, or Fighting, Quakers. He figures prominently in the novel *Hugh Wynne, Free Quaker,* by S. Weir Mitchell, who called him "one of the most public-spirited and brilliant men of his time." Chemistry became the hereditary profession of the Wetherill family, and by 1830 three of Samuel's descendants, the brothers Charles, John, and William Wetherill, were managing a prosperous pharmacy on North Second Street in Philadelphia. At a time when pharmacology depended so largely on botanical preparations, this would have been a likely meeting place for Rafinesque and the Wetherills. In separate establishments the brothers also had a chemical laboratory, an oil of vitriol works, a lead pipe factory, and, most remunerative of all, a white lead factory that produced the pigment for paint.

Work was divided up among the three, and to Charles fell the task of managing the very successful factory that produced and ground white lead. Though no one at the time would have recognized the hazard, his having to breathe in a lead compound through the dust in the factory made him too ill to continue by the time he reached the age of thirty-eight. On January 1, 1837, the brothers appraised the value of their holdings at $180,320, of which Charles presumably was entitled to a third on his retirement. With more than $60,000 to draw against, he had the resources to make a mark in the world. In Rafinesque he had a friend eager to help him.[5]

Charles Wetherill aspired to be remembered by posterity as a cosmologist. Since 1835 he had been working on what he called "the electric theory of the solar system"; meanwhile, perhaps to prepare the world for a revolution in astronomical theory, he reprinted—with Rafinesque's job printer, H. Probasco—an eighteenth-century work "to demonstrate that I am not singular in my views, but have even the support of a very learned English Author." The book was the 1750 *Original Theory or New Hypothesis of the Universe*, by Thomas Wright of Durham, which is known, when known at all, as a precursor of and possible influence on Immanuel Kant's conception of the cosmos. With eleven pages of notes provided by Rafinesque, the reprint was issued in 1836 under the title *The Universe and the Stars* as "printed for Charles Wetherill."

Wetherill did not live to complete his own book, but annotating Thomas Wright brought Rafinesque's mind back to a science he had neglected since 1820, when he published, in Lexington's *Western Review and Miscellaneous Magazine*, an essay titled "Enquiries on the Galaxy or Milky-Way." The result now was a small book of 135 pages on *Celestial Wonders and Philosophy*, which was printed in 1838 "for the Central University of Illinois." Rafinesque claimed originality only for the metaphysical implications of the "celestial religion" based on his summaries of the discoveries of a group of eminent astronomers, all of whom are credited in his book. Inconsequential as the book may be (and even Rafinesque apologized for its having been "hastily written in a single week"), what cries out for explanation is how it could have been the first publication of the University of Illinois, which did not come into existence until 1867, twenty-nine years later.[6]

The anomaly is explained by an eight-page pamphlet[7] issued June 1, 1837, by the Philadelphia Land Company of Aurora, whose office happened to share space with the Divitial Institution of North America in Rafinesque's residence at 172 Vine Street and whose president was Charles Wetherill, its secretary C. S. Rafinesque. Account books had been opened at that office for the public to subscribe shares at sixty dollars each. For his sixty dollars a subscriber was assured a farm of forty acres, a city lot of one-fourth acre, and a town lot of one acre. Although at this date the company was unsure of the exact location of its proposed colony, its organizers had firmly in mind what they intended to build there.[8]

As soon as $96,000 had been collected, the company expected to purchase 100 square miles or 64,000 acres, which would yield 1,600 forty-acre farms and provide space for Agathopolis, a shining new city surrounded by four satellite towns named Industry, Honesty, Benevolence, and Tolerance. The company promised to provide free public schools in each of the towns and a free college in Agathopolis. Later the paper college was enhanced and became a paper university, now envisioned to consist of five colleges: a college of agriculture and labor, a college of arts and sciences, a college of teachers and languages, a college of medicine,

and—as the fifth college—an "eleutherium," an institution to provide free education to adults on a variety of subjects; in short, what a later age would call a university's extension service. All this was shaping up very much to resemble the land-grant college plan actually devised in the 1857 legislation proposed by Sen. Justin Smith Morrill two decades later.

In fact, it was the linkage of education with available cheap land that inclined the Aurora Company to favor Illinois. The Aurora pamphlet ends by remarking that shareholders would meet July 10 to decide whether to buy land in Ohio, Indiana, Illinois, Michigan, or Wisconsin. We do not know when they did decide, but the company's interests must have been widely known because John Plumbe, Jr., a land speculator in Dubuque—then still part of Wisconsin Territory—wrote in the local newspaper that he was trying to get Professor Rafinesque to "examine the merits of Wisconsin."[9] However, in Philadelphia just then at least three speculators were pushing the sale of Illinois real estate, which they had acquired at bargain prices from insolvent soldiers who had received bounty lands for their service in the War of 1812 but had greater need for cash. Lying in the center of the state, these "bounty lands" probably were what the Aurora Company decided on, for Rafinesque told A. P. de Candolle[10] that he was going to deposit specimens of his new plant species in the museum of a university to be erected in Bloomington. And he had told Torrey[11] earlier that he planned to go to Illinois during the summer of 1838 to lay out a city and five colleges. But he never went, probably because of the death of Wetherill some time that year.

To encourage education, each state that was carved out of the old Northwest Territory was supposed to apply the proceeds from the sale of every section sixteen (one square mile) of the thirty-six sections in each township to purposes of education. And Illinois was especially blessed; unlike any of the other states, its enabling act provided, in addition to the section-sixteen benefit, that 3 percent of the proceeds from the sale of all public lands should go "for the encouragement of learning, of which one-sixth part shall be exclusively bestowed on a college or university."[12] No doubt this provision also encouraged the Aurora Company to favor Illinois. Rafinesque and his friends may not have known, however, that the state legislature had already borrowed from the college fund to avoid raising taxes, and that dating from 1833, when the first bill was introduced to fund a state university, the populist and ill-educated legislators had consistently rejected the idea. One Illinois legislator boasted that he had been "born in a briar thicket, rocked in a hog trough, and had never had his genius cramped by the pestilential air of a college."[13]

It is unlikely that during the depths of the depression many Philadelphians wished to invest even $60 in Illinois land, though payments could be made in quarterly installments. Yet, at the date of the Aurora pamphlet, "$16,000 were subscribed at the outset by the promoters of this Colony," surely much, if not all of it, by Charles Wetherill. It is unknown whether any of this subscription had been paid in before the scheme failed and, if so, whether any funds actually were invested in land. If they were, they were dissipated like other speculative investments in land that had helped to bring on the financial panic in the first place. Gen. Asahel Gridley reminisced about selling $20,000 to $30,000 worth of Bloomington town lots to Philadelphians in 1836, some going as high as $150 each. "After the crash came, they would not bring over $5 apiece," he said, and "the parties in the East let them go for taxes."[14]

After Wetherill's death something substantial must have remained of the Aurora assets, because the Eleutherium of Knowledge continued to publish Rafinesque's writings. The first of these titles was another that harked back to his interests in Kentucky, when he had reviewed *An Easy Entrance into the Sacred Language, Being a Concise Hebrew Grammar without Points*,[15] by Martin Ruter, with whom he shared the platform when both were awarded the honorary M.A. degree at the 1823 commencement of Transylvania University. Now, published in 1838 for both the Eleutherium and the Central University of Illinois, Rafinesque's 259-page book on the *Genius and Spirit of the Hebrew Bible* made use of Ruter's guidance and drew on scores of other authorities, principal among them, according to the author, the French Fabre d'Olivet (1767–1825) and the German Johann Gottfried von Herder (1744–1803). The book investigates some of the cipher meanings of Hebrew roots developed in the Cabala (in keeping with Rafinesque's continued fascination with secret codes), in general supports the deistic "celestial religion" advocated in *The World or Instability*, and—what is perhaps most of interest—elaborates a method for transliterating Hebrew by means of roman letters printed from left to right, a system for which Rafinesque claimed originality.

The next year, 1839, two bulky pamphlets were published by the Eleutherium, but their title pages are silent about the Central University of Illinois. The first of these was the *American Manual of the Mulberry Trees*, written as a kind of companion volume to Rafinesque's *American Manual of the Grape Vines* (1830), the first intended to encourage the making of silk, the second the making of wine. To these he expected to add a third, to encourage the making of sugar from beet roots and maple sap; but it appears that this task was never accomplished. The second Rafinesque pamphlet published by the Eleutherium in 1839 was his *Improvements of Universities, Colleges, and Other Seats of Learning*, a long essay he had prepared in 1833 for the *Atlantic Journal*, which never had space enough to include it. Written before the Central University of Illinois had been conceived, this pamphlet does tell us something about what Rafinesque would have aimed for had he been able to start that university. He opposed charging tuition and believed the total cost of higher education was a responsibility of the state. He advocated the use of competitive examinations for the selection of professors. He was not so radical as to suggest coeducation, but he did proclaim the "absolute need of female Colleges as auxiliaries to the Universities, in order to teach *one half of mankind*" (17), and he argued that boys would benefit from being taught by learned women as well as by men. Finally, he proposed to scrap the fixed curriculum in favor of allowing students, in consultation with their advisors, to select their own course of studies, among which would be innovative courses in such areas as agriculture. In more ways than one, Rafinesque truly was in advance of his time, as he himself so often claimed.

The last publication known to come from Rafinesque's pen, *The Good Book, and Amenities of Nature,* also was "Printed for the Eleutherium of Knowledge." Released in January 1840, this was intended to be "issued quarterly or semi-annually" and was the usual *omnium gatherum* of essays on Rafinesque's various interests. According to the introduction, one of its features would be the inclusion of letters by "eminent men, or writers on historical and natural sciences." Among those named, no other letters have been found written to Rafinesque by Adriano Balbi, A. P. de Candolle, Stephen Elliott, William Elford Leach, Johann Jacob Roemer, and Edouard de Verneuil. For this reason alone it is to be regretted that only one number of *The Good Book* ever appeared; that number did not contain any of the letters of his correspondents.

In the letters which *are* extant and date from the last few years of Rafinesque's life, much discussion is devoted to the New York State Natural History Survey, which was conducted between 1836 and 1843. Even while telling Torrey that he did not have the slightest notion of how to apply to become a member of the survey, Rafinesque was writing letters of application to New York's governor; and Torrey himself, when he got the post of botanist for the survey, turned to his friend Asa Gray, not Rafinesque, for assistance. Meanwhile, despite all he had going with Wetherill in their attempt to found a colony and a university and, as far as we know, the continuing obligations of the Divitial Bank, Rafinesque had time to compose a four-page broadside outlining the benefits of natural history surveys and explaining how they ought to be conducted. The one example known is titled *First Scientific Circular for 1837*—though another, never found, was issued in 1838—and addressed to all the governors, lieutenant governors, and speakers of the legislatures of all the states and territories of the United States.

With or without public patronage, Rafinesque continued his scientific and historical studies with remarkable vigor. Throughout 1839 he pursued his intellectual disputes with John Torrey in their exchange of letters. Despite their earlier differences, Rafinesque also now resumed his correspondence with William

Swainson in the form of long professional letters. With John Howard Payne he schemed to find commercial use for some of the Indian lore he had accumulated. But, knowing as we do that the end was not far off for him, we seek clues in his letters for any premonition he himself might have had.

A petty quarrel over borrowed books exhibited in his letters to William Wagner in the spring of 1840 suggests an unusually testy temperament, but it can be matched by his earlier irascibility. He must have believed he would live long enough to enjoy the pension he petitioned for in a letter of June 1, 1840, to Louis Philippe, King of the French. It is not until the last letter known to have come from his hand, that of August 5, 1840, to Elias Durand, that there is any hint of the cancer which was to cause his death two months later. There he mentioned having been "unwell for 2 months of a dyspeptic complaint"—unaware that it was cancer of the stomach—which he expected to relieve by a visit to the shore. We have no way of knowing whether he went there or not, but a month later fellow botanist John Leonard Riddell, then visiting Philadelphia, jotted in his diary that he had called on Rafinesque and found him "sick a bed. Fear he will not recover."[16] This was only two weeks before the naturalist's death. It is likely that he never left his bed after that, for by September 11 he was complaining of a painful bedsore on his left side. A week later he was dead.

III

Even today, the best-known "facts" of Rafinesque's life concern his leaving of it. The story began thirty-seven years later, and it is worth examining as the embryology of successful myth making. As Pennell has told us, a person signing himself "H. H.," and never otherwise identified, published in the *Philadelphia Ledger* (May 5, 1877) a letter to the editor "to revive the memory" of C. S. Rafinesque who had "selected a garret for his labors, and abode in Race Street between Third and Fourth." Something about the words "garret" and "Race Street"—neither of which is true—captures the imagination of nearly everyone who has written about Rafinesque since. Almost nothing in the account of H. H. is correct, including his assertion that Dr. James Mease and Mr. Bringhurst, the undertaker, had to lower the corpse by ropes to the back yard and sneak it away from creditors. But H. H. sounds plausible, and he claims that Mr. Bringhurst himself "assured me of these facts relating to the burial." The ever-charitable Francis Pennell thought that if Mr. Bringhurst said so—founder of the firm of undertakers still prominent in Philadelphia—then the story "seems likely true, at least in essentials."

After the passage of another six years, Thomas Meehan—botanist, publisher, seedsman—took up the story: "When I was a young man in Philadelphia, thirty years ago, some of Rafinesque's contemporaries were still living. His chief home was here, and here in a dingy garret, with scarcely a loaf of bread to eat, he worked for science, as he understood it, to the last. He died on a cot with hardly a rag to cover him, and without a solitary friend to stand by him in his last hours."[17] After another eight years, when he published Rafinesque's will that had been recovered from the city archives, Meehan's heated imagination pictured the death scene—again in a garret on Race Street—where the naturalist died "with nothing but a hard cot and pillow for furniture, and no living soul to close his eyes."[18]

Meehan could not understand why the body had been let down with ropes, but the biographer Richard Ellsworth Call, picking up where Meehan left off, explained that the "landlord refused permission of burial; he hoped to find a market for the body in a medical school, and thus obtain the rental Rafinesque could not pay when living," though Call's only cited source for this additional insight was Meehan's own article, where nothing at all like this appears. Call, too, was firm in his conviction that Rafinesque had "lived in the most abject poverty on Race Street, Philadelphia, in a garret." For Call, and for his successors aiming to rehabilitate the reputation of Rafinesque, it was essential that the neglected and despised genius live in poverty. Except as the result of a religious vow, poverty seldom is seen as elevating, and in Rafinesque's case "abject" always accompanies asseverations of his poverty to assure us that he was humiliated by it. It was also essential to the myth that he live out his miserable existence in a garret, though that place of residence sounds less wretched and more bohemian when expressed in French as "dans une mansarde," as it is in the untranslated original of Reynaud's genealogy. Some writers have declared, entirely gratuitously, that Rafinesque was buried in a potter's field. Call concluded that "he died without a word to cheer him," adding for additional pathos, "without a tear shed for him."[19]

Now, as far as the truth can be ascertained at so great a remove from the events, let us see what actually happened just before, and after, the death of Rafinesque.

IV

Readers of S. S. Haldeman's obituary notice of Rafinesque's zoological writings learned from it that their author had died of cancer of the stomach and liver. More than a century passed before it was learned that Haldeman had written with the authority of an autopsy to guide him. This information was brought to light by Francis Pennell in 1949—after his Transylvania essay had been published—in a short article pointing out where the printed autopsy report could be found. Apparently it did not occur to Pennell that the remains of few paupers living in garrets were autopsied in 1840. Nor, for that matter, were paupers likely to be visited daily by one and sometimes two physicians during their illness.

In the published "Case of Cancer of the Stomach and Liver,"[20] Dr. Edward Hallowell (1809–1860), who wrote the report, says that he was called for consultation on the case by Dr. William Ashmead (1801–1888). Doctors Hallowell and Ashmead were both highly regarded practitioners, both of them medical graduates of the University of Pennsylvania, both of them members of the College of Physicians of Philadelphia. They may have been personally acquainted with Rafinesque as well, because both also were members of the Academy of Natural Sciences, as he was. Dr. James Mease is nowhere mentioned in the autopsy report, though we know that he did serve afterward as executor of Rafinesque's estate. Because he had long been a personal friend of Rafinesque, it may be that Mease preferred having another physician take charge of the case, although, as we shall see, he was a frequent visitor to the sickroom.

Be that as it may, Dr. Hallowell began his report by stating that the patient "had throughout life enjoyed almost uninterrupted health until the winter of 1839–40," when he began to experience severe constipation, which continued into spring. On June 15 he first had nausea, with occasional vomiting. It was not until early September that the patient "observed, for the first time, a tumor in the right hypochondriac region, when he also noticed yellowness of the skin; since then he has complained of much debility, and for the last week has been confined almost entirely to his room." It may be inferred that Dr. Ashmead had been attending Rafinesque for some time, for only on September 10 had he called in his colleague, Dr. Hallowell, who then comments on the patient's condition on each of his almost daily visits. Though junior to Ashmead by eight years, Hallowell specialized in pathology and may have had more experience with this sort of disease, because it was he who prescribed a light diet and tried a "Hydragogue" to extract water or serum from the patient's distended abdomen. His only other prescription on the first visit was hyoscyamine, a powerful sedative, which he then continued daily. On the first visit, too, he measured the five-and-a-half-by-four-inch protruding tumor and noted the decubitus on the patient's back resulting from immobility in bed. He also recorded the customary vital signs, appearance of the urine, etc.

Rafinesque took an active part in his own treatment, refusing the next day to take any calomel "internally in any shape, even in the smallest doses"—which was true to his conviction that the "heroic" treatments of conventional physicians were more likely to kill than to cure. The bedsore continued to bother him, and it was reported that he "wishes his back bathed with a decoction of bark, which he thinks would afford him relief." So, far from spending his last days in solitude, there was someone in the house at least to bathe his back. Although Dr.

The Last Days of Rafinesque

Hallowell had not stated even a tentative diagnosis, he continued what would be considered conservative treatment, probably with little expectation of saving the life of the patient. On the third visit he had nothing to add to the therapeutic regimen. He skipped seeing his patient on September 13, and when he returned the next day found a racing pulse of 108. The doctor prescribed the bitter tonic quassia and more of the sedative. When he visited on the fifteenth, the patient was weaker and had had little sleep; he had been "much troubled with hiccup," and he was unable to take the quassia. Dr. Hallowell now increased the dosage of the hyoscyamine and added morphine as well. On the sixteenth the patient had been having nosebleed, vomiting, a copious flow of urine "highly tinged with bile"; but his pulse had dropped and he slept better. For some reason Dr. Hallowell did not see his patient on the seventeenth—though it may be that Dr. Ashmead did—and when he returned on the eighteenth at 10 A.M. the patient was "evidently sinking"; his respiration was labored, his ankles were swelling, and his abdomen was even more distended. To afford nourishment, Dr. Hallowell merely prescribed oyster broth, weak wine and water, and "injections of milk"—all of which implies that someone had to have been in attendance all along during Rafinesque's last hours, even if his physician was not. Dr. Hallowell finally recorded under the date September 18: "Died at 9 P.M."

His report goes on to say that he and Dr. Ashmead performed the autopsy "thirteen hours after death"—which would place it at the reasonable hour of 10 A.M. the next day—and this explains, as Francis Pennell pointed out in his account of the autopsy, why the body had been taken to another room when undertaker Bringhurst came to prepare it for burial. Anatomical specifics of what the two physicians found during the autopsy will be reserved for the discussion of Rafinesque's posthumous adventures that follow in "Who's Buried in Rafinesque's Tomb?"; it remains to be said now that the autopsy report concludes with observations on the problems of detecting cancers of the stomach and of the liver, and cites several French authorities on the diagnosis of this disease.

Most people died at home in 1840; hospitals were used more as pesthouses for the confinement of those having contagious diseases—especially if they were indigent—than as treatment centers of last resort for solvent, respectable citizens. There is every reason to conclude that in his last illness Rafinesque had the best treatment available in a city that led the nation in the quality of its medical care.

Except for outstanding public figures, the newspapers of the time devoted little space to deaths. The only death notice so far discovered is one that appeared September 23, 1840, in the *Philadelphia Public Ledger*. There—under the heading "DIED"—Rafinesque's death was mentioned along with those of six others. The notices for two of them, who had died more recently, included the dates and times of their funerals. That for Rafinesque, whose body must have been interred by the time the notice appeared, merely stated: "On the 18th inst., Professor C. S. RAFINESQUE, for many years a resident of this city, and the author of several scientific and literary works."

It seems unlikely that the notice was reprinted elsewhere, but it may have been—on January 22, 1841, these lines were written by a Bostonian to Jules Rafinesque in France: "I sent you a long time since a paper containing a notice of the death of your uncle in Philadelphia, but I have not been able till now to obtain any decisive information as to the state of his affairs, though I have been in constant correspondence with a gentleman who resides there." This letter, which is missing the last page containing the writer's signature, has been kept in the Rafinesque family in Paris, but all efforts to identify its writer have failed.[21] He must have been well acquainted with Rafinesque's nephew, for he addressed him "My dear Jules." He went on to say that he had obtained a copy of the will of Jules's uncle and because of its bulk would summarize the contents. In the remaining pages of the letter he did this, but there is no evidence that Jules ever took any action to press his and his sister's claims on the estate, probably because his American friend remarked that shipping to France the entire text of the will would alone cost more in postage "than the whole is worth"—a judgment that must have been provided by his acquaintance in Philadelphia and, as we shall see, undoubtedly was true by the date of the Bostonian's letter.

V

Thomas Meehan first published Rafinesque's will in 1891 in the *Philadelphia Public Ledger,* where he gave credit to one of the newspaper's reporters, Millard Reeves, for digging it out of the city archives. Either Reeves or Meehan "improved" the diction of the text to the extent that this version is untrustworthy; it also is marred by silent deletions. However, when Richard Ellsworth Call decided to print the will in his Rafinesque biography he obtained a certified copy from the Register of Wills, who took it from Will Book 14 (1840), where it is will number 214 beginning on page 394. This is the form of the will used ever since by students of Rafinesque as a prime document for biographical information. But even this represents the text as copied over by a clerk from Rafinesque's own holograph and is only as good as its copyist was accurate. Call's copy also contained, immediately following the will and its codicils, a sworn statement by Samuel Hood and James Henry Horn that, being personally acquainted with the handwriting and signature of C. S. Rafinesque, the foregoing will and its codicils were truly those of the said C. S. Rafinesque. Hood and Horn swore their statement on November 16, 1840, by which time, as will be seen, the probate court already knew that the will had been repudiated by its maker, and so did Samuel Hood, who had taken Rafinesque's deathbed dictation. It is hard to understand now why an invalid will was witnessed at all.

Finally, in 1980 I also wrote to the Register of Wills for a copy. What I received, typed out on stiff legal paper and bearing the great seal of the City of Philadelphia, had the aspect of such solemn probity that it was hard not to accept it as unalloyed truth, though of course the typist who prepared this document might have introduced another layer of error. Nor could the clerk at city hall understand my motive when I went there insisting that I, personally, be allowed to search of the Philadelphia city archives—I already had "a legal copy," he explained. After extensive persuasion, considerable effort, and much good luck I did, however, come across the original will in Rafinesque's handwriting which Hood and Horn had witnessed, and from which the Will Book copy had been made. It is a wonder the holograph copy still existed after all these years, since the legal copy now was that in the Will Book. Yet, there it was in a separate, crumbling manila folder, written out in Rafinesque's characteristic chirography which I had become so familiar with as a result of poring over thousands of pages of his manuscripts and letters. (After careful comparison of the two, I am happy to report that the nineteenth-century handwritten transcript in the Will Book is accurate; my twentieth-century typewritten copy of that had introduced some minor errors.) Of greater importance were the two sheets of paper I found laid in loosely behind the holograph will. These sheets had never been copied into the Will Book, and hence they had remained unknown to previous searchers.

One of the two sheets is Rafinesque's deathbed revocation of all earlier wills; the other appears to be notes taken by a clerk of the probate court at the time the second will was proved. Herewith, for the first time, they are given verbatim:

> In the name of God Amen. The last will and testament of Constantine S. Rafinesque citizen of the United States whereby he revokes all former wills. I therefore the said Constantine S. Rafinesque being of sound mind and memory but sick of body do hereby express my forgiveness of all my enemies as in duty bound to do—especially Brulté and his associates whose persecution has been a main cause of my present illness. I hereby appoint Dr Samuel Mease and Samuel Palmer and Dr Samuel Betton of Germantown my Executors to whose management and care I refer all my affairs, and if one or more of them refuses to act it is my will the Court should appoint a substitute or substitutes. In witness whereof I hereby set my hand and seal this

Signed sealed and declared in) 18th day of Sep. 1840—
the presence of us—)
The above was prepared at the dictation)
of C. S. Rafinesque, in our presence.)
He became so weak when the will pro-)
ceeded thus far that he decided that)
it should be closed so far and was)
unable to sign it from bodily weakness.)
)
      Sam Hood)
        her)
    Lucia + Lyon)
      mark)

What one immediately notices is that all the complicated provisions of the 1833 will and its codicils have been swept away, and the executors are left to manage the estate according to their best judgment. Of the three executors named, Samuel Betton was one of Rafinesque's oldest friends, nothing is known about Samuel Palmer, and one wonders whether "Samuel" Mease shouldn't be James Mease. Did Samuel Hood make a mistake? Evidently this was the central issue when the will was proved. There the matter did get cleared up and in the process revealed more about Rafinesque's last day alive.

The second document found behind the deathbed will was obviously scribbled down in haste by a clerk or more likely by someone participating in the hearing and trying to write at the same time. It is as follows:

M^r. Hood sworn, on the day of the date I was call'd to go to M^r. R. found him very ill, I asked him (R) if he ~~wanted~~ wished me to make his Will, he requested me to make his Will. I wrote the form of the Will he then dictated to me the Will, I wrote it as dictated ~~to me~~. I put more formal language down[;] to the word substitute, I introduced another word, he became so weak he found it difficult [to] articulate, he then requested me to stop. I drew up signed sealed &c, I then asked him if he was able to sign it, by this time he was unable to sign it in consequence of his weakness, I waited about 20 minutes probably ½ an hour, to see if he would be able to sign, I found him still unable to sign it at that time, I then wrote the memorandum under signed sealed and declared &c and signed my name as a Witness, and got the woman of the house to witness it, she could not write, she made her mark, she was present during the whole time. Lucy Lyon sworn[,] says on friday last 18

The Last Days of Rafinesque

Sept., at 1 o'clock P.M. he Rafinesque [told me] to get a sheet of paper & pen & ink, he then spoke of Doc^r. Mease, Doc^r. Betton & Saml. Palmer as his Executors, he then said something about this Gentleman now present (M^r. Hood) and M^r. Horn, I was very much worried, he was not able to sign, it was from weakness, he told M^r. Hood to call next day <at 12 o'clock> and he could finish it the next day, he died next day 10 minu[te]^s before 9: he was to[o] weak to sign it at the time he was dictating—he had Gold in the room at the time he died[;] he continued weak until next day—when M^r. hood left there was about 4 o'clock.

Objected by M^r. Hirsh—are you acquainted with Doc^r. J Mease, I called on him at the request of (R.) to attend him, D^r. Mease frequently called at & before his il[l]ness, D^r. Mease was intamate [*sic*] with R, I know of no other Person who the[y] call Doc^r. Mease but Doc^r. James Mease, I have have [*sic*] no doubt but this is the same Doc^r. Mease—the[y] both requested me to make my mark, I thought he would die

Septr. 23. 1840

Unless the illiterate Lucy Lyon was confused about the date of an event that happened less than a week earlier (or the person who took down her testimony misunderstood), Rafinesque did not die September 18, as commonly believed, but before 9 A.M. on September 19. But this raises another problem we have no means to resolve: how could two experienced physicians perform an autopsy only an hour after death without being aware of it? Therefore it seems likely that the date she stated is in error.

Lucy Lyon's testimony does reassure us that the faithful Dr. Mease was present during Rafinesque's last illness. Her reference to gold in the bedchamber may only mean the French medal Rafinesque prized so highly, or it might be a sum worthy of the evil machinations of "Brulté and his associates" mentioned in Rafinesque's dictation. There is no way of knowing what remained of the assets of the Aurora Company or the status of the Divitial Bank. Yet, in his last days Rafinesque must have been engaged in financial affairs larger than his usual trafficking in natural history specimens; when Elias Durand wrote to Torrey that "our poor Rafinesque is dead," he added that "the poor man is dead[,] broken hearted by all the miseries he has experienced by his financial undertakings."[22]

VI

We know Peter Brulté only as a fruit seller on North Second Street whose residence was somewhere in the Northern Liberties of Philadelphia; he was also one of the Philadelphia tradesmen who had signed a petition in 1836 requesting a charter from the state for the Divitial Bank, so he had been associated with Rafinesque's financial affairs for some time. Whatever he was up to regarding Rafinesque's possessions, he helps to tip the balance toward the belief that Lucy Lyon erred about the date of Rafinesque's death. According to Philadelphia records of the administration of wills (Administration Book P [1840], file 285, p. 123) Brulté scurried down to city hall and "On the 19th day of September AD. 1840 Letters of Administration on the Estate of Constantine S. Rafinesque deceased were granted unto Peter Brulta [*sic*] of the Northern Liberties." Surely this could have happened no sooner than the day following the death.

In this instance it is possible even to recover the document where Brulté— lying through his teeth—"did depose, declare, and say" on "his solemn oath" on the nineteenth that "C. S. Rafinesque died without a Will." There is also a bond of the same date for four hundred dollars to cover the actions of the administrator, signed by Brulté and also by Francis Timmins and Joseph A. Clay, about whom nothing is known but who probably were the "associates" Rafinesque had in mind. In his declaration, Brulté estimated the value of the estate he was going to administer at less than two hundred dollars, but that probably was to keep the bond as small as possible rather than a realistic valuation.

Whatever Brulté was trying to do with Rafinesque's assets, he had about ten weeks to carry it out; but we have no information on what took place during this two and a half months, which surely was long enough for him to loot much of whatever wealth remained. It was not until November 28 that James Mease was properly sworn as executor.[23] Since the court was aware five days after his death that Rafinesque did not die intestate, it appears that an inordinate amount of time was left to pass before a proper executor was appointed. The only explanation I can think of is the fact that three different candidates refused the job of executor. But during this ten-week period, administrator Brulté seems to have had continued access to Rafinesque's estate. After executor James Mease finally took over we hear no more of administrator Peter Brulté. Knowing now, as we do, that on his deathbed Rafinesque gave his executor full liberty to follow his own judgment, there is no longer any reason to reproach Dr. Mease for failing to carry out the almost impossible provisions of the voided 1833 will and its codicils.

The inventory Mease must have filed when he closed the estate can no longer be found; we have to rely on what Meehan printed in the *Public Ledger*, which probably is as inaccurate as his transcript of the will printed in the same place and, even if accurate, surely represents only what was left after Brulté took everything he could lay hands on. According to Meehan, the total value of the estate was only $132.42, of which the cash portion is represented by this confused account: "Cash taken from one of two rouleaux of half-eagles found on the person of [the]

The Last Days of Rafinesque

deceased at the time of his death and $1 from his purse, $6." A half-eagle was worth five dollars; so what happened to the rest of those that made up a "roll" of them? "To cash received from Joseph S. Clay [*sic*—Brulté's associate?—] from one of the said rouleaux in the evening for funeral expenses, and 62 cents, $25.62." Whatever this may mean, Lucy Lyon was right—there was gold in the room, two rolls of $5 gold pieces, as well as the medal from France. The medal itself was valued at the Philadelphia mint for $16.55 and sold to Charles Poulson for that sum, as well as some manuscript notebooks for $5.00.[24] Other items sold were boxes and bottles, sheets of paper, clothes, and a number of copies of the *Medical Flora*.

On the debit side, such expenses are mentioned as "arranging the library" and "moving it for sale," making a catalogue of minerals and shells, and cartage for "eight loads of books, minerals and shells to Lord's auction and thence"—apparently the remainder that did not sell—"to Freeman's." No mention is made of the money realized from the sale of any of these items, and it may be noted that no mention is made at all of the disposition of Rafinesque's huge herbarium, which he had valued at $1,800 in 1832, though sixteen years after his death Durand reported that he paid $45 for it.[25] It has often been observed with derision that the estate ended up indebted to the executor in the amount of $13.43, but Meehan, like other myth makers later on, wanted to underscore the poverty of the neglected genius. In view of the bulk of the items sent for auction, Dr. Mease must have been entitled to more than the $12.14 he allowed himself as executor's commission. This is said to be 10 percent, and that would be approximately correct for the total valuation of the estate before the sale. We can only conclude that the various items Meehan lists may be correct, but he has not told the whole story by any means.

It happens that the burial costs at least can be checked in an independent source, and, unless they figure under "other expenses," they also are nowhere mentioned in Meehan's account. As Pennell first reported, the Bringhurst Company billed Dr. Mease $7 for a walnut coffin, $3 for the winding sheet, and $6 for a rented hearse to bury Rafinesque in Ronaldson's Cemetery.[26] This was an unpretentious funeral; some funerals on the same page of the Bringhurst account book cost ten times as much. Invariably, though, the more expensive funerals included many yards of black crepe to drape over the house of the deceased, numerous rented carriages for the mourners, and such items as black gloves presented to each mourner as a memento mori. There is no record of any religious service in connection with Rafinesque's burial, nor any honorarium for a clergyman.

Through most of his adult life, Rafinesque had consoled himself in the face of obstacles and adversity with the thought that he was ahead of his time. It is easy to attribute this attitude to his overweening egotism, which was undeniably a flaw of his character. Yet he really was ahead of his time in many ways, and in his thoughts on funeral practices he was very much ahead of his time. His 1833 will stated the wish that his "body if possible to be burnt rather than buried as I do not want to contamine the Earth by decay, nor be a cause of disease to other men." This request, far from being overlooked, as Pennell thought, was too advanced to be carried out. The first cremation in Philadelphia did not take place until 1888, forty-eight years later.

Notes

1. Though as usual he preferred to make exchanges with other botanists whenever possible. His one attempt to explore the flora of the southeastern states, in 1833, having been frustrated by bad weather, he now wrote to Lewis Gibbes in Charleston (Feb. 15, 1836; Library of Congress), offering to buy southern plants and mentioning that he would like to finance the exploration of that region by someone else. In the *American Journal of Science* 29 (1836): 393, he also offered to buy "at a fair rate, according to value" plants from the southern states and assured any collectors willing to visit those states that he would purchase specimens from them up to the value of one hundred dollars. He succeeded to some extent, for by November 1837 he had received "parcels of plants from Florida, Alabama & Louisiana" (Rafinesque to Charles Wilkins Short; Nov. 7, 1837; Filson Club History Society).

2. Rafinesque to John Torrey, Dec. 15, 1822, New York Botanical Garden Library.

3. Rafinesque to John Torrey, May 21, 1834, New York Botanical Garden Library.

4. Rafinesque to John Torrey, Apr. 18, 1837, New York Botanical Garden Library. The last letter known to come from his hand was dated from this address also, when he was already suffering from the cancer that took his life. See Rafinesque to Elias Durand, Aug. 5, 1840, New York Botanical Garden Library.

5. Samuel Wetherill, the founder, is in the DAB. The little that is known about Charles Wetherill's wealth comes from Miriam Hussey, *From Merchants to "Colour Men," Five Generations of Samuel Wetherill's White Lead Business* (Philadelphia, 1956).

6. Winton U. Solberg, writing the official early history of *The University of Illinois, 1867–1894* (Urbana, 1968), tried to account for Rafinesque's scheme before taking up the story promised by his title. He concluded that Rafinesque's death in 1840 "spared the author obloquy in a state which was just completing the frontier phase of its development and was not receptive to educational innovation" (19).

7. A list of publications in the endpapers of Rafinesque's *American Manual of the Mulberry Trees* (1839) states that five numbers of *New Aurora* appeared in 1837. The one cited here is called a second edition and is the only one that has been found.

8. *Plan of the Philadelphia Land Company of Aurora* (Philadelphia, 1837). Although the educational plans of the Aurora Company have been stressed here, Rafinesque and his

associates had additional utopian expectations for the colony. Concord would be assured because "no Drunkards nor Lawyers shall be allowed to . . . settle in Aurora." To avoid the cost of fences, livestock would be kept in pens at night and watched over by herdsmen in daytime. When land was cleared, all sugar maple trees would be spared. Bank notes were to be banned, and the circulating medium of exchange among settlers would be certificates of deposit issued by the savings bank of the colony.

9. *Dubuque Iowa News,* Nov. 18, 1837.

10. Rafinesque to Augustin-Pyramus de Candolle, June 1, 1838, Gray Herbarium Library. This letter, given to Mrs. Asa Gray as a souvenir, was reproduced in facsimile in Call's biography, but entirely ignored by Call himself. Rafinesque's wish to deposit holotypes of all his new plant species in the university's museum was an innovation, because not until the twentieth century did the deposit of type specimens become a regular requirement of plant taxonomy.

11. Rafinesque to John Torrey, Mar. 20, 1838, New York Botanical Garden Library. Knowing as we do that he was fated to die in eighteen months, it is sorrowful to see Rafinesque telling Torrey here that "I shall again botanize in the West & mean to settle there between 1845 & 1850 giving all my Collections & Books to [the new university], receiving Land for it which I shall dedicate to perpetual benefits."

12. R. Carlyle Buley, *The Old Northwest: Pioneer Period, 1815–1840* (Bloomington, Ind., 1951), II: 363.

13. Charles Henry Rammelkamp, *Illinois College: A Centennial History, 1829–1929* (New Haven, 1928), 65. Among the several ironic parallels between the imaginary Central University of Illinois at Agathopolis and the real University of Illinois at Urbana is their shared "classical" place names. The name of the former is pseudo Greek, that of the latter spurious Latin.

14. *History of McLean County, Illinois* (Chicago, 1879), 343. Bloomington is the county seat of McLean County.

15. The review appeared in the *Cincinnati Literary Gazette* 1 (May 22, 1824): 161–62.

16. John L. Riddell, manuscript "Personal Journal," XIX: 23, Tulane Univ.

17. Thomas Meehan, "C. S. Rafinesque," *Botanical Gazette* 8 (Feb. 1883): 177.

18. Thomas Meehan, "Rafinesque," *Philadelphia Public Ledger,* Feb. 19, 1891.

19. Richard Ellsworth Call, *The Life and Writings of Rafinesque* (Louisville, 1895), 55–57.

20. The autopsy report was printed in the Philadelphia *Medical Examiner* 3 (Sept. 19, 1840): 597–99. Despite its printed publication date, this number actually was issued the following month.

21. In the 1840s, many ambitious American physicians capped their professional training by a year or two of experience in the lecture halls, clinics, and dissection rooms of Paris. If the unknown Bostonian were a physician, this might explain how he became

acquainted with Jules. The directors of both the Harvard Medical Library and the Harvard University Archives have kindly checked the handwriting of this letter against manuscripts of several dozen Bostonians known to have studied medicine in Paris during the period. No reasonable match was found.

22. Elias Durand to John Torrey, Nov. 25, 1840, New York Botanical Garden Library.

23. The Will Book shows that Samuel Betton, Jacob Green, and the French consul in Philadelphia—all of whom had been named in the 1833 will as possible executors—refused to serve.

24. A note pasted inside the cover of one of Rafinesque's manuscript notebooks at the Smithsonian Institution is the receipt Mease gave Poulson for this purchase. The medal is now in the Mütter Museum of the College of Physicians of Philadelphia. It is pictured here as fig. 1.5.

25. Francis W. Pennell, "How Durand Acquired Rafinesque's Herbarium," *Bartonia* No. 23 (1944–45): 43–46. The evidence is from a letter Durand wrote to Charles Wilkins Short where he said the sale "was forced upon me by Dr Mease, who had no money of the deceased in hands to pay 45 dollars storage." This letter is the source of the often repeated derogation of Rafinesque's skill in preserving botanical specimens, for Durand remarks that the herbarium "was truly a dunghill"—and no wonder, because "it had remained three years in a stable garret, where a numerous colony of rats found it a convenient hiding and breeding place." This $45 could not appear in Meehan's inventory, since three years elapsed before Durand made the purchase. However, we have no way of knowing why the herbarium that Rafinesque had valued at $1,800 was put in storage at all.

26. "Bringhurst & Co. Accounts Ledger—Feb. 1824 to April 1849," p. 82, Historical Society of Pennsylvania.

Rafinesque Portraits

CHARLES BOEWE

SO MUCH CONFUSION attends the three portraits said to be likenesses of Rafinesque that the scattered facts known about them have been gathered here. At the outset it may be remarked that one picture certainly is genuine, as it was drawn from life and had the approval of the sitter. This is the engraved frontispiece of *Analyse de la Nature*, dated 1810 by the artist, thus representing Rafinesque in his twenty-seventh year (fig. 4.1). Another, an oil painting reproduced in Call's biography (fig. 4.2), where it is attributed to the Kentucky artist Matthew Jouett, probably is not of Rafinesque, and the attribution also is doubtful. The third (frontispiece of this book), an enamel miniature, likely is Rafinesque but probably was not painted from life. There are problems about its attribution.

Call first drew attention to the emblematic elements in the background[1] of the 1810 portrait but passed over without notice both the names of the artist and of the engraver. The latter's name appears as Vaincher, who as "P. V." was one of several engravers contributing plates to *Caratteri* (1810) from Rafinesque's own drawings of fishes and plants. The artist's name on the 1810 picture is signed Falopi, who, like Vaincher, probably was a Sicilian. Nothing more is known about him, and his work has been detected in no other Rafinesque publication.

This portrait, with its oval border and pedestal, was abstracted from the stipple background and published as a line engraving by an engraver signed "REA" in *Potter's American Monthly*.[2] Elsewhere in *Potter's* there is a reference to Jeremiah Rea as the engraver of portraits for the magazine. Beneath the portrait appear the words

Fig. 4.1. *Frontispiece of* Analyse de la Nature *(1815).*

Fig. 4.2. *The alleged Jouett portrait. Courtesy of Transylvania University Library.*

Fig. 4.3. *Detail of* Potter's *magazine engraving. U.S. National Archives.*

Rafinesque Portraits

"Your Friend C. S. Rafinesque" that the magazine acknowledged as the facsimile of an autograph clipped from a letter. Rea's skill as an engraver may be judged from fig. 4.3 reproduced here.[3] Call averred that the *Potter's* engraving had been made from a "painting owned by Doctor William Kent Gilbert"—thus cluttering the record with a nonexistent painting because he misread the magazine's acknowledgment to Gilbert for the opportunity to engrave his "rare portrait of Rafinesque," which was none other than the frontispiece of *Analyse de la Nature*, a book somewhat rare to be sure even then.

Call helped to obscure the record further when he published the so-called Jouett portrait and included under it the same handwritten words "Your Friend C. S. Rafinesque." This autograph gives the portrait the spurious appearance of having been signed by the sitter, but the words were in fact lifted without comment from the *Potter's* engraving, where of course their source in a letter had been acknowledged. To his credit, it must be observed that Call had misgivings about the portrait he attributed to Jouett. He printed the reproduction only because both it and the *Analyse* frontispiece had "been seen by persons who knew Rafinesque, but opinion on them is very equally divided." If it suggests anything, this divided opinion implies that neither picture is a very good likeness, for it is almost impossible to believe both represent the same man.

Then in the possession of the Wisconsin State Historical Society, this oil painting, measuring about ten by fourteen inches, is attributed to Jouett only in the caption by Call, who adds that the painting is on black walnut, and that it was obtained by the society's collector, Lyman C. Draper, about 1876, either in Kentucky or Philadelphia. Call gives no source for this information.

More than two decades later, however, the superintendent of the society wrote to Transylvania's dean, confirming that in 1876 Draper had purchased the portrait; but in Cincinnati, from Robert Clarke & Co. He quoted from the 1876 Clarke catalogue as follows: "4371 Rafinesque (C. S.), Portrait of Rafinesque, painted by himself, in oil colors, on cedar panel, in gilt frame, in fine preservation; this portrait was painted for and presented to Charles Paulsen, Esq.,[4] of Philadelphia, by Rafinesque, in gratitude for his having translated one of his works from the French." The letter continues that the society paid the price stated in the catalog, without mentioning what that price was; then adds that in the writer's opinion the painting is on black walnut rather than cedar.[5] A copy of the Clarke catalogue recently turned up and shows that the picture was priced at thirty dollars.

Although no writer on Matthew Harris Jouett (1787–1827) has chosen to include the painting among that artist's *oeuvre*,[6] the notion that it could be a self-portrait of either Rafinesque or Jouett is even more troublesome. The way the eyes engage the viewer suggests that it well may be a self-portrait; but then it certainly was not by Rafinesque, for none of his attempts at portraiture shows this degree of skill, nor did he ever paint with oils as far as we know. If Call's attribution is correct it would have to be a self-portrait of Jouett. Most Jouett portraits of others are in three-quarters view, as is this one, with the sitter looking to his own left. In a known self-portrait, Jouett represents himself looking to the right, as in this one. This pose was necessary because of his use of a mirror, as would be true for any other right-handed painter. But since the picture Call believed to be Rafinesque does not show a person resembling that known to he Jouett, it seems improbable that this painting is an unknown Jouett self-portrait.[7]

The third portrait, unknown until 1934, is an enamel miniature, two and a quarter inches in diameter, mounted in a massive gilded frame. One item in what has been described as the "greatest private collection of early American miniatures in this country," the painting was offered for sale by Parke-Bernet Galleries upon the death of its owner, Erskine Hewitt, in 1938. At that time several friends of Transylvania University made a frantic effort to scrape together the $175 price, so it could join the so-called Jouett portrait, which had been on permanent loan to Transylvania for some time. They succeeded, and these two somewhat dubious but fascinating Rafinesque artifacts are at Transylvania today.

Both the friends of Transylvania and Parke-Bernet acted in good faith, but investigation of the pedigree of the miniature shows a tangled history much like that of the painting on wood. The Parke-Bernet catalog traced ownership of the miniature from Hewitt back to Hiram Burlington, from him to Effingham Schieffelin, and from him to Jacob Schieffelin. This provenance, which is not questioned, derives from two undocumented articles that provide additional information which *is* subject to question, part of it being the attribution of the painting to William Birch (1755–1834).

An anonymous article[8] at the time of the Hewitt sale referred, without any shadow of doubt, to "William Birch's portrait of Rafinesque the naturalist." It turns out that this calm assurance of certainty derives from an equally undocumented article[9] that reproduces the miniature, calling it "one of really exceptional charm." The article provides no reasons for the assertion that the picture is "that of the botanist, Constantine Samuel Rafinesque, by William Birch." The article goes on to allege that the first owner of the painting, Jacob Schieffelin, was an "associate" of Rafinesque, and it shows that the painting passed through the Anderson Galleries at the time of the Burlington sale in 1934 before reaching Parke-Bernet. And there the trail ends.

Rafinesque Portraits

If the "documented" provenance ends in a fog of doubt, can connoisseurship dispel the darkness? Examination of the painting today reveals that William Birch's name does appear, but only on the back of the wooden frame, in faded brown ink, and might have been put there by anybody through whose hands the picture passed. Neither Birch's name nor that of Rafinesque appears anywhere on the painting itself. To be sure, the surface of the painting once bore a serial number—10453—which has since been removed but may be seen in early reproductions. It seems improbable, however, that anyone would have decreased the value of the painting by removing the artist's signature, if it ever appeared there.

There are several other problems with this attribution to William Birch. In the first place, few Birch miniatures contain background details as this one does. In the second place, if Birch painted it he almost surely did not paint from life, for, given the travels of the two men that can be traced, he could have had Rafinesque as a sitter only after the latter's return to Philadelphia in 1826. By the 1820s Birch was something of a society painter, whose work was hardly within Rafinesque's financial range. Moreover, as early as 1825 Rafinesque was described by an acquaintance as "rather bald . . . and withal is rather corpulent."[10] Of course, the painting might have been made by Birch, or someone else, from the 1810 Falopi print, which it does resemble. And it could have been commissioned by Jacob Schieffelin (1757–1835), a New York manufacturing chemist who was wealthy enough to engage Birch's talent. However, nothing in Rafinesque's published or unpublished writing has ever established any connection with the Schieffelin family, who are still in business—now as wine importers—and whose family papers also have revealed no connection.

With such uncertainty about artists' renderings of the appearance of Rafinesque, one turns to the written accounts of those who knew him personally—only to find the confusion deepened. R. E. Call printed those descriptions he was able to gather, which at least enable us to form some judgment of Rafinesque's appearance during the Transylvania years, 1819–26, when he was in his late thirties and early forties.

At that time he appeared to a girl[11] then ten or twelve years of age as "small and slender, with delicate and refined hands and small feet. His features were good and his eyes handsome and dark, or apparently so from the long, dark eyelashes. His hair, which he wore long, was dark and silky." A man[12] who had been a student during Rafinesque's tenure believed that "he had a full suit of hair and black eyes," but, comparing an adaptation of the Falopi print with the "Jouett" painting, judged the painting "a better likeness." Another student[13] recalled Rafinesque as "a man of low stature, not more than 5 feet 10 inches in height, strongly built, and capable of great physical force; his head rather larger than usual, square shoulders but not stooped, dark grey eyes, and dark hair." Yet another person[14] in Lexington, identified only as one whose home Rafinesque frequented, reported that "as I recall the old man he was a small, peculiar looking Italian, with a large, rather bald, head and stooping figure. . . ."

About all such conflicting testimony establishes is that during middle age Rafinesque had a robust appearance but retained a physical delicacy, had dark eyes, had a head large for his body, and had dark hair which was tending to go bald.[15] Meager as they are, these details seem consonant with both the Falopi drawing and the "Birch" miniature.

The tendency to corpulence first attested in 1825 must have increased. Young Joseph Henry remembered Rafinesque from the Rensselaer School trip on the Erie Canal as "a short man stoutly formed" and also noticed the tendency to baldness.[16] And during Rafinesque's brief tenure as teacher at the Franklin Institute, 1826–27, William Kite, one of his students, later had these recollections:

> Mr. Kite described Rafinesque as a corpulent man, with a queer French accent, and said he sometimes became very angry with the class. When he appeared to lecture, his odd manners and dress attracted the boys, who laughed and made fun of them, and his lot seemed not to have been an ideal one. Rafinesque, said Mr. Kite, was very large about the waist, and wore wide Dutch pantaloons of a peculiar pattern, and never wore suspenders. As he proceeded with a lecture, and warmed up to his subject, he became excited, threw off his coat, his vest worked up to make room for the surging bulk of flesh. . . .[17]

This clearly is no longer the dapper young man we have in any of the three pictures.

Finally, for what it is worth, Rafinesque appears to have lost the corpulence during his last three years, when the cancer that caused his death may already have started to take its toll. A newspaper writer remarked that "he was, about the year 1837, a little, dried up, 'muffy' looking old man, resembling an antiquated Frenchman . . . he became one of the most neglected among God's creatures—a poor philosopher."[18]

Perhaps in composite these several sketches by pen and brush will give us some conception of Rafinesque's appearance. Until more evidence appears we are left with the conclusion that the visual images we have are those of the young man, probably not more than thirty years of age. Among these, the miniature is artistically the best picture and may have been painted by William Birch, but, if so, probably was not painted from life. The painting attributed to

Rafinesque Portraits

Jouett is questionable both as to painter and subject but cannot be dismissed out of hand; almost certainly, though, it is not a self-portrait by Rafinesque.[19] And the Falopi frontispiece to *Analyse de la Nature*, esthetically the poorest of the three, remains the only one of unimpeachable integrity and may have served as the source of the miniature.

Notes

1. Richard Ellsworth Call, *The Life and Writings of Rafinesque* (Louisville, 1895), 68.

2. Theodore L. Chase, "Constantine Schmaltz Rafinesque," *Potter's American Monthly* 7 (1876): 97–101.

3. This portion of the *Potter's* magazine engraving, including the full oval border but with the pedestal removed, was published to illustrate the Rafinesque chapter in William Jay Youmans, *Pioneers of Science in America* (New York, 1896), 182–95. There originated an error (193), since repeated, that the likeness had come from the obverse of the gold medal awarded Rafinesque in 1832 by the Société de Géographie.

4. This would be Charles Poulson, who did in fact translate Rafinesque's *Monograph of the Fluviatile Bivalve Shells* in 1832, by which time Rafinesque could not have looked like this—even to himself.

5. Joseph Schafer to Thomas Macartney, Madison, Dec. 3, 1923, Transylvania Univ. Archives. Much of the information given in this letter is included, with no source stated, in Elizabeth M. Simpson, *Bluegrass Houses and Their Traditions* (Lexington, 1932), 90–91.

6. Samuel Woodson Price, *The Old Masters of the Bluegrass* (Louisville, 1902), a Filson Club publication as was Call's biography, lists more than 300 Jouett paintings without mentioning a portrait of Rafinesque. A *Catalogue of All Known Paintings by Matthew Harris Jouett*, by Mrs. William H. Martin (Louisville, 1939), lists 529 Jouett pictures, about a fourth of which are now known to be by other artists, and also does not mention the alleged Rafinesque portrait. In 1980 Transylvania University mounted a Jouett exhibition that included three of his portraits owned by the university, as well as many from other institutions and from private owners, without feeling a need to include the so-called Rafinesque portrait, though this was in its possession; and the catalogue of this exhibition, by William Barrow Floyd, *Matthew Harris Jouett: Portraitist of the Ante-Bellum South* (Lexington, 1980), did not allude to the portrait. Thus, Call's attribution has never been supported by serious students of Jouett's work.

7. The only point of similarity between the "Rafinesque" picture and the Jouett self-portrait is in the treatment of the hair. However, this is more than offset by the fact that Jouett pictures himself with a head rather small for the massive body depicted; both of the other likenesses of Rafinesque show the reverse, a large head for a small body.

8. Anonymous, "Historic Americana on Miniature & Canvas: Hewitt Collection," *Art News* 37 (1938): 16.

9. Helen Comstock, "American Historical Portraits: The Burlington Collection," *The Connoisseur* 94 (1934): 31–37. The record has been further confused by the publication of an excellent, enlarged, color reproduction of the miniature in a popular, undocumented article (Peggy Robbins, "The Oddest of Characters," *American Heritage* 36 [1985], 58–63), where, along with other misleading information, this painting also is "said to be by Matthew Jouett"! In the color reproduction used in *Natural History* (October 1996) to illustrate David Oestreicher's article, the image has been flipped, with the figure's hand pointing to the left. This surely will lead some scholar to conclude that Rafinesque was a southpaw.

10. Edmund Porter quoted in *Pennsylvania Magazine of History and Biography* 16 (1892): 249.

11. This was Frances Paca Dallam, who, in old age, reported her memories to Call through her daughter, Johanna Peter (Call, *Life and Writings of Rafinesque*, 63).

12. George W. Jones (ibid., 64), who adds that he never saw Rafinesque as well dressed as the figure in the so-called portrait. Elsewhere the fashion-conscious Jones described the dress of a young gentleman when he was a student at Transylvania, an account that would do very well as a description of the clothes in the miniature, a picture Jones could not have known. He wrote of "Canton-crepe trousers, buff colored buck-skin boots, dark blue or black swallowtail coat with brass buttons, which were sometimes flat and sometimes bullet shaped, white waist-coat, shirt ruffled at the bosom and sleeves, very stiff and high-standing collars, and white or black silk broad stock." He adds that he himself had his collars starched so stiff "that they could draw blood from my ears" (John Carl Parish, *George Wallace Jones* [Iowa City, 1912], 260).

13. Belvard J. Peters (Call, *Life and Writings of Rafinesque*, 66).

14. Ibid., 65.

15. One may entertain some doubt about the literal correspondence of Audubon's "Eccentric Naturalist" with Rafinesque, since Audubon varnished his stories as he varnished his oil paintings. Yet the two men were acquainted, and Audubon's trained eye should give weight to his written description, even though he never chose to record Rafinesque's appearance with his brush. If the Eccentric Naturalist truly is Rafinesque, and if Audubon did not exaggerate for literary effect (which is unlikely), he confirms only the observations of others that Rafinesque had a forehead "broad and prominent." Most of his description is devoted to the odd clothing of his visitor in the summer of 1818. He also mentions the man's long beard and "lank black hair"—neither characteristic appearing in any of the pictures of Rafinesque—but goes on to acknowledge that these resulted from the naturalist's extensive rambles in the backwoods. See his sketch included here as chapter 19.

16. *The Papers of Joseph Henry*, ed. by Nathan Reingold (Washington, 1972), I: 152.

17. Edwin C. Jellett, *Personal Recollections of William Kite* (Germantown, 1901), 27–28.

18. Probably a paragraph appearing originally in a Philadelphia newspaper, this is signed only "Jacob" and was reprinted in the Morrisania, N.Y., *Historical Magazine* 6 (1869): 244–45, where its source was attributed to the *Salem Gazette*. The writer could have been "an aged amateur naturalist, William Jacob, a lithographer I was informed, who regularly frequented the library of the Academy" of Natural Sciences, who "told me he had known and met Rafinesque" (Henry W. Fowler, ANSP Monograph No. 7, *A Study of the fishes of the Southern Piedmont and Coastal Plain* [Philadelphia, 1945], 15). On the occasion of the Rafinesque centennial, Francis Pennell reported—in a footnote comment transferred from the reprint of his article to this place—that Mr. Jacob was "an exceedingly gaunt and tremulous old man, who occasionally visited the herbarium of the Academy of Natural Sciences when I first knew it in 1906, and who had surprised the curator of those days by recounting botanical bits that Rafinesque had told him."

19. All three portraits have been reproduced many times, but it bears mention that if one wishes to examine the "Jouett" portrait in color it may be found in Aurèle La Rocque, "A Rafinesque Portrait," *Sterkiana* No. 16 (Dec. 1964): 1–2, where, however, nothing is added about its history. Another adaptation of the Falopi engraving, by Elsa S. Ames, may be found in the form of a drawing in *Taxon* 22 (1973): 228.

Part II

The Naturalist

Rafinesque among the Field Naturalists

CHARLES BOEWE

Preponderance of Europeans

ANY ROSTER of the early naturalists who catalogued the flora and fauna of the region that lies west of the Allegheny Mountains and east of the Mississippi River has to include such names as John James Audubon, Charles Lucien Bonaparte, Charles Alexandre Lesueur, André and François Michaux, C. S. Rafinesque, and Louis Pierre Vieillot (all of whom were French, either by birth or culture); John Bradbury and Thomas Nuttall (who were English); Alexander Wilson and William Maclure (Scottish); Gerard Troost (Dutch); and Frederick Pursh (German). Few native-born Americans can be added to the list, Titian Peale and Thomas Say being the only major figures who readily come to mind.

Most of these, but not all, were field naturalists, though they had to maintain professional connections with settled acquaintances on the East Coast, for that was where such publication as was possible usually took place. Others, if not working in the field, had agents who supplied them with specimens from the same region. Not all the sedentary naturalists, as Rafinesque liked to call them, were native-born and several had European antecedents—G. W. Featherstonhaugh (English) and Elias Durand (French) being examples of the former; Henry Muhlenberg and L. D. von Schweinitz of the latter, as their Germanic names imply—but all of them had established positions in society which, by and large, provided them a living. Muhlenberg and Schweinitz were clergymen; Benjamin Smith Barton and John Torrey, though educated as physicians, made their livings as professors, as did L. C.

Charles Boewe

Beck and Amos Eaton; William Darlington, Richard Harlan, and David Hosack were practicing physicians; Durand was a pharmacist; Isaac Lea was a publisher; Stephen Elliott was a banker; Zaccheus Collins was a merchant; George Ord inherited money, and Featherstonhaugh married it.

Though less numerous, there also were resident naturalists living beyond the mountains, of course. Among them, for example, were the physicians Daniel Drake, Samuel R. Hildreth, and John Locke. In Ohio, Hildreth contributed descriptive work in zoology from his base in Marietta, as did Locke in botany from Cincinnati. Whether in Cincinnati, Lexington, or Louisville, Drake pursued his topographical studies of the influence of environment on diseases. Most notable among the botanists was Charles Wilkins Short, briefly Rafinesque's colleague at Transylvania during the winter of 1825–26. And the field versus settled dichotomy is by no means absolute. Even such an ornithologist as Bonaparte, whom Rafinesque admired, clung to the cozy collegiality of the Academy of Natural Sciences and the American Philosophical Society in Philadelphia, and let others— among them Audubon—bring him dead birds to classify and name.

The competition and sometime rivalry that developed between the field naturalists and the more sedentary ones often is accounted for by the plausible explanation that only the settled naturalists, who also controlled most of the publication media, had at their disposal the library resources, museums, herbaria, and other reference collections necessary to classify and order the raw information coming in from the field, especially as new knowledge swelled to flood tide later in the nineteenth century. From this point of view, the finest field naturalist was such a botanist as the diffident Charles Wright, who flourished around the middle of the century. Wright collected throughout the American Southwest, in Cuba, and even in the Far East—then meekly shipped his plants to Asa Gray to publish, satisfied with the small glory of seeing the title *Plantae Wrightianae* (1852–53) on Gray's books. But many of the earlier field naturalists, and especially Rafinesque, thought otherwise.

For one thing, Rafinesque tried to eke out his living by his collecting work. Turning over his collections to others would be tantamount to depleting his capital. At a time when the nation could support few academic posts in the natural sciences, and republican sentiment was opposed to public patronage, some of the field naturalists went to extraordinary lengths to sustain themselves while getting on with their work.

Lesueur and Say turned "communist" in New Harmony. Their hope, like those of the other Owenites, was to find leisure for research when not making their required contributions to the welfare of the colony. Audubon turned celebrity in Europe. Dressed in buckskin he became the toast of London; if he were alive today we would see him touting his bird book on TV talk shows and sitting at Oprah's knee. Nuttall turned bitter, especially after he inherited a small estate in England that made his part-time residence there mandatory. Wilson died young, and his fiancée buried him. After losing his plant collections to Pursh, Bradbury gave up natural history in dejection and drudged in a cotton mill to feed his wife and eight children. Pursh himself died a drunkard in Canada. And conventional wisdom has it that Rafinesque simply went mad.

He suffered himself to be "laughed at as a mad Botanist," Rafinesque said, in order "to be a pioneer of science." Why he and his fellow field naturalists so gladly chose "the gloom of solitary forests," to "be burnt and steamed by a hot sun at noon, or drenched by rain," while "gnats dance before the eyes" and ticks "stick to you in crowds, filling your skin with pimples and sores" has not been answered, perhaps because seldom asked. In retrospect, "the pleasures of a botanical exploration fully compensate," he wrote, for the hardships he and the others endured.[1] To give the stay-at-homes a hint of these joys he had to resort to erotic imagery. "Nature is a beautiful and modest woman," he told a lecture audience in Lexington, "concealed under many Veils, some of which she throws aside occasionally or allow[s] them to be removed by those who deserve such a high favor."[2]

Yet one wonders why so many footloose and impecunious young Europeans should aspire to disrobe demure Nature in the American backwoods. Unlike most of us today, few of them had been educated for the professions they chose to follow. Of the botanists, only the Michauxes (father and son) and Pursh had anything resembling formal education in botany. Nuttall, the greatest traveler of them all, was trained to be a printer. Even his biographer was at a loss to explain how Nuttall could anticipate the pleasures of botanical exploration in the New World and go on to alienate his family by rashly undertaking it.[3] Rafinesque came from a merchant family and was intended by his apprenticeship to follow that occupation; all the botany and zoology he knew came from his undirected reading.

One clue to their motivation seems to lie in the foreign birth or background of so many of the early field naturalists in America, and in this respect Rafinesque was typical rather than exceptional. Less likely than the native-born to occupy fixed places in the American social order, they possessed at the same time a greater European sense of the potential rewards of terrestrial exploration. Though Rafinesque's name is especially associated with botany in the history of science, in truth he viewed himself as an explorer first, naturalist second. "A traveler from the cradle," he wrote as a motto for his autobiography, "I'll be one to the grave." It is not much too emphatic to say that he made himself a naturalist in order to be an explorer. In his own language, he was ever a voyageur; though the French language lacks the distinction English makes between travel on land or water, the French word does carry a richer connotation of adventure than our paler word traveler.[4]

We need to recall that in the early decades of the nineteenth century, Europe was still in the expansionist period that began with the discovery of the New World itself. The United States, physically on the margin of Europe's expansion, would not be in a position to participate in global exploration until the Wilkes Expedition of 1835–42, though the impulse to contribute a continental portion was present as early as Jefferson's wish (1793) to send an overland expedition to the Pacific, which resulted a decade later in Lewis and Clark's epochal trek. As exemplified by Jefferson's hopes for the Lewis and Clark journey, by the time Americans began to participate in terrestrial exploration the element of scientific research had been added to the process of discovery. Then, as the untouched regions of the globe shrank until only those surrounding the poles remained terra incognita, explorers equipped for the task—as not all of them were—had opened before them vast fields of endeavor in the uncharted flora and fauna of geographical areas otherwise sufficiently well known. This Second Great Age of Discovery, as one historian has called it, has been seen as a scientific revolution.[5] Its lure was reflected in the lives of all of the early field naturalists. Rafinesque's career, which merited much of the censure it received during its course and provokes dispute even today, is illustrative of an approach to science long since superseded by a more prosaic and less impetuous manner of discovery.

Adventure as a Motive

Though most of the details of Rafinesque's early life are unknown to us, his autobiography is explicit about some of his earliest interests. In addition to the standard subjects taught him by tutors, "I learnt many other things by myself," he wrote, "as I was greedy for reading any book I could get; but I liked travels above all. Those of Leguat, Le Vaillant, Cook, &c., gave me infinite pleasure: I wanted to become a traveler like them and I became such. They increased also my taste for natural history, in which Pluche and Bomare became my guides and manuals."[6]

Left unmentioned in his published autobiographical writings was a French author who had a profound influence on Rafinesque, and very likely—either directly or indirectly—on many other early naturalists as well. This was Rousseau's friend and disciple, Bernardin de Saint-Pierre, whose novel *Paul et Virginie* Rafinesque wept over at the age of 12. The novel's vision of prelapsarian innocence in a state of nature contributed to a romantic ethos that was by no means confined

Rafinesque among the Field Naturalists

to literature and art. It helped to establish a mind-set in Rafinesque which caused him to exclaim that "every pure Botanist is a good man, a happy man, and a religious man! He lives with God in his wide temple not made by hands."[7] Moreover, odd as it seems, even though it was fiction, the novel *Paul et Virginie* was intended to be the fourth volume of Bernardin's projected multi-volume *Études de la Nature* (3 vols., 1784), which originally was planned as nothing less than a complete natural history in the manner of Aristotle. However, its author was wise enough to give up the larger enterprise when he saw that an exhaustive study of the strawberry plant alone would take a lifetime. Of the three volumes of the *Harmonies de la Nature* (1796) that Bernardin later published, Rafinesque wrote that "they would be my library in the solitude" of the wilderness "if ever I was to be deprived of all books."[8]

The little we know about the naturalist's father, François G. A. Rafinesque (1750–1793), appears in Reynaud's genealogical note (p. 74), but it is enough to suggest that following in the paternal footsteps promised to be the life of adventure the son was seeking.[9] International commerce at that time was fraught with physical as well as financial danger. In 1802, when the Philadelphia merchant John D. Clifford—who later became Rafinesque's best friend in Kentucky—had shipped as supercargo to Livorno, where he first met the young Rafinesque, he referred to the voyage in his letters as a "merchant adventure." As well he might, for, preparing to return his ship the *Philadelphia* to home port, he wrote to his father and brother about the need to guard against the Barbary pirates. To calm their apprehension, he said he would sail "in company with the armed merchant ship *Serpent* of Baltimore," expecting as a result that "there will be little danger from the Tripolitans"—i.e., in modern context, the Libyans.[10]

It was through the Clifford contact that Rafinesque came, aged nineteen, to Philadelphia for his apprenticeship. Probably the same connection enabled him, three years later, to work for Abraham Gibbs in Palermo,[11] where, he said, he made a small fortune in trade himself, which enabled him then to devote his full attention to natural history. And it was Clifford again—since transplanted to Lexington, Kentucky, where he became a trustee of Transylvania—who wangled for Rafinesque the only university post he ever held. Had Clifford lived longer, Rafinesque mused, "he might have become for me the Cliffort" who was the patron of Linnaeus.[12]

Here Rafinesque's analogy was not very apt. He might better have drawn a parallel between himself and Linnaeus's traveling collector, Pehr Kalm. Though he liked to think of himself as the Linnaeus of his age, as the lawgiver for natural history, the truth is that Rafinesque never was comfortable plodding along in the careful, systematic way that had brought about Linnaeus's monumental *Genera Plantarum* and *Species Plantarum*.[13] He wanted always to be in the vanguard of exploration, tingling with the thrill of discovery.

After ten years there, Sicily proved too small an arena for Rafinesque's vaulting ambition; he proposed that Sir Joseph Banks send him as a naturalist to explore Australia. Nothing came of the venture, just as nothing had come of his earlier hope that Jefferson would send him on some expedition to explore beyond the Mississippi. On his return to the United States in 1815, he was shipwrecked off the coast of Connecticut—losing his savings, collections, and books—and henceforth had to support his travels as best he could, all means of patronage having failed.[14]

It was three years later, in the summer of 1818, that he first made his way across the Alleghenies and down the Ohio as far as Shawneetown, in Illinois Territory—what he called "my great Western tour of 2,000 miles"—and, returning, visited John D. Clifford in Lexington, where Clifford pulled strings to enable him to teach there the following year. Without disparaging the hardships of such a trip, we cannot help noting that it was a mere stroll in comparison with Thomas Nuttall's several journeys. Yet, in a special way, the trip was indeed a great one for Rafinesque. At the falls of the Ohio, just downstream from Louisville, he had an experience comparable to Darwin's in the Galápagos. In a sense, the experience for both men

Fig. 5.1. Falls of the Ohio, a focal point of Rafinesque's field research. Here, a fossil madrepore he dug out of the Ordovician limestone reminded him of the cap worn by Marianne, the symbolic female figure who stands for France. He named it Turbinolia phrygia *because this "petit espèce ressemblant à un bonnet phrygian renversé." Courtesy of the Filson Historical Society.*

was a secular epiphany—for it changed their lives. For Rafinesque, the consequences were immediate; Darwin had to wait another twenty-four years before he could publish *The Origin of Species* (1859), where his theory of natural selection unraveled such puzzles as those posed by the Galápagos tortoises and finches.[15]

Rafinesque's account of the western trip is sketchy in his autobiography, *A Life of Travels* (53–58), but as Pennell has shown, pp. 10–13, in a remarkable series of letters sent back East while he was on the road, Rafinesque disclosed how excited he became with the joy of discovery during this trip. One letter, datelined "Falls of the Ohio, 20 July 1818," poured out a torrent of discoveries (mostly in conchology and ichthyology), and the writer observed that "fossils are numberless in the valley of the Ohio." Returning from the barrens of Western Kentucky and the prairies of Southern Illinois eight weeks later, he again wrote from the falls of the Ohio that his discoveries had "continued to exceed" his "most sanguine expectations." And he concluded his last letter written on this trip by exclaiming: "The total of New Animals discovered this year by me is 583! of Plants 125!"[16]

Three consequences for Rafinesque's career resulted from his great Western tour which had come to a radiant focus at the falls of the Ohio. The first was his teaching position at Transylvania University, which lasted six stormy years following the death of his protector Clifford. The second was his publication program to make known the Western discoveries, which resulted in such a flood of papers that it was one of the causes—though probably not the most important one—for his contributions being banned by the editor of the *American Journal of Science*, who said that single-handedly Rafinesque could have filled the whole journal. When other manuscripts were rejected about the same time by the Academy of Natural Sciences of Philadelphia, Rafinesque thereafter turned to French-language journals in Europe, to obscure American media such as newspapers and literary journals, and finally to self-publication in pamphlets and books having almost no circulation—all of which led most of his peers to conclude that they had no obligation to read him at all. And the third consequence—which could have been most important of all, had Rafinesque been more reflective and less impetuous—was the confirmation of his views on biological variation.

Consequences of Ohio Valley Field Research

The seven years Rafinesque spent in the Trans-Allegheny West provided collections he continued to exploit the rest of his life. When he left Kentucky in the spring of 1826, it took forty crates to hold the plunder he shipped back to Philadelphia. These years also caused Rafinesque to confront fundamental biological problems, any one of which, had he pursued it as exhaustively as he searched for new species, might have brought him more lasting fame than the exploit of devising 6,700 Latin plant names. The issue of biological variation was one of these. Among other factors that he glimpsed in the course of collecting living plants, animals, and fossils was the importance of plant geography[17]and the significance of index fossils as a means of dating sedimentary geological strata.[18] But Rafinesque never could pause long to think about the meaning of a discovery; he plunged headlong after fresh adventures—in the field whenever possible, from other sources when unavoidable.

During the last decade of his life, he often alluded to the impermanence of species (even genera) as an explanation for his having found life forms in the West undetected by others. Thus, writing of the lopseed plant *(Phryma leptostachya)* that grows in Kentucky, for which other botanists distinguished but a single species within the genus, he described three distinct species, adding that if others choose to consider them "mere varieties," then at least "they afford a fine illustration of incipient species forming under our eyes in our woods."[19]

In recent years, Rafinesque's views on the mutability of species have brought him considerable praise as an early evolutionist; but, in truth, as Elmer D. Merrill has pointed out, he was largely reflecting the views of the eighteenth-century scientist Michel Adanson.[20] Far from having anticipated Darwin, as some have claimed, he more closely resembled Charles Darwin's own grandfather, Erasmus Darwin, who also recognized variation in plants. Rafinesque sought, unsuccessfully, to honor Erasmus Darwin (not Charles) by naming a plant genus for him. After the *Origin of Species* occasioned a furor, Charles Darwin searched earlier literature for quotations to support his own views and printed them in "An Historical Sketch of the Progress of Opinion on the Origin of Species," prefixed to his third and later editions. There he quoted from Rafinesque's *New Flora of North America* (I: 6) that "all species might have been varieties once, and many varieties are gradually becoming species"; this citation has given rise to the notion that Rafinesque "anticipated Darwin." But if mutability were the only issue in Darwinism, then a line of precursors stretched back as far as Aristotle, as Darwin himself well knew.

Rafinesque seems never to have considered it a problem that he had no way to account for the origin of new species, which Charles Darwin found in natural selection, and Rafinesque's conception of historical time—still the Biblical one—was far too confined. Yet, the most mature expression of Rafinesque's opinion about species is worth noting, and it is almost totally unknown because it appears not in any of his technical publications but as a long footnote to a poem. The book-length poem, *The World or Instability* (1836), is itself in the tradition of Erasmus

Rafinesque among the Field Naturalists

Darwin's *The Botanic Garden* (1791) and was written as a philosophical meditation on mutability. There Rafinesque declared:

> A species is the collection of all the individuals acquiring distinct forms and colors, and all the deviations that can breed together. They are abstract terms of our own; Nature only acknowledges individuals,[21] and var[ies] them constantly; so as to produce new species now and then, particularly among plants. Genera vary also, but so slowly, as not to be easily perceived. It is probable that new genera are also forming, and that all our generic and specific form[s] of animals and plants have been produced by successive deviations from the original types discovered among the fossils of the former earth.[22]

The most serious thought Rafinesque ever gave to the principles involved in classification must have been prompted by the problems forced on him by the collections he made in the vicinity of the falls of the Ohio. Since he was known already as a notorious "splitter" of botanical genera, the way he was compelled by the evidence to treat certain mollusks found there suggests the degree of judgment he was capable of but too often failed to exercise.

In the letters to Collins he had mentioned his discovery of dozens of species of a fluviatile bivalve to which he there gave the manuscript generic name *Potamila*.[23] These bivalve mussel shells, found in great abundance along the Ohio and especially at the falls, presented "infinite anomalies in form and structure," he wrote, causing him later to doubt whether some of them should be fitted into the genus *Potamila* after all. He was, in short, presented with the kind of classification problem Darwin found in the variation of beaks among the Galápagos finches. Specimens of both creatures exhibited a graduated series without any sharp breaks between species. If arranged in a series of some length, the specimens at either extreme were clearly of different species but at no point in the series was it possible to say the first species stopped and the second one began. In his earliest professional publication on the subject, Rafinesque attempted to classify his specimens provisionally by accommodating all of these shells under eight subgenera of the established genus Unio,[24] and the genus *Potamila*, though his own invention, was abandoned for the time being. After pondering the problem yet another year, what he finally decided was "to give the name *Unio* . . . to all my new species, in conformity with the views of naturalists," for the elaboration even of more subgenera "would render the definition of the species prolix." Despite the fact that the genus *Unio* would "then consist of more than seventy species," he was unable to find justification in the evidence for dividing it.[25] Of course, in his final analysis of the Ohio River mollusks he did find it necessary to call on twelve additional genera

Charles Boewe

and sixteen additional subgenera. As a result, Cain concludes that "nearly every mussel in his samples did indeed belong to a different species; and whereas at the time his introduction of numerous genera instead of the all-embracing *Unio* was merely irritating, it is now seen as justified."

Yet, his lust for discovery seldom afforded Rafinesque leisure for such extended examination of a problem. It is not generally known that during the last year of his life, when he probably was already suffering from the cancer which caused his death, he read the recently published *Journal of Researches* (1839) of the voyage of the *Beagle*, still exploring vicariously as he could no longer do on foot. In six pages of cramped notes he made on the book, his halting writing recorded that "Charles Darwin . . . was a naturalist," for, to be sure, this Darwin was unknown in Philadelphia at that time; Rafinesque went on to note that "seldom Engl[ish] travelers" are "so well informed." He jotted down Darwin's references to many of his own varied interests—ballooning spiders, a new octopus, lizards, frogs, fossils, and tortoises—and he exclaimed with delight that in the Galápagos Darwin had found "26 land birds all new!"

Finally he succumbed to old habits. From Darwin's words and woodcut illustration he derived the description of a fungus he himself had never seen—"sessile, globular, bright yellow, on beech trees . . . outer skin with small circular pits," etc.— and resoundingly declared it to be a new species of a new genus named *Endelmis edulis* Raf.[26] The passion for discovery pursued him to the grave.

Notes

1. See Pennell, pp. 47–48, for the full quotation of this often cited Rafinesque panegyric on the joys and sorrows of research in the field.

2. Rafinesque, unpublished MS, "Lecture on Knowledge" (Nov. 7, 1820), American Philosophical Society.

3. Jeannette E. Graustein, *Thomas Nuttall* (Cambridge, Mass., 1967), who asks, "How did an English youth before he reached his majority become committed in his heart to so unlikely a project" (12) as Western exploration? "His writings give no clue," she adds.

4. Rafinesque's *A Life of Travels* was written first in a shortened form, in 1833 in French, and published long after his death as *Précis ou Abrégé des Voyages, Travaux, et Recherches de C. S. Rafinesque,* ed. Charles Boewe, Georges Reynaud, and Beverly Seaton (Amsterdam, 1987). The Original French form was sent to France for publication in the *Bulletin de la Société de Géographie,* failing which the author hoped it would be published in the *Annales des Voyages.* Thus he saw geographers and explorers as the appropriate audience for the story of his life, not naturalists.

Rafinesque among the Field Naturalists

5. William H. Goetzmann, "Paradigm Lost," *The Sciences in the American Context: New Perspectives,* ed. Nathan Reingold (Washington, D.C., 1979), 21–34. Goetzmann's title implies that the natural history research of the "Second Great Age of Discovery" is a paradigm in the sense explicated by Thomas S. Kuhn in *The Structure of Scientific Revolutions,* 2d ed. (Chicago, 1970).

6. Rafinesque, *A Life of Travels,* 8. François Leguat (1637–1735) was a French traveler to India; François Le Vaillant (1753–1824) explored Africa and described African birds; James Cook (1728–1779) was the British navigator who explored the coasts of Australia, discovered the Hawaiian Islands, and died there; and Jan Janszoon Struys (d. 1694)—included in the French but dropped from Rafinesque's English version of his autobiography—traveled through Greece, Muscovy, Tartary, Persia, India, and Africa. Such was the company of adventurers the youthful Rafinesque hoped to join. Nöel Antoine Pluche (1688–1761) was the author of an eight-volume illustrated compendium of natural history; J. C. Valmont de Bomare (1731–1807) published a five-volume *Dictionnaire Raisonné, Universel d'Histoire Naturelle.* See Cain, pp. 134–37, for additional literary sources that probably influenced Rafinesque.

7. Rafinesque, *New Flora of North America,* I: 15.

8. Rafinesque, unpublished MS, "Ideas on Harmonies of Nature," American Philosophical Society. Undated, this MS appears to have been written at the time of Bernardin's death, therefore, ca. 1814.

9. Nor was such a life incompatible with botanical interests. As Reynaud points out, it was the elder Rafinesque's navigator, Pierre Blancard, who had introduced the chrysanthemum to Europe. Blancard discovered this ornamental on one of his voyages to the Far East.

10. John D. Clifford to Thomas and John Clifford, Leghorn, Mar. 1, 1802, Clifford Family Papers, Historical Society of Pennsylvania. On a Mediterranean adventure two years earlier, Clifford turned a 30 percent profit on muslins in the same city but, unable to sell nankeens at any price, took the *Philadelphia* to Palermo in company of the *Martha,* which, he pointed out, carried sixteen guns. To explain his commercial problems to his partners in Philadelphia, he conveyed such international intelligence as the Prussians had withdrawn from Italy, Genoa was blockaded by the British, and Bonaparte had defeated the Austrians (John D. Clifford to the Clifford Brothers firm, Leghorn, June 29, 1800, Historical Society of Pennsylvania).

11. Gibbs, though a British subject, was U.S. consul at Palermo during Rafinesque's Sicilian residence. As banker to Lord Nelson, he had come to Sicily when the court of the Kingdom of the Two Sicilies had to flee there for safety under British protection. Clifford's letters mention him as a merchant in Naples as early as 1802. He shot himself to death in 1816, leaving a note that those he trusted were unworthy of the trust. Happily, C. S. Rafinesque had returned to the United States the previous year.

12. Rafinesque, *A Life of Travels,* 114. Clifford's adventures did not end when he turned his attention toward Kentucky. In 1803 he was mugged by a gang of vicious whites while he was passing through the Choctaw Nation in Alabama. In 1804 he loaded fifty tons of cotton on the brigantine *Louisiana* and sailed with it from the mouth of the Cumberland River down the Ohio, through the Mississippi, and on to Liverpool. He died suddenly in 1820, in Lexington.

13. Grotesque as it may appear, his early book, *Analyse de la Nature* (Palermo, 1815), whose epigraph was "La Nature est mon guide, et Linnéus mon maître," was intended to be the Rafinesquian equivalent of Linnaeus's *Systema Naturae;* his *Principes fondamentaux de somiologie* (Palermo, 1814) of Linnaeus's *Philosophia Botanica;* and even his *Good Book and Amenities of Nature,* the periodical launched in the last year of his life that saw but a single issue, echoed the Linnaean title *Amoenitates Academicae.* While wishing to emulate the Linnaean product, he deplored the Linnaean method in botany, the so-called sexual system.

14. The proposal to Banks is mentioned in *A Life of Travels,* 41. Rafinesque met Jefferson in Washington and wrote to him on November 27, 1804, stating his intention to visit Ohio and Kentucky the following spring (which did not occur), adding, "I wish I could go still farther and across the Mississip[p]i into the unexplored region of Louisiana." The offer of employment in Sicily arrived suddenly, for Rafinesque was gone before Jefferson's ambiguous reply, written two weeks later, reached him. See Edwin M. Betts, "The Correspondence between Constantine Samuel Rafinesque and Thomas Jefferson," *Proceedings of the American Philosophical Society* 87 (1944): 368–80.

15. To be sure, Gertrude Himmelfarb shows in *Darwin and the Darwinian Revolution* (Garden City, 1959) that Charles Darwin failed to appreciate the significance of the biological puzzles he had turned up in the Galápagos until he reached home and had the help of specialists to sort out his collections.

16. Unpublished, these letters exist in manuscript at the American Philosophical Society. Collins's docket on the back indicates that most of them were read to the Academy of Natural Sciences of Philadelphia. At the same time, Rafinesque must also have copied out the substance of them and sent it to New York, where his discoveries were summarized in a series of articles in the *American Monthly Magazine and Critical Review,* which Pennell neglected to mention. This magazine also had been faithfully reporting the activities of the New York Lyceum of Natural History, of which Rafinesque was a founding member.

17. One example is Cosmonist VIII, "On the Botany of the Western Limestone Region," *Kentucky Gazette* Apr. 4, 1822, where Rafinesque also touched on the idea of plant succession.

18. With Clifford, Rafinesque undertook to produce "a general Descriptive History of all the fossil remains of Kentucky"—an effort that ended with Clifford's death (C. S.

Rafinesque among the Field Naturalists

Rafinesque to Samuel Burnside, Lexington, Oct. 2, 1820, American Antiquarian Society). The need for this was expressed in his review of *Observations on the Geology of the United States,* by William Maclure, where Rafinesque wrote that "we must especially collect and describe all the organic remains of our soil, if we ever want to speculate with the smallest degree of probability, on the formation, respective age, and history of our strata" (*American Monthly Magazine and Critical Review,* 3 [May 1818]: 41–44).

19. Rafinesque, *New Flora of North America,* II [1837]: 37–38.

20. Elmer D. Merrill, *Index Rafinesquianus: The Plant Names Published by C. S. Rafinesque with Reductions, and a Consideration of His Methods, Objectives, and Attainments* (Jamaica Plain, Mass., 1949), 47–48. Adanson, *Families des plantes* (Paris, 1763).

21. As early as 1814 he began to doubt that the categories of biological nomenclature corresponded with reality in nature: "La Nature n'a peut être créé que des Individus ou tout au plus des Espèces, toutes les autres Dénominations ne sont que des notions idéals inventées par nôtre imagination, pour nous faciliter la connaissance des objets"—possibly Nature creates only Individuals or at the most Species, all the other Designations being only ideal notions invented by our imagination, to facilitate our knowledge of objects (*Principes fondamentaux de somiologie,* 13).

22. Rafinesque, *The World or Instability* (Philadelphia, 1836), 229. The source commonly cited for Rafinesque's evolutionary conviction is his letter to John Torrey printed in *Herbarium Rafinesquianum* (Philadelphia, 1836), 11–12, where he wrote that "I shall soon come out with my avowed principles about G[enera] and Sp[ecies]"; namely, that "Species and perhaps Genera also, are forming in organized beings by gradual deviations." Neither here nor elsewhere did he ever speculate on the cause of these gradual deviations. He concluded with the hope that "If I cannot perform this give me credit for it." Rafinesque did not "perform" it, nor did Torrey give him credit.

23. When he conveyed much the same information to the New York Lyceum of Natural History in a letter dated from "Falls of the Ohio, 20th July 1818" the ending of the name was changed—whether by Rafinesque or the editor we cannot tell—and the genus was designated *Potamilus,* one of three "new genera" (*American Monthly Magazine and Critical Review* 3 [Sept. 1818]: 355). Only recently, under "Opinion 1665," the generic name "Potamilus Rafinesque, 1818 (Mollusca, Bivalvia)" (*Bulletin of Zoological Nomenclature,* 49 [Mar. 26, 1992]: 81–82) was officially conserved by the world's zoologists.

24. Rafinesque, "Prodrome de 70 Nouveaux Genres d'Animaux Découverts dans l'Intérieur des États-Unis d'Amérique, durant l'Année 1818," *Journal de Physique, de Chimie, et d'Histoire Naturelle* 88 (1819): 417–29.

25. Rafinesque, *A Monograph of the Flaviatile Bivalve Shells of the River Ohio* (Philadelphia, 1832), 10–11, trans. by C. A. Poulson from the *Annales Générales des Sciences Physiques* 5 (1820):

287–322. Moreover, eleven years later he added two additional species to the genus *Unio*, described from specimens brought to Philadelphia from the Ganges River in India (*Continuation of a Monograph of the Bivalve Shells* [Philadelphia, 1831], 7).

26. These notes probably are unknown because they appear beneath an incongruous title. They are in a MS notebook: "Book 43d or Z.Y.; Materials for the History of the American Nations . . . By C. S. Rafinesque, 1839," American Philosophical Society. Later editions of Darwin's book note that the fungus was described by J. M. Berkeley and named *Cyttaria darwinii*.

Rafinesque on Classification

Arthur J. Cain
(1921–1999)

Rafinesque had the benefit of a polyglot early environment, a good acquaintance with French culture and science, much commercial experience, and wide travels as a naturalist. In his appreciation of the necessity for a natural system of classification, his familiarity with European scientific literature put him ahead of most of the older American botanists, who were convinced Linnaeanists. He corresponded with many other scientists (for example, Cuvier, Bory de St. Vincent, and Persoon), and sent specimens to several of them. He had no university education (on which he congratulated himself) but was extremely widely read in all sorts of literature, and thought himself capable of taking up any subject whatever and improving it.

I. Intellectual Influences

His formative years were at a time of great intellectual ferment and political change. The French Revolution broke out when he was six; in the same year George Washington became the first president of the recently formed United States. In 1792 France declared war on Austria and Prussia; in 1793 Louis XVI was executed; in 1795 Napoleon Bonaparte came to the fore with his "whiff of grapeshot" and began his victories in 1796 by defeating the Austrians. In 1798 he became master of Egypt, when Rafinesque was fifteen. In 1803 the United States bought Louisiana and New Orleans from the French. In 1804 Napoleon was crowned emperor, in 1812 he was forced to retreat from Moscow, and the Americans were defeated by British forces at Queenston Heights, stopping the further invasion of Canada. Napoleon lost the battle of Waterloo on June 18, 1815. On December 10, 1817, the State of Mississippi was added to the Union; Illinois became a state on December 3, 1818. Equally tumultuous were the histories of Turkey, India, and South America at this period.

But intellectually the period was one of the most fascinating in modern history; and since the pace of scientific change was not so great as it is today, we must go back well before Rafinesque's birth for some of the great books and controversies that were still exerting their influence in his young days and later—some even up to the present day. As Rafinesque was interested in everything, one must mention the preamble to the constitution of the United States (1787), Clarkson's *Essay on Slavery* (1786), Jeremy Bentham's *Introduction to the Principles of Morals and Legislation* (1789), Thomas Paine's *The Rights of Man* (1791–92), Condorcet's *Tableau of the Progress of the Human Spirit* (1793), Malthus's *Essay on the Principle of Population* (1798), Bentham's *Civil and Penal Legislation* (1802), the Code Napoléon (1809), and Robert Owen's *A New View of Society* (1813).

In science pure and applied there was a plethora of inventions, books, papers, and discoveries, of which a few of the notable books can be mentioned. In the inorganic sciences, the following are outstanding:

1784	R. J. Haüy, *Attempt at a Theory of the Structure of Crystals*
1786	W. Herschel, *Catalogue of Nebulae*
1788	J. Hutton, *New Theory of the Earth*
1789	A. L. Lavoisier, *Elementary Treatise of Chemistry*
1796	P. Laplace, *Exposition of the System of the World*
1797	N. de Saussure, *Chemical Researches on Vegetation*
	J. L. Lagrange, *Theory of Analytical Functions*
1807	J. L. Gay-Lussac, *Observations on Magnetism*
1808–27	J. Dalton, *New System of Chemical Philosophy*
1809	K. F. Gauss, *Theory of the Movement of Celestial Bodies*
1810	J. W. Goethe, *Theory of Colors*
1812	H. Davy, *Elements of Chemical Philosophy*
1814	J. J. Berzelius, *Theory of Chemical Proportions and the Chemical Action of Electricity*

But of course it was in natural history that the most relevant works appeared. Linnaeus's *Philosophia botanica* (1751) and *Critica botanica* (1737) were fairly

Rafinesque on Classification

widely accepted when Rafinesque was young, but the whole basis of Linnaeus's practice in classification, accepted in Germany, Scandinavia, and Britain, had been questioned in France right from his first promulgation of it. The French tradition stemmed from Tournefort; perhaps its ablest exponent was A. L. de Jussieu, in his *Genera plantarum* (1789), which set out a natural arrangement as against Linnaeus's artificial one.[1] As Stafleu points out (xxv) in his excellent introduction to the Cramer reprint (1969) of Jussieu's *Genera plantarum*, both Linnaeus and A. L. de Jussieu, and indeed his brother Bernard de Jussieu and Adanson, all achieved within their very different classification systems a high degree of naturalness when judged by the percentage of genera still placed in the same families today. It was at the higher levels that they differed greatly, and if Linnaeus's system, based on the numbers of stamens and pistils, is confessedly artificial (he also published a list of natural groups), the higher classification of A. L. de Jussieu's natural system looks rather rigidly artificial at the present day. However, a natural system was less easy to use as a key for identification, which was the great virtue of Linnaeus's. Rafinesque had no doubt about the necessity for a natural classification, but apparently also no doubt that he could provide trenchant definitions for every entity in it, which others had found more difficult.

Rafinesque's earlier years came at a wonderful time for exploration, and for botanical and zoological discovery. New species and genera were pouring in, and naturalists first tried to cram them into Linnaeus's classification. Then (when this proved obviously impossible) there was an outburst of genus making; in this Rafinesque was following a trend of his own time.

The journeys of Captain Cook (1768–79), P. S. Pallas in Siberia (1768–69), James Bruce (1772) in Abyssinia and nearby in Africa, George Vancouver (1790) in northwestern North America, and Mungo Park (1795), to name only a few, had opened up vast tracts and made hitherto unsuspected faunas and floras available for investigation, those of Australia being the most extraordinary. There was an enormous output of local and regional natural histories, monographs, descriptions of private and public collections, new classifications, and speculative theories. Moreover, from about 1796, Cuvier's lectures and writings in Paris virtually founded modern comparative anatomy, and greatly upset Linnaeus's higher classifications. And Lamarck's *Philosophie zoologique*, with its doctrine of evolution, clean contrary to Cuvier's ideas, appeared in 1809.

It was a time of questioning everything, of controversy over everything, from the exact classification of barnacles to the very foundations of society. Adanson, now widely hailed as the remote progenitor of numerical taxonomy, had not only adopted a totally different approach to the natural classification of plants, but had even produced a phonetic spelling for use in French, and published his books in it—probably the best thing he could have done to antagonize his compatriots generally. Rafinesque especially mentions Adanson as among those who influenced him scientifically.

The greatest single influence on him, however, was Linnaeus. On the title pages of *Précis des découvertes et travaux somiologiques* and the *Analyse de la nature* respectively, he proclaims that "He has chosen the genius of Linnaeus for his guide" and "Nature is my guide, and Linnaeus my master"; the *Principes fondamentaux de somiologie* is dedicated to Linnaeus as founder of the methodical study of organized bodies (and to Buffon as their delineator), and in the introductions he explicitly claims to be the successor and perfecter of Linnaeus (see especially the *Précis*). It is clear that he was attracted by Linnaeus as the universal classifier, the genius who had given laws for definitions and nomenclature to the world. Conscious of his own genius, Rafinesque resolved to complete the edifice of which Linnaeus had laid the foundations. And, since the glorious sentiments of the French Revolution, the American Revolution, and the English Utilitarians were obviously right, he would be as much a benefactor to mankind in his exposition of the structure of universal knowledge—an essential basis for the enlightenment of Mankind—as in his inventions and commercial innovations.

The style of some of Rafinesque's works may well make us wonder what use they were intended for; and several authors have commented unfavorably on the lack of definition of many of his names (e.g., Jordan 1887), the brevity of the descriptions (Merrill 1948) and Rafinesque's hastiness and lack of thoroughness (Pennell 1945). More than one, including people who knew him personally, declared roundly that he was mad at least in his later years. Whether this is so or not, I think some criticisms of his earlier work rest upon a misunderstanding of the genre in which he was working. He must be judged by the usages, standards, and knowledge of his time, not of the present, and it is not always clear that his later critics (justifiably irritated though they may be) have appreciated this point.

II. Rafinesque's Scientific Usages

French Vernacular Names

A major point in understanding Rafinesque arises from the usages of the period in French scientific circles. It was standard practice with Cuvier, Lamarck, Latreille, Duméril, Blainville, and others to give generic and other group names in French—which required the production of a vast number of new names, formed from the official taxonomic names—for hitherto unknown animals and plants.

Rafinesque on Classification

Thus Latreille (1817) gives the aquatic pentamerous carnivorous beetles as forming a tribe, "Hydrocanthares" (*Hydrocanthari* Latreille) and the first genus is "Les Dytisques" *Dytiscus,* Geoffroy. (This practice was carried so far by some authors that genera were proposed and named only in the French form, thus creating later a nomenclatural difficulty.) French being a Romance language, equivalents could most often be made with ease, suiting the genius of the language, although Cuvier's "Les Koala" (1817, I: 184) looks a little odd. Since French had become the universal diplomatic language with the conquests of Napoleon, and had already enjoyed for several centuries a primacy in culture, it is possible that some patriotic French people were even contemplating the eventual supersession of the Latin international scientific nomenclature, standardized by Linnaeus, with French equivalents, and preparing for it by a popular nomenclature that could eventually take its place.[2] Certainly their efforts were much more practicable than the corresponding Victorian popularizations in England; one does not feel that the conversion of *Trifolium subterraneum* into "Subterranean Trefoil" helped Victorian schoolchildren very much, still less *Claytonia perfoliata* into the "Perfoliate Claytonia," while the transmogrification of two unfortunate freshwater mussels, *Unio pictorum* and *U. tumidus,* into "Thin Painter's Union" (why thin, anyway?) and the "Tumid Union" was downright grotesque. (Engelmann [1896: 303] notes also a work by Freiherr K. von Meidinger [1787] on a German systematic nomenclature for the Linnaean system.)

Rafinesque adopted this procedure wholeheartedly and provided near-transliterations into French of all his names, however uncouth, but he insisted on retaining the Latin names as well.[3] It is only fair to emphasize that the practice was a normal one, although particular products of his must have set many teeth on edge.

Families

A further point explains a good deal more of the strangeness of his names. Among taxonomists in general, even when the word *family* was used to denote particular groups, it was a recent innovation, not provided for in Linnaeus's rules, and employed (often irregularly and sporadically) with no fixed termination. Thus Latreille (1817) gives for the first order of insects "Les Myriapodes" (*Mitosata* Fabricius), for the first *famille* "Chilognathes" (*Chilognatha* Latreille) and for the second "Les Chilopodes" (*Chilopoda* Latreille), for the second order "Les Thysanoures" (no Latin name) and for its two families "Des Lepismènes" (*Lepismenae* Latreille) and "Des Podurelles" (*Podurellae* Latreille). Lindley (1836) says in the preface to the second edition of his *Natural System of Botany,*

Arthur J. Cain

I have ventured upon a reformation of the nomenclature of the natural system, by making all the names of divisions of the same value end in the same way. The orders are here distinguished by ending in *aceae*, the suborders in *eae*, the alliances in *aes*, and the groups in *osae*.[4] To some it may seem that such alterations are fanciful, but I think it will be found that many advantages and conveniences will attend the establishment of uniformity in these matters. I fear, however, that I have in some cases been obliged to offend against the laws of construction in order to carry this into effect; but I trust it will be found that I have done so only in cases of inevitable necessity.

Swainson (1836: 230) gave the endings of tribes as *es*, families as *idae* or *adae* and subfamilies as *inae*, remarking that "This plan of designating the groups in question has been so extensively employed, more especially in ornithology, that it will now be adhered to by all who desire to establish a fixed nomenclature. It is not so material that the names of the higher groups should have definite terminations, because they are comparatively so few, and are so well known, that the change would not be productive of any real advantage." But I note that in Stephens's *British Entomology*, vol. I (1828), family names in *idae* were already in fully consistent use.

The peculiar endings (as they now seem) of Rafinesque's groupings above the genus, are not, then, a piece of eccentricity but are conformable to the usages of the school of systematists he followed. But although Lindley apologized for departing from the strict classical rules of combination and inflection of words, Rafinesque had no inhibitions whatever in this matter.

Definitions

Rafinesque modeled himself on Linnaeus, and Linnaeus's own definitions have been criticized for brevity and occasional unintelligibility (e.g., by Dodge [1952, 1959] on the Testacea). In the tenth edition of Linnaeus's *Systema Naturae*, each group below the Imperium Naturae (the Kingdoms and the Classes) that receives special treatment in a prefatory essay is defined in as few words and using as few characters as possible. Each definition (and this applies equally to the actual definitions of the greater groups) is therefore comparative, serving only to differentiate each group from related ones. It follows that the definitions are relative to the knowledge of the time. Linnaeus himself was quite explicit on this. When new species were discovered, the names (i.e., definitions) of old ones in a genus would have to be altered; and this is why he thought it impossible to give more than a generic definition to a species which is the only one in its genus—how could one know what were only specific characters until other species had been discovered? They would share with the formerly unique species various characteristics which could then be recognized as generic, while others which turned out to be limited to the first species could only be specific to it. (Linnaeus, indeed, claimed that with a good definition one needed no illustration, illustrated books being appallingly expensive; but he mitigated the bleakness of this attitude by often referring to the illustrations of others immediately after a specific definition).

Rafinesque, therefore, thought himself fully entitled to make his definitions as short as he could, provided that in the then state of knowledge they differentiated the forms he was defining. A long and prolix description merely confused the reader and was too great a burden on the memory. The definitions of Lamarck in the *Natural History of Animals without Vertebrae* (1815–22)—especially the generic ones—were often a good deal longer than Linnaeus's, but he had far more species and genera to separate, and the technique was still the same, although Lamarck made much more use than did Linnaeus of informal comment and description in his "Observations," appended to many genera and higher groups.

Nomenclatural Rules

Nothing can have given more offense to his contemporaries than Rafinesque's incredible proliferation of (usually undefined) generic names, unless it was his alteration or rejection of so many already in use. Yet in this he claimed to be merely applying consistently or improving Linnaeus's own legislation, and it must be admitted that there was truth in this claim. The fact that the "improvement" was really a grotesque and tasteless exaggeration of Linnaeus's practice to the point of downright caricature would not have occurred to Rafinesque.

The points in Linnaeus's practice and legislation that Rafinesque specially seized on are:

Hybrid names formed from both Greek and Latin are to be rejected. (*Philosophia botanica*, aphorism 223)

Generic names ending in *-oides* are to be rejected. (*Phil. bot.* 226)

Generic names with a similar sound are liable to give rise to confusion. (*Phil. bot.* 228)

Generic names which are not derived from Greek or Latin are to be rejected (*Phil. bot.* 229). In the *Critica botanica* under this aphorism Linnaeus comments that an unbroken series of several consonants is not good.

Generic names formed from names of good botanists are to be retained as an act of piety (*Phil. bot.* 238) but (*Critica botanica*, 238, comment 3) they should be made easier to pronounce; thus *Barreliera* becomes *Barleria*. And (comment 5) over-long names are to be cut down; thus *Gundelsheimera* becomes *Gundelia*.

Arthur J. Cain

The terminations of generic names, and the pronunciation are to be
made as easy as possible (*Phil. bot.* 248) (but not altered to become confused
with others, *Critica botanica,* under 298); but unfamiliar endings can be
altered under certain (specified) circumstances because the Romans
followed this practice in adopting Greek words into Latin.

Generic names (in addition to those formed from botanists' names) are
to be avoided if too long, hard to pronounce, or unpleasant *(nauseabunda)*
(*Phil. bot.* 299). Over-long names are those with more than 12 letters; thus
Kalophyllodendron is shortened to *Calophyllum, Titanoceratophyton* is replaced by
Isis. Difficult names have too many consonants, for example *Acrochordodendros*
and *Alectorolophus.* Unpleasant names are unlike proper botanical names and
uncouth (*Critica botanica,* under 239); thus *Caraxeron* is to be replaced,
Potamogeiton shortened to *Potamogeton.*

In several of his rules, however (*Phil. bot.* 239, 241), Linnaeus provided for
the retention of names if, although they did not conform to his rules, they were
harmless, or familiar, or used in pharmacy. Also, if a name had to be changed, a
new one should be made which is reminiscent of the old one (*Critica botanica* under
aphorism 242). And (*Phil. bot.* 293) if a name is good, it is not to be altered to an
even better one.

Rafinesque, however, paid little or no attention to those of Linnaeus's apho-
risms which would help toward nomenclatural stability; and he "improved" on his
others by determining that names which were too short should also be rejected.
Moreover, he contracted not only generic names derived from modern personal
names, but any generic (or higher) names, in such a way that their meaning by
derivation was totally obscured. Now a modern personal surname would have no
Greek or Latin derivation; Linnaeus surely thought that the contraction (above) of
Gundelsheimera to *Gundelia* would obscure nothing; and it is to be noted that when
he contracted *Kalophyllodendron* (beautiful leaf tree) to *Calophyllum* (beautiful leaf)
he preserved a meaning. Indeed, in his examples in the *Philosophia botanica* and the
Critica botanica he preserves a meaning in all such shortened forms, when he does
not substitute a completely different name. Linnaeus was very concerned to avoid
criticism from classical scholars, as well he might be in his day when Latin was the
universal scholarly language. Not a good linguist himself, he was anxious to avoid
apparent solecisms, and to prevent other scholars from laughing at botanists and
botany. Several of his explanatory comments in the *Critica botanica* make this concern explicit. In Rafinesque's time, the classics were still a major part of a good
education; yet he wholly ignored Linnaeus's practice and shortened many of his
and others' generic and higher names to the point of total incomprehensibility.
His friend William Swainson (1840) commented that Rafinesque's names were

euphonious and classically derived; but Swainson, who pushed the quinary system to outrageous lengths (see, e.g., Cain 1984) was himself a rather strange personality. Merrill (1949) surprisingly said that Rafinesque was "very adept in selecting short euphonious generic names for the new entities that he proposed," but points out that he often wrecked the meaning, so that for those who are intrigued with the meanings of generic names, most of which are derived from classical sources, Rafinesque's work presents many puzzles.

This judgment is an anachronism. When a word has a meaning, it is far more easily remembered, and Linnaeus was most emphatic on the necessity of remembering genera (Cain 1958). He was equally emphatic that each name should wherever possible convey a distinct *idea* of the characteristics of the genus. What could be better than *Helianthus*, Sun Flower, which brings the color and shape of the flower immediately to mind (*Critica botanica*, under aphorism 290)? Only a true botanist, therefore, can bestow a good generic name, since only he understands the genera (*Critica botanica*, comment under aphorism 218). No sane person, he says (aphorism 220), introduces primitives as generic names, because primitives are words of no known significance. (Linnaeus would allow, however, ancient classical names known to denote plants, aphorism 241.) Generic names borrowed from a different science or craft have meanings appropriate to their origin and must be rejected as ridiculous because of their meanings (*Critica botanica*, under aphorism 231, where Linnaeus gives a characteristically coarse, not to say obscene, example, to laugh the practice out of court).[5] And, as is often the case, if one can't think of a truly expressive name and uses a commemorative one from some botanist or patron of botany, one can always "discover a link by which to connect the name with the plant, and indeed there will be such charm in the association that it will never fade from . . . memory" (Hort's translation, *Critica botanica*, under aphorism 238). Linnaeus then proceeds to give some hilarious examples of how to do this, scoring vigorously off some of his opponents in the process.[6]

A name had to have meaning (one way or another) for Linnaeus. Rafinesque often so contracted his constituent words that very few could work out what they meant; Merrill gives one of the few examples actually explained by Rafinesque, *Diodeilis* "abridged" from *Diodontocheilis*. But what Merrill obviously regarded as matter for a mild antiquarian curiosity in a few botanists so inclined, was to Linnaeus the whole soul and spirit of naming. Take one more example from Rafinesque—his dicotyledonous class *Eltrogynia*. He says himself it is from two Greek words signifying "free women." The correct form of the first syllables must therefore be *eleuthero*, found as the first element of various compound words in classical Greek. By no known linguistic process could this have contracted to *eltro,* and no one without Rafinesque's explanation could have expanded that on any principles of grammar or phonetic change to *eleuthero.* It cannot be explained by Rafinesque's knowledge of modern Greek, in which language there would be merely a change of the *u* to *v.* Moreover, with all such barbaric contractions, the corrigibility of the word would be lost. A misspelling of "eleutherogyne" would be easily detected by anyone who knew its derivation from well-known Greek words; but the new form would have to be learned, as what Linnaeus would have called a primitive and detested. Rafinesque knew well what he was doing; but to justify his practice by saying only that a meaning "is not absolutely necessary," as he said in his own Rule 15, was hardly sufficient.

By requiring the shortening of all long names, and the lengthening of short ones, Rafinesque was able to change a huge number of generic names. By splitting the Linnaean genera and those of others—a rather more defensible process— he was able to introduce hundreds of names of his own coining. The resulting upset in established botanical nomenclature is well surveyed by Merrill (1949) in the prefatory pages of his *Index Rafinesquianus.* I do not know of any corresponding survey for the zoological names,[7] but the results would be exactly the same. One demand of Linnaeus that Rafinesque quite threw overboard was aphorism 241, with its commentary in the *Critica botanica,* which begins in Hort's translation:

> 241. The generic names, Greek or Latin, given by the Fathers of Botany, if they are good, should be retained, as also should those which are most famil-iar, or which are officinal [used in medicine].
>
> With this principle of ours agrees that of Ray: "There should be as little innovation as possible, nor should names be altered which are received in com-mon use, or commonly employed in the writings of physicians, while the con-fusing and erroneous implications which they give them should be avoided. Even those moderately acquainted with the history of plants know what obscu-rity and bewilderment the multiplication of names has introduced."

Nothing, I believe, in Jussieu, Candolle, Cuvier, Lamarck, or any other writer of the period suggests that they would not endorse these excellent remarks.

Encyclopedic Endeavor

Linnaeus himself set out to classify in detail all three kingdoms of nature—animal, vegetable, and mineral—and it was not to be expected that Rafinesque should do less. There was a long tradition, going back to the early years of the eighteenth

Rafinesque on Classification

century, of encyclopedias, dictionaries, lexicons, handbooks, and the like, summarizing and classifying all knowledge, in English, French, and Latin. (There is no evidence whatever that Rafinesque was influenced by the more magical encyclopaedic treatises of the sixteenth or seventeenth centuries.) I suspect that it was not formal taxonomic practice alone that influenced him. Many compendia existed that were little more than memoriae technicae or introductions to the vocabulary of subjects, and he could regard himself as producing one of these classified verbal inventories.

One of the most celebrated of such authors was Benjamin Martin (1704– 1782), whose *Philosophical Grammar; being a View of the Present State of Experimented Physiology, or Natural Philosophy,* etc. (1735) was not only divided into four parts, Somatology, Cosmology, Aerology, and Geology, highly reminiscent of Rafinesque's nomenclature, but in 1749, the year of the seventh English edition, was translated into French. The French edition was reprinted in 1764 and again in 1777. When we remember the remarkable prestige of English works and ideas in France in the eighteenth century (e.g., Hazard 1953) and the role they played in preparing for the Revolution, and add to that Rafinesque's admiration for French science, it seems probable that this style of compendium, wholly compatible with Linnaeus's own practice, also influenced him in his presentation. In Martin's *Philosophical Grammar,* the inevitable aridity of the style is somewhat relieved by its being cast into dialogue form (it is not surprising that this one of his productions was chosen for translation), but the speeches of the instructor, B, apart from their beginnings and endings, are simply large chunks of erudition which could have been lifted directly from his other works.

Of these, Martin's *Bibliotheca Technologica; or, a Philological Library of Literary Arts and Sciences* (1st ed. 1737, 2nd 1740) in 25 sections, also divides "Physiology or Natural Philosophy" into four parts, Somatology (nature of matter in general), Uranology (constitution of the heavens), Aerology (the atmosphere "and the various Meteors thereof"), and Geology ("which takes a View of the Earth and Sea, with all their various Productions"). Uranology includes Heliography, Selenography, Planetography, Cometography, and Astrography; Aerology comprises Aerography, Anemography, and Meteorography. Geology is divided into Geography and Hydrography, the former into Geography in the strict sense, Mineralogy, Phytology, and Zoology. Zoology has as its parts Anthropography, Zoography ("of Beasts or Brutes," i.e., of mammals), Ornithography, Ichthyography, Entomography, Herpetography, and Zoophytography in that order, reptiles being taken in a very general sense to include "worms, snails, caterpillars, etc."

Arthur J. Cain

Similarly, Anatomy, "the Art which teaches the true knowledge of the *Human Body* principally, but of any *Animal Body* in general," is composed of Osteology, divided into Osteogeny, Osteography, and Synosteology or Synosteography, and Sarcology, divided into Myology, Splanchnology, and Angiology or Angiography. This last is further subdivided into "(1.) *Neurology*, or the Doctrine of the *Nerves;* (2.) *Arteriology*, of the *Arteries;* and (3.) *Phlebotology*, which treats of the Veins." (It is noticeable that although the study of the eye, heart, etc., is detailed under Splanchnology, Martin was apparently too early to revel in cardiology, ophthalmology, otology, laryngology, and a number of other terms in very necessary use in present–day medicine.)

Under some of the subdivisions in the *Grammar*, especially of groups of living things, are what are really mere lists of names. "Of the *first Species* [i.e., insects with a single metamorphosis—species in the logical, not zoological, sense] are the *Libellae* or *Pertae* produced from an Insect with 6 Feet; the *Cimices Silvestres*, having the Figure of St. Andrew's Cross on their Backs. The *Locusts*. The *Gryllo–Talpa*, or Mole Cricket. *Crickets* of all Kinds. The *Grasshopper*. The *Blatta*. The *Ephemera*, which lives but a Day. The *Water Scorpion*. *Water Flies* of several sorts, &c." The reader presumably feels he has been given some idea of the contents of such a group, and is provided with names to recognize if he proceeds to consult a natural history to further his knowledge. The use of such a list to a beginner would be considerable; even a more advanced student might use it as a sort of checklist.

Even in Ephraim Chambers's highly influential *Cyclopaedia: or, an Universal Dictionary of Arts and Sciences* (1728, editions in 1738, 1739, 1741, 1746), which stimulated Diderot and d'Alembert's famous *Encyclopédie ou Dictionnaire raisonné des sciences, des arts et des metiers*, 35 vols. (1751–80), and is arranged alphabetically, a preface explains how knowledge is divided into its parts and what those parts are, with lists of all the terms to be looked up under each section, so that the reader can appreciate the words of the title page "The whole intended as a Course of ancient and modern Learning." Chambers saw his work as a preservative against narrow-mindedness and fanaticism (1738, I: xxv), a means of "encreasing our sensibility, to the making our faculties more subtile and adequate, and giving us a more exquisite perception of things that occur; and thus enabling us to judge clearly, pronounce boldly, conclude readily, distinguish accurately, and to apprehend the manner and reasons of our decisions. . . ." Diderot and d'Alembert also saw their work as highly educative, and a major influence in their war against authority, obscurantism, and tyranny. I am fairly sure that Rafinesque used the format of Benjamin Martin's works, and I suggest that he did so in the spirit of Chambers's remarks just quoted, and that of the *Encyclopédie*.

Rafinesque on Classification

Such encyclopaedic surveys, then, acting as lexicons of technical terms and names, and reasoned inventories of subjects, even partially of groups (see, e.g., the article Animal in Chambers's *Cyclopaedia*, which gives a synopsis of John Ray's classification among much other information) were a well-established genre of general and scientific letters.

Within professional zoology of the most modern description, Duméril's *Analytical Zoology, or Natural Method of Classifying Animals, Made More Easy by Synoptic Tables* (1804) is described by its author (xvii) as forming as a whole a single vast synoptic table showing all known genera of animals (but using dichotomous division). Many years later (1830) the same author's *Elements of the Natural Sciences* (originally entitled *Elementary Treatise of Natural History*, 1804) would give a conspectus of all Nature for the benefit of young students, beginning as always with a division of all natural bodies into inorganic and living or organized. Rafinesque, if somewhat old-fashioned in his actual practice, could feel that he was contributing to a solid tradition and the best modern practice. In particular, his *Analyse de la nature* gives a "Tableau of the Universe," which in essence is an expansion and modernization of the *Introitus* and *Imperium Naturae*, the celebrated introductory (and exhortatory) essays at the beginning of the tenth edition of Linnaeus's *Systema Naturæ*, volume I, and indeed of some of Linnaeus's orations. Rafinesque's method, with clear comparative definitions of the greatest possible terseness, was that of the greatest systematist of all time, and of the authorities (French) of his own day. The lists he gives under the groups in the *Analyse de la nature* are wholly creditable. It is very probable, moreover, that Rafinesque had a far more exalted view of encyclopedic surveys than his actual practice suggests.

Conclusions

Rafinesque's practice, then, if he is read in his context, is not quite as grotesque as it has appeared, granted that in some respects he was old-fashioned in his own day. Moreover, he could regard himself as having some precedents to quote for what he was doing. Linnaeus *had* contracted names out of grammatical recognition, and by his new rules caused a major break in generic (and specific) nomenclature. Lamarck and others *had* created many new genera and the Linnaean genera were clearly insufficient; A. L. de Jussieu *had* emphasized the necessity for a natural classification, and Linnaeus himself had said that the man who could attain to it would be for him the Great Apollo. Linnaeus, also, had insisted in the preface to the *Critica botanica* that one must have rigorous laws of nomenclature, and observe them rigorously, to achieve a universal system and clarify the ideas of genera. "Nor is there any need for one to be afraid about assigning a new name to a plant which has been already named by some collector: for the name which he gave is bound to be untrue in so far as he did not understand genera . . ." (*Critica botanica,* under aphorism 210, Hort's translation). But Linnaeus himself had not been wholly consistent.

Nevertheless, one must admit that Rafinesque's development of Linnaeus's rules would be taken at the time as wholly contradicting the intentions of them, and producing a chaos in generic nomenclature; moreover, his practice, so characteristic of an idiosyncratic and isolated worker, of publishing without defining, or giving definitions so short that they were useless, was indisputably bad. And his criticisms of others were unjustifiably severe even in that outspoken age. It is not surprising that many of his contemporaries rejected his work completely.

In a few taxonomic groups, notably the North American freshwater bivalves, his practice is now seen to be well in advance of his age. Merrill (1949: 63) suggests that in such fields "this is due to the fact that he was there the pioneer explorer, and the situation was not confused by much work of earlier specialists. . . ." Rafinesque could generate his own confusion in any group, and this cannot be the reason. I suggest that in the amazingly species-rich fauna of the Ohio River, Rafinesque found material that thoroughly suited his extraordinary temperament—nearly every mussel in his samples did indeed belong to a different species; and whereas at the time his introduction of numerous genera instead of the all-embracing *Unio* was merely irritating, it is now seen as justified. It is certainly true that he saw distinctions and groupings, in taxonomy and biogeography, that were ignored by his contemporaries and that he met with harsh and unjustified treatment at their hands. But one is as much inclined to sympathize with them as with him. Had he paid more attention to the sobriety of method, nomenclature, and definition of his great contemporaries, Cuvier, Lamarck, Candolle, and their predecessors, he would not have alienated his own acquaintances so thoroughly.

Notes

1. [A simple explanation of the distinction between these two systems of classification appears later, pp. 205–6. **Ed.**]

2. [Writing to Persoon (*Précis des découvertes* [Palermo, 1814], 6) about his plan to publish a series of works that would advance the science of classification and nomenclature, Rafinesque said, "J'imprime maintenant ici et en *français,* langue que je désire

substituer a la latine"—I now print here and in *French*, the language I desire to substitute for Latin—his pamphlet titled *Principes fondamentaux de Somiologie* (Palermo, 1814). Though his magazine, *Specchio delle Scienze*, had been produced in Italian, for local consumption, the following year when he issued his *Analyse de la Nature*, which was intended to introduce him to the savants of the Continent, that too, though printed in Sicily, was in the French language. **Ed.**]

3. [Early in his career Rafinesque also made one unsuccessful experiment of providing English equivalents for his Latin binomials. In a letter of September 1, 1807, to Samuel Latham Mitchill, which the latter printed in his *Medical Repository*, Rafinesque translated *Cerastium velutinum* as "velvet mouse-ear," but having no equivalent for the generic element of *Callitriche terrestre* all he could make of it was "terrestrial callitriche." Other examples are *Aster leucanthemus*, which he rendered as "white star," but had to leave *Smilax heterophylla* as "heterophyllous sarsaparilla." He gave "semi-hearted mock-plantain" as the English equivalent of *Alisma subcordata;* "triflorous garlic" as that of *Allium triflorum;* and "uniflorous euphorby" for *Euphorbia uniflora*. Thereafter he was content to list English common names only when they were already known, and no longer tried to devise new ones from Latin. **Ed.**]

4. [Rafinesque: "[T]he uncouth latin plural of EAE, against which I protest as a barbarism, and shall ever use instead the elegant termination of IDES both singular and plural, adapted to the Greek, Latin, English, French and nearly all languages. . . ." *The Good Book* (1840), 39. **Ed.**]

5. [In this aphorism, Linnaeus gives examples of generic names used by anatomists and pathologists that, he says, "would make even the dullest bottle-washer laugh" if employed in botanical nomenclature: Auriculae, Tunicae, Umbilici, Clitoridis, and Priapi. Nevertheless, at one time or another, genera of flowering plants were designated *Auricula* by Tournefort, *Tunica* by Hall, and *Umbilicus* by Candolle. Linnaeus himself, in a fit of forgetfulness—dare we use the verb botanists usually employ in this context?—erected the genus *Clitoria* in 1737. The name of the wayward son of Dionysus and Aphrodite probably never struck botanists as appropriate for a flowering plant; but Lamarck, with a marine worm to classify, aptly named it *Priapulus*. And Linnaeus himself must have known that as early as 1564 the Dutch botanist Hadrianus Junius had the happy inspiration to give the genus of the stinkhorn fungus the emblematic name *Phallus*. **Ed.**]

6. [Rafinesque did the same. See an example in chap. 17, n. 6.]

7. [There is no index of Rafinesque's zoological names comparable to Merrill's for the botanical ones. The next chapter, however, shows how the zoological names often were derived by the same process and gives some indication of the extent of their present acceptance. **Ed.**]

Arthur J. Cain

References

Cain, A. J. 1958. Logic and Memory in Linnaeus's System of Taxonomy. *Proceedings of the Linnean Society of London* (169 Session): 144–63.

———. 1984. Islands and Evolution: Theory and Opinion in Darwin's Earlier Years. *Biological Journal of the Linnean Society of London* 21: 5–27.

———. 1992. The *Methodus* of Linnaeus. *Archives of Natural History* 19: 231–50.

———. 1995. Linnaeus's Natural and Artificial Arrangements of Plants. *Botanical Journal of the Linnean Society* 117: 73–133.

Candolle, A. P. de. 1813. *Théorie élémentaire de la botanique ou exposition des principes de la classification naturelle et de l'art de décrire et d'étudier les végetaux.* Paris: Déterville.

Chambers, E. 1738. *Cyclopaedia: Or, an Universal Dictionary of Arts and Sciences; Containing an Explication of the Terms, and an Account of the Things Signified Thereby, in the Several Arts, both Liberal and Mechanical; and the Several Sciences, Human and Divine: the Figures, Kinds, Properties, Productions, Preparations, and Uses of Things Natural and Artificial: the Rise, Progress, and State of Things Ecclesiastical, Civil, Military, and Commercial: with the Several Systems, Sects, Opinions. &c among Philosophers, Divines, Mathematicians, Physicians, Antiquaries, Critics &c. The whole Intended as a Course of Antient and Modern Learning . . .* 2nd ed., 2 vols. London: Midwinter, etc.

Cuvier, M. le Chevalier. 1817. *Le Règne animal distribué d'après son organization . . .* Vol. I. Paris: Déterville. Reprint, 1969. Brussels: Impression Anastaltique Culture et Civilisation.

Dodge, H. 1952. A Historical Review of the Mollusks of Linnaeus. Part I. The Classes Loricata and Pelecypoda. *Bulletin of the American Museum of Natural History* 100: 1–263.

———. 1959. A Historical Review of the Mollusks of Linnaeus. Part 7. Certain Species of the Genus *Turbo* of the Class Gastropoda. *Bulletin of the American Museum of Natural History* 118: 207–58.

Duméril, A. M. C. 1804. *Zoologie analytique, ou Méthode naturelle de classification des animaux, rendue plus facile à l'aide de tableaux synoptiques.* Paris: Allais.

———. 1830. *Élemens des sciences naturelles.* 4th ed., 2 vols. Paris: Déterville.

Engelmann, W. 1896. *Bibliotheca historico-naturalis. Verzeichniss der Bücher über Naturgeschichte welche in Deutschland, Scandinavien, Holland, England, Frankreich, Italien und Spanien in den Jahren 1700–1846 erschienen sind.* Vol. I Leipzig: Engelmann. Reprint 1960, Weinheim: H. R. Engelmann, & Codicote, England: Wheldon & Wesley (Historiae naturalis classics, vol. 14).

Hazard, P. 1953. *The European Mind, 1680–1715.* Harmondsworth, England: Penguin Books Ltd. (Pelican Book).

Jordan, D. S. 1887. Note on the "Analyse de la Nature" of Rafinesque. *Proceedings of the U.S. National Museum* 1887: 480–81.

Rafinesque on Classification

Jussieu, A. L. de. 1789. *Genera plantarum secundum ordines naturales disposita* . . . Paris: Herissant and Barrois. Reprint 1964, with introduction by F. Stafleu; J. Cramer: Weinheim (Historiae naturalis classica, vol. 35).

Lamarck, J. B. 1801. *Systême des animaux sans vertèbres* . . . Paris: Déterville.

———. 1809. *Philosophie zoologique, ou exposition des considérations relatives a l'histoire naturelle des Animaux* . . . Paris: Dentu. Reprint 1960, Weinheim: H. R. Engelmann, & Codicote, England: Wheldon & Wesley (Historiae naturalis classica, vol. 10).

———. 1815–22. *Histoire naturelle des animaux sans vertèbres* . . . 7 vols. Paris: Verdière.

Latreille, [P. A.] 1817. *Les Crustacés, les arachnides et les insectes.* Vol. 3 in the *Règne animal* of Cuvier (see above). Paris: Déterville. Reprint 1969. Brussels: Impression Anastaltique Culture et Civilisation.

Lindley, J. 1836. *A Natural System of Botany; or, a Systematic View of the Organization, Natural Affinities, and Geographical Distribution, of the Whole Vegetable Kingdom* . . . 2nd ed. London: Longman, Rees, Orme, Brown, Green, and Longman.

Linnaeus, C. 1737. *Critica botanica in quo nomina plantarum generica, specifica, and variantia examini subjiciuntur, selectiora confirmantur, indigna rejicuntur: simulque doctrina circa denominationem plantarum traditur. Seu Fundamentorum Botanicorum pars 4.* Lugdunum Batavorum [Leiden]: Wishoff. Trans. Sir Arthur Hort, London: Ray Society, 1938.

———. 1751. *Philosophia botanica in qua explicantur fundamenta botanica, cum definitionibus partium, exemplis terminorum, observationibus rariorum, adjectis figuris aeneis.* Stockholmiae: Kiesewetter.

———. 1758. *Systema Naturae per Regna tria Naturae, secundum classes, ordines, genera, species, cum characteribus, differentiis, synonymis, locis. Tomus 1. Editio decima, reformata.* Holmiae: Salvius. Facsimile reprint 1956, London: British Museum (Natural History).

Martin, B. 1737. *Bibliotheca Technologica: or, a Philological Library of Literary Arts and Sciences* . . . London: Noon.

———. 1738 *The Philosophical Grammar; Being a View of the Present State of Experimented Philosophy in Four Parts . . . The Second Edition, with Alterations, Corrections, and Very Large Additions by Way of Notes.* London: Noon.

Merrill, E. D. 1948. Foreword to Rafinesque, *Précis des découvertes et travaux somiologiques* 1814. Reprint 1948, by Peter Smith. Wakefield, Mass.: Murray Printing Company.

———. 1949. *Index Rafinesquianus. The Plant Names Published by C. S. Rafinesque with Reductions, and a Consideration of his Methods, Objectives, and Attainments.* Jamaica Plain, Mass.: Arnold Arboretum of Harvard Univ.

Pennell, F. W. 1942. The Life and Work of Rafinesque. *Rafinesque Memorial Papers, October 31, 1940,* ed. L. A. Brown. *Transylvania College Bulletin* 15: 10–70.

Rafinesque, C. S. 1814a. *Précis des découvertes et travaux somiologiques.* Palerme: Aux dépens de l'Auteur.

Arthur J. Cain

————. 1814b. *Principes fondamentaux de Somiologie ou les lois de la nomenclature et de la classification de l'empire organique ou des animaux et des végétaux contenant les Règles essentielles de l'art de leur imposer des noms immuables et de les classer méthodiquement.* Palerme: Aux dépens de l'Auteur.

————. 1815. *Analyse de la nature ou tableau de l'univers et des corps organisés.* Palerme: Aux dépens de l'Auteur.

Stephens, J. F. 1828. *Illustrations of British Entomology; or, a Synopsis of Indigenous Insects: Containing Their Generic and Specific Distinctions . . . Mandibulata.* Vol. 1. London: Baldwin and Cradock.

Swainson, W. 1836. *On the Natural History and Classification of Birds.* Vol. 1. London: Longman, Reese, Orme, Brown, Green & Longman, and Taylor.

Swainson, W. 1840. *Taxidermy with the Biography of Zoologists.* London: Longman & Co.

The Historical Background and Literary Sources of Rafinesque's Mammalian Taxonomy

CHARLES BOEWE

THE SUCCESSFUL COMPLETION of the Lewis and Clark expedition across the Louisiana Purchase opened an unprecedented arena of discovery for American naturalists. Meriwether Lewis was a close observer who had been personally coached by Jefferson and advised by the American Philosophical Society on what to look for, but he was not a professional naturalist. William Clark was a skilled map maker and he produced respectable drawings of natural objects, but he had only a layman's knowledge of natural history. The explorers' steps were soon followed, however, by a series of professional naturalists. Some of them, like Thomas Nuttall, were supported by private Eastern philanthropy; others, like Edwin James, Titian Peale, and Thomas Say, were participants in subsequent federal expeditions.

Rafinesque was a naturalist who very much wanted to be part of the scientific exploration of the West, but various circumstances prevented his ever crossing the Mississippi River. Failing to get a berth on any expedition sponsored by the federal government and never able to find private financial support for his own western travels, Rafinesque had gone back to Europe in 1805 after his first two years in the United States, and remained in Sicily for the following decade.

On his return to the United States in 1815, he was pleased to find the Lewis and Clark journals in print—that first flawed version edited by Biddle and Allen (1814)—and he was soon to make good use of them. The lure of Louisiana was so strong that he took a bold—and surely ill-advised—step to exploit its riches. From the published account of an obscure French traveler, Claude Cesar Robin, he detailed a small portion of the flora of the newly admitted state of Louisiana in a book titled *Florula Ludoviciana,* implying that a known part along the Gulf Coast could stand for the whole until the rest of the state—later the entire territory of Louisiana, stretching all the way to the Pacific Northwest—could be explored botanically. When this book appeared in 1817, he incurred the ill-will of botanists that has lingered to this day because he named not only plants he never saw but also split off new genera from established ones on the basis of some real, but many fancied, distinctions he detected in the descriptions of Robin, an avowed amateur botanist.

The process seems to have been little recognized, but many of Rafinesque's contributions in zoology came about in much the same way. In addition, S. S. Haldeman, writing his obituary "Notice of the Zoological Writings of the Late C. S. Rafinesque,"[1] sniffed that "he was very credulous, which led him to believe the exaggerated accounts of the vulgar; and to write essays and found 'species,' upon grounds which should be beneath the notice of any naturalist." And such was the decline in Rafinesque's reputation that when Elliott Coues came to annotate the second published version of the Lewis and Clark journals in 1893 and had to deal with the animals Rafinesque had "discovered" in their pages, he grudgingly referred to Rafinesque as "that rattle-headed genius."[2] The question naturally arises why Rafinesque was always suspect when the taxonomic conclusions of such sedentary naturalists as George Ord and Richard Harlan—neither of whom ever stepped foot outside Philadelphia—were considered deserving of respect. For instance, in 1845, while he was working to improve the text of Audubon's *American Quadrupeds,* John Bachman queried a correspondent in New York for a copy of the magazine where, he said, Rafinesque "established the genus Lynx." In this, Bachman was as wrong as Rafinesque himself, both of them being unaware that Robert Kerr had segregated *Lynx* from *Felis* in 1792. Bachman went on to say that he wished Rafinesque "had never written, but since he has we are compelled to look at what he has done, if it is only to throw it off again."[3] It is true enough that today zoological names are conventional labels not required to be descriptive,[4] but nineteenth-century naturalists did take a conceptualist view and did agonize over the selection of appropriate Latin names to express what they considered to be a salient and easily recognizable feature of the object they were naming. Moreover, though not yet formally codified, there was a decent respect for priority as well as the practical consideration that names familiar by custom should not be wantonly changed unless there were compelling reasons for the change.

An instructive example of the problem these naturalists faced occurs where Lewis and Clark refer to the "sheep" they saw in the Rocky Mountains but

Rafinesque's Mammalian Taxonomy

elsewhere speak of the "bighorn" without any qualifier. This ambiguity caused Coues to sputter that, while the bighorn *is* indeed a sheep, Lewis and Clark's "sheep" is in fact a goat. It happens that in this case the Philadelphia naturalists had more than a written description to guide their decision. A skin given by Lewis to Peale's Museum was examined by George Ord, who then exhibited one of the detached horns at the Academy of Natural Sciences of Philadelphia. Rafinesque, coming down from New York where he was then living, happened to be present at this meeting. Though Ord knew perfectly well that the bighorn is a sheep, such was the power of Lewis's written words that he was compelled to believe the goat skin also was that of a sheep, though from a species of sheep having very unusual horns. In May 1817, he published a substantial account of the animal in the academy's journal,[5] where he quoted the Lewis and Clark notes, described the quality of the hair on the skin, gave a life-size cut of the horn (about four inches long; the animal was undoubtedly a young one, Ord wrote, for the skin measured "three feet from the insertion of the tail to the neck"), and named the creature *Ovis montana,* a name he had published earlier (1815) in his contribution to the second American edition of Guthrie's *Geography.* Back in New York, Rafinesque, writing to Ord on other business, concluded his letter with the curt comment: "Your *Ovis montana* is not a Sheep, I told you so when you exhibited a horn at the Academy."[6] Meanwhile, not having seen any scrap of the animal, A. G. Desmarest, in Paris, writing in Cuvier's *Dictionnaire des sciences naturelles* (1816–30), located the creature among the goats in the *Capra* genus—which was good enough reason for Richard Harlan next to make it out as *Capra montana* in his 1824 *Fauna Americana.* Harlan, who was accused by some of his contemporaries of plagiarizing most of his book from Desmarest anyway, was not one to let observation stand in the way, though he, too, was a member of the Academy of Natural Sciences.

Aside from his unpublished letter to Ord, Rafinesque became embroiled in this brouhaha when he wrote a review of the first number of the academy's journal, where Ord's article had appeared. Rafinesque was still smarting over the academy's rejection of two articles of his own—the action which caused fellow botanist William Baldwin to congratulate the Academicians for having "sufficient independence to reject the wild effusions of a literary madman"[7]—and devoted most of his space to picking apart the botanical essays in the journal. But he laconically remarked of an "Account of the *Ovis montana* by Mr. Ord" that he wrongly "calls by that name the white wild sheep of the rocky mountains, which has been called *Mazama dorsata* by Rafinesque, since it belongs to that genus rather than to the *Ovis,* having solid horns not spiral."[8] Indeed, nine months earlier Rafinesque had published in New York a description of *Mazama dorsata.*[9] Rafinesque failed to mention, though, that he—as Harlan would do later—had plucked his characters for the genus *Mazama* from the same Desmarest, who had penned them in classifying a Central American swamp deer, or bracket. Rafinesque did go on to say that he was assigning Ord's specimen to a new genus of his own, to *Oreamnos*. Going back another two months, to a Rafinesque article about new mammals published by Desmarest,[10] we do find the source of Mazama revealed—along with an even more startling revelation—that Desmarest's own information had come, not from any specimen, but merely from a description in the published *Travels* (1800) in Paraguay of Félix de Azara. The outcome of all this, more through luck than through logic, is that today the authoritative list of *Mammal Species of the World*, published by Honacki and others,[11] credits Rafinesque with the genus *Oreamnos* for the mountain goat and also with the genus *Mazama* for the bracket. While Rafinesque undoubtedly examined a horn from the first, he almost as surely never laid eyes on any part of the latter, whose range barely reaches into Mexico.

In the course of two decades Rafinesque proposed more than 150 new mammalian genera, but when the circumstances of their naming are looked at closely the twelve recognized today nearly all depend on luck, not on Rafinesque's sagacity. His position is somewhat different in other branches of zoology. The first to make a systematic study of the fishes of the Ohio River, in 1820 he published *Ichthyologia Ohiensis,* a slim book that ever since has served as a baseline to understand the fauna of that river, and because he was the first in the field it has netted him the distinction of establishing generic names for many of the freshwater fishes of the whole Mississippi drainage basin. Because of the contiguous waterways including the Missouri River, perhaps it can be said therefore that Rafinesque did manage to add a mite to our knowledge of the Louisiana Territory. His publications on the nomenclature of freshwater mollusks probably should have stood up as well, but they failed to, largely because they were mostly ignored by the domineering Isaac Lea. A few of his crustacea have prevailed; he is credited with a bird or two, and also a few snakes and salamanders. His contributions in entomology are fewer, and oddly, for all his onomastic zeal, he never named a butterfly.

Among mammalian orders, Rafinesque's talent was most widely devoted to bats—a fact that comes as no surprise to those who accept at face value Audubon's best-known story about him, given here in Audubon's own words, p. 369. Whether Rafinesque's specimens were captured in Audubon's house or elsewhere, he did publish very acceptable descriptions of nine Kentucky bats; however, having classified them under the Linnaean genus *Vespertilio,* he lost credit for all when that genus was broken up. On the other hand, he presently does get credit for no fewer than five bat genera: *Tadarida* (free-tailed bat), *Nycticeius* (evening bat), and *Eptescius* (big brown bat)—all of which he observed—but also for *Eidolon* and *Vampyrum*, which he probably did not. As an indication of the cosmopolitanism of Rafinesque's bat studies, it may be noted that his free-tailed bat was observed in Sicily and published there (in French),[12] his evening bat was observed in Kentucky but published in Paris,[13] and his big brown bat also was observed in Kentucky and was published there[14]—where Rafinesque volunteered the observation that "it often comes in[to] the house at night" but failed to mention whether or not he meant Audubon's house.

The circumstances of Rafinesque's other two prevailing bat genera—*Eidolon* and *Vampyrum*—are quite different, however, and go far to explain why so many of his proposed genera have been forgotten. These appear in his book, *Analyse de la Nature* (1815), which he published in French, just before his departure from Sicily. *Analyse de la Nature*, where Rafinesque assigned to his own credit no fewer than seventy new mammalian genera, was projected as nothing less than a complete schematic natural history, covering both the plant and animal kingdoms, written in imitation of Linnaeus's *Systema naturae.* It is not surprising, therefore, that Rafinesque had to fill gaps in his personal experience by inventing generic names, hoping perhaps that someday appropriate taxa would be identified and his proposed names adopted for them (which has, in fact, happened in a number of instances). Though nearly all the new mammalian names are *nomina nuda*, nevertheless this book is the source for four of Rafinesque's currently recognized genera: the two bats; *Muntiacus,* an Asian deer he almost certainly never saw; and *Addax,* a North African antelope he possibly could have seen in a Palermo menagerie.

Returning with Rafinesque to the United States, where he remained the rest of his life (and was naturalized in 1832), we have yet to deal with his naming of three other mammals. *Geomys,* his genus for the Eastern pocket gopher, a name that has prevailed, may have been based on personal observation, but in his brief description Rafinesque cited the names of four prior authorities who may just as well have been the source of his knowledge.[15] The other two mammalian names, those for the prairie dog and the mule deer, which also have prevailed, represent in the first instance a kind of poetic justice, but both have elements of comic confusion.

When the Lewis and Clark expedition, even before it left the wilderness, managed to ship on ahead a living prairie dog for display at Peale's Museum, Philadelphians took the inquisitive little animal to heart much as Americans in our own time dote on the panda. Everyone crowded Independence Hall to see it; while its Indian name, "wish-ton-wish," was good enough for the title of a novel by James Fenimore Cooper, the stolid Academicians vied with each other to stiffen the new rodent under a proper Latin cognomen. Like the blind philosopher who described the elephant as "very like a rope" from the evidence afforded by its tail, the redoubtable George Ord, even with a living prairie dog in front of him, got it confused with the written account of William Clark's "burrowing squirrel" and decided both were marmots, naming the former *Arctomys columbianus* and the latter *Arctomys ludoviciana*. Two years later,[16] unencumbered by the living specimen, which had not survived, Rafinesque studied Lewis's description of the "barking squirrel" and devised *Cynomys socialis* for the gregarious prairie dog by splitting off the burrowing squirrel as *Anisonyx brachiura*—a name that also has not survived. Nevertheless, the nomenclature of the two quite distinct animals remained hopelessly entangled until the end of the century, when Rafinesque's *Cynomys* was restored as the genus and Ord's *ludoviciana* was taken away from the burrowing squirrel and applied as the specific epithet for the prairie dog. It seems a mark of appropriate justice that Rafinesque, who chose to throw in his lot among the Americans, should finally have the credit for establishing the generic name of a mammal so distinctly American.

Now to conclude with how the mule deer—a very American kind of deer—got its scientific name. Every deer hunter knows that, in distinction to the Eastern white-tailed deer, the Western mule deer has a longer tail tipped with black and it has long ears that resemble a mule's. Trappers and mountain men had been calling it the mule deer long before Lewis and Clark penetrated the wilderness. It was from one of these—one of the vulgar, as the proper Philadelphian Haldeman would say—that Rafinesque first learned of the mule deer. This man was said to be named Charles LeRaye, about whom nothing is known with certainty (though lack of knowledge has not forestalled speculation that he was a Frenchman of noble blood). Reported to have been a hostage of the Sioux Indians from 1801 to 1805, he figures in a narrative of his captivity published in Boston in 1812. Since the title of the book mentioned Louisiana, Rafinesque scanned this account[17] as carefully as he had those of other Western travelers. From it he derived a description of the mule deer, which he placed quite reasonably in the Linnaean genus *Cervus*, to which he then attached the elegant specific epithet of the Asian wild ass, the onager, which is *hemionus*.[18] Even if Rafinesque's eyewitness description of the mule deer was a secondhand one, it seems to be based on valid information—except for the uncomfortable likelihood that "LeRaye" himself was a fabrication. Such are the geographical inaccuracies and wild improbabilities of Indian practices reported by him that his narrative itself is now thought to be a hoax.[19]

Rafinesque's Mammalian Taxonomy

Today most North American deer are no longer classified in the genus *Cervus* but rather in *Odocoileus,* and we owe this genus to Rafinesque as well. If he could know his *Odocoileus hemionus* now designates the mule deer—and at that a deer whose range never extended east of the Mississippi River—Rafinesque might be gratified, but he surely would be astonished. *He* thought he was naming a "dwarfish oxen," long extinct, when he devised *Odocoileus.*

It came about this way. In 1832 Rafinesque was visiting a dentist friend in Baltimore, a Dr. Horace Hayden who dabbled in geology and paleontology as a hobby. Not long before, Dr. Hayden had visited a cave near Carlisle, Pennsylvania. What would strike a dentist's attention more than a couple of large teeth? These teeth had been found in the cave by a Mr. Wardel, "who had broken them from a jawbone sticking out of the lime rock," according to Rafinesque.[20] Hayden obtained the teeth, which he took to be fossils. Receiving the teeth in turn from Hayden as a gift, Rafinesque first dashed off a letter to the paleontologist Alexandre Brongniart in Paris, telling him about the discovery and including a sketch; then he dashed off in person to visit the cave himself. He confided to the French paleontologist that this find surely was a new genus, which he said, "I propose to name *Coilenodon* shortened from Coiloselenodon."[21] Though disappointed at finding nothing more in the cave when he went there, he published an account of the episode in his own magazine, the *Atlantic Journal,* giving as a woodcut a flipped copy of the sketch he had sent to Paris. In his magazine he said that because the teeth were hollow he was naming the creature they came

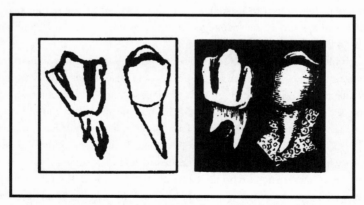

Fig. 7.1. Teeth of Odocoileus speleus. *Woodcut on the right.*

149

from *Odocoileus speleus*—Odocoileus, "teeth well hollowed," and speleus, "belonging to a cave." He further stated his belief that *Odocoileus speleus* was an extinct oxlike animal about the size of a goat. These two pictures, crude as they are, are the only visual evidence we have to authenticate any of Rafinesque's published mammalian nomenclature.

No logical process can get us from the Pennsylvania cave to the plains of Missouri, but after several years and the publication of many conflicting opinions about appropriate names for American deer, C. Hart Merriam resurrected "Odocoileus" from the *Atlantic Journal* in 1898 as "The Earliest Generic Name for the North American Deer."[22] Once that genus was accepted by zoologists for the deer found across all of North America, the mule deer fell into place as *Odocoileus hemionus*—thus conveying to Rafinesque's credit the full binomial, but by a process he never could have imagined.

Perhaps Merriam was more concerned about priority of publication than many of his colleagues, but he cannot have studied Rafinesque's article with much care. He flatly stated that the "name was based on the second or third . . . left upper premolar of the Virginia deer," a dental analysis hardly warranted by Rafinesque's crude woodcut. And Merriam expressed a conviction not substantiated by Rafinesque's own hesitant opinion when he wrote that "there is no room for doubt as to the animal to which the tooth belonged, for Rafinesque described it in detail"—as indeed Rafinesque did *not*. And finally Merriam alleged that Rafinesque "published natural-size figures of both outer and inner faces of the tooth." What Merriam took to be two faces of the same tooth were, in fact, sketches intended to represent *two different teeth*, one of which is shown in the woodcut as "united to its socket and the fragment of a jaw," according to Rafinesque. Of course, since zoological names are now merely conventional, perhaps it no longer matters.

Finally, as if Rafinesque had not done enough to complicate mammalian taxonomy, it may be mentioned that there are several allusions in his writings to a new whale, which he published in 1815. Though searches had been made, both in Europe and in the United States, none of his bibliographers ever found this publication, and the whale did not get listed among published names for mammals. At last, in 1987, a copy of the Palermo *Il Portafoglio* turned up—in Palermo, of course—where there appear not one but two Rafinesque articles about a strange cetacean that had washed up on the beach there.[23] Adopting the Linnaean genus *Balaena*, Rafinesque named the mammal *"Balena gastritis,"* a binomial that now has been entered in the Smithsonian's database of published cetacean names. Perhaps that name, too, one day will be resurrected.

Rafinesque's Mammalian Taxonomy

Notes

1. *American Journal of Science and Arts* 42 (Jan.–Mar. 1842): 280–91.

2. *History of the Expedition under the Command of Lewis and Clark,* III: 856. This edition has extensive annotations on natural history by Elliott Coues. The earliest edition, that used by Rafinesque, Ord, and others in Philadelphia, was begun by Nicholas Biddle and completed by Paul Allen (1814). When Reuben Gold Thwaites published a re-edited version of the journals (1904–5) he included many of Coues's annotations. A new edition has recently appeared: *The Journals of the Lewis and Clark Expedition,* ed. Gary E. Moulton (Lincoln: Univ. of Nebraska Press, 1983–2001).

3. John Bachman to Victor G. Audubon, Charleston, Nov. 24, 1845, Charleston Museum Archives.

4. George Gaylord Simpson, *Principles of Animal Taxonomy* (New York, 1961), 34 passim.

5. "Rocky-Mountain Sheep: *Ovis montana,*" *Journal of the Academy of Natural Sciences of Philadelphia* 1 (May 1817): 8–12. Ord's names for animals discovered by Lewis and Clark usually are cited from his contribution to the second American edition of William Guthrie's *A New Geographical, Historical and Commercial Grammar* (Philadelphia, 1815)—which does, in fact, establish his priority for the bare names. Less well known, but important to resolve these tangled issues, is a reprint of the zoological portions of this book which the bookseller Samuel N. Rhoads issued as a pamphlet (Haddonfield, N. J., 1894); in an extensive appendix, Rhoads printed Ord's second thoughts "taken from Mr. Ord's private, annotated copy" of Guthrie.

6. C. S. Rafinesque to George Ord, New York, Oct. 1, 1817, Academy of Natural Sciences of Philadelphia.

7. For this and other contemporary slurs on Rafinesque, see Ronald L. Stuckey's essay included here in the next chapter.

8. "[Review of] *Journal of the Academy of Natural Sciences of Philadelphia,*" *American Monthly Magazine and Critical Review* 3 (Aug. 1818): 269–74.

9. "Descriptions of Seven New Genera of North American Quadrupeds," *American Monthly Magazine and Critical Review* 2 (Nov. 1817): 44–46.

10. "New Species of Mammifers, Noticed in . . . Methodical Picture of the Mammiferes, by D. Desmarets [*sic*] . . . ," *American Monthly Magazine and Critical Review* 1 (Sept. 1817): 361–63.

11. James H. Honacki, Kenneth E. Kinman, and James W. Koeppl, eds. *Mammal Species of the World* (Lawrence, Kans., 1982). Compiled for the Checklist Committee of the American Society of Mammalogists, this volume is taken here to be the authoritative reference for genera and species currently recognized. Though to be sure, systematists still disagree; E. Raymond Hall, in *The Mammals of North America* (New York, 1981) insists on *Dama*

as the genus for the mule deer, though he does credit Rafinesque with *hemionus* for its specific epithet.

12. *Précis des Découvertes et Travaux Somiologiques* (Palermo, 1814).

13. "Prodrome de 70 nouveaux genres d'animaux découverts dans l'intérieur des États-Unis d'Amérique, durant l'année 1818," *Journal de Physique, de Chemie et d'Histoire Naturelle, et des Arts* 88 (June 1819): 417–29.

14. *Annals of Nature* (Lexington, Ky., 1820). Though sometimes regarded as a periodical and catalogued as such—Rafinesque subtitled it a "first annual number," but never succeeded in printing a second one—this is actually a sixteen-page pamphlet.

15. "Descriptions of Seven New Genera of North American Quadrupeds," *American Monthly Magazine and Critical Review* 2 (Nov. 1817): 44–46. Rafinesque remarked that the animal "has been called Georgia Hamster, by Milledge, Mitchill, Anderson," and "Mease."

16. Ibid. Rhoads (William Guthrie's *A New Geographical, Historical and Commercial Grammar* [pamphlet], appendix, 18) points out that Ord merely adopted a genus established by Pallas in the eighteenth century. Coues, in a rambling footnote, traces the changing nomenclature through Harlan, Bachman, Baird, Allen, Merriam, and his own technical publications. It is noteworthy that Harlan (*Fauna Americana,* [Philadelphia, 1835, 308) chose this animal in particular to trounce Rafinesque, correcting his spelling and sneering that because of his "insulated situation" at Transylvania University in Kentucky, as well as his "almost utter ignorance of the labours of other naturalists," he had been "seduced" into "grievous errors, and occasioned much confusion in natural history." Of course, Rafinesque had been in New York—not Kentucky—when he published on the prairie dog, and in both places he had reference materials equal to if not superior to those in Philadelphia. Anyway, Rafinesque had his revenge in 1832 when he exposed paleontologist Harlan's great "discovery" of the extinct Allegheny rhinoceros as being based on nothing more than an adventitious piece of sandstone, not a jaw bone as Harlan considered the rock.

17. It was published in [Jervis Cutler's] anonymous *Topographical Description of the State of Ohio, Indiana Territory and Louisiana* . . . (Boston, 1812). Cutler called LeRaye a "gentleman, who is a native of Canada"; but, if not as vulgar as Haldeman believed Rafinesque's informants to be, the dubious LeRaye certainly was no naturalist. Nevertheless, I. M. Cowan, in the most comprehensive study yet of the mule deer, *The Deer of North America*, ed. Walter P. Taylor (Harrisburg, 1956), treated him as though he had been a zoologist out on a collecting expedition, saying (342) that Rafinesque's description of the mule deer came from "field notes made by Charles LeRaye," and noting sadly that "no type specimen was designated"—as if LeRaye, if there really was such a person, should have tucked away in a museum the deer he and his Sioux captors probably ate.

18. "Extracts from the Journal of Mr. Charles LeRaye, Relating to Some New Quadrupeds of the Missouri Region," *American Monthly Magazine and Critical Review* 1 (Oct. 1817): 435–37.

Rafinesque's Mammalian Taxonomy

19. See Clyde D. Dollar, "The Journal of Charles LeRaye: Authentic Or Not?" *South Dakota Historical Collections* 41 (1982): 67–191.

20. "Description of Some of the Fossil Teeth Found in a Cave in Pennsylvania," *Atlantic Journal* 1 (Autumn 1832): 109.

21. C. S. Rafinesque to Alexandre Brongniart, Philadelphia, June 25, 1832, Muséum National d'Histoire Naturelle, Paris. Rafinesque wrote: "je propose de nommer *Coilenodon* abrégé de Coiloselenodon." It is hard to follow whatever logic is involved in going from Coiloselenodon to Coilenodon to Odocoileus, although the root for hollow is retained.

22. "The Earliest Generic Name for the North American Deer, with Descriptions of Five New Species and Subspecies," *Proceedings of the Biological Society of Washington* 12 (Apr. 30, 1898): 99–104.

23. See "The Fugitive Publications of Rafinesque," chap. 18.

Opinions of Rafinesque Expressed by His American Botanical Contemporaries

RONALD L. STUCKEY

RAFINESQUE WAS WITHOUT DOUBT North America's most versatile and controversial naturalist. Active during the first half of the nineteenth century, he was an eccentric and erratic individual of remarkable mental ability and indefatigable energy whose works generated much interest and came under continuing scrutiny, evaluation, and re-evaluation from his contemporaries and their immediate successors. Little study has focused on the reasons for which Rafinesque was largely rejected by his contemporaries. My intention here is to provide a view of Rafinesque and his botanical work derived from statements by those botanists who knew him personally or who had some knowledge of his work. These private comments on his character and work have survived in the many letters that were exchanged among the various botanists active in eastern North America during the first half of the nineteenth century.

The North American Botanical Scene, 1800–1840

Prior to 1840, botanical activity in North America centered in Philadelphia and the surrounding area, where several individuals began to compile comprehensive floras of the continent. Benjamin Smith Barton (1766–1815), professor of materia medica, natural history, and botany in the University of Pennsylvania and author of the first North American textbook of botany, projected such a flora, but himself did

Rafinesque and His Botanical Contemporaries

little toward its preparation. To aid in this effort, in 1807 Barton hired Frederick Pursh (1774–1820), a German-born horticulturist, and, following Pursh's early defection, in 1808 Barton engaged the services of Thomas Nuttall (1786–1859), an English naturalist. Each employee worked more diligently for himself than for Barton, and each published his own flora, the first being Pursh's *Flora Americae Septentrionalis* (1814) and the second Nuttall's *The Genera of North American Plants . . .* (1818).

Seventy miles west of Philadelphia in Lancaster, Pennsylvania, the Lutheran minister Gotthilf Henry Ernst Muhlenberg (1753–1815) prepared a *Flora Lancastriensis* (1785) and an *Index Florae Lancastriensis* (1793), based on herbarium specimens from the area. He exchanged specimens and information with other North American botanists, principally Stephen Elliott (1771–1830) of Charleston, South Carolina, who wrote a remarkable *Sketch of the Botany of South-Carolina and Georgia* (1816–24), and Dr. William Baldwin (1779–1819) of Wilmington, Delaware, and later of Georgia. Baldwin, an ardent botanist, early studied the flora of the southeastern United States and became an authority on sedges, but because of ill health and an early death, left most of his work in manuscript, some of which was published by later botanists.

Muhlenberg held the strong conviction that knowledge of the entire North American flora could best be assembled through the production of local floras by botanists working in their own neighborhoods. Toward that end, in addition to his own local flora, he published a *Catalogus Plantarum Americae Septentrionalis . . .* (1813) derived from information sent him by his correspondents. In the years following his death, Muhlenberg's idea had support from several botanists. Foremost among them in the early 1820s was the Rev. Lewis David von Schweinitz (1780–1834), a Moravian clergyman first at Salem, North Carolina, and later at Bethlehem, Pennsylvania, whose work in phanerogamic botany consisted of monographic treatments, a local flora, descriptions of new species, and a 357-page unpublished manuscript, *Synopsis Plantarum Americanum* (undated), containing the names, descriptions, and habitat notes on the then known flowering plants and ferns of the continent. Others who shared similar viewpoints were two of Barton's students, William Darlington, M.D. (1782–1863), of West Chester, Pennsylvania, who wrote *Florula Cestrica* (1826) and *Flora Cestrica* (1837), and Charles Wilkins Short, M.D. (1794–1863), in Lexington and later Louisville, Kentucky, who prepared a *Florula Lexingtoniensis . . .* (1828–29). With his colleagues Robert Peter, M.D. (1805–1894), and Henry A. Griswold (d. 1873), Short also compiled *A Catalogue of the Native Phaenogamous Plants, and Ferns of Kentucky* (1833), followed by four supplements (1834–40). Farther west at St. Louis, the German-born physician George Engelmann (1809–1884) took up the study of botany and was its principal western figure from the 1840s through the 1860s.

Ronald L. Stuckey

Meanwhile, Amos Eaton (1776–1842), pioneer educator in natural history and senior professor in the Rensselaer School at Troy, New York, popularized the study of botany through his *Manual of Botany for the Northern States* (1817). It saw eight editions (1817–40), the last co-authored with one of his students, John Wright (1811–1846). Eaton trained many students who subsequently contributed to the knowledge of floristic botany in eastern North America. Among them were Lewis C. Beck (1798–1853) of New York, Douglass Houghton (1809–1845) of Michigan, Robert Peter (1805–1894) in Kentucky, and John L. Riddell (1807–1865) in Ohio and later Louisiana. Eaton's foremost student was John Torrey (1796–1873), professor of chemistry at several schools in and near New York City. Torrey first learned botany, 1811 to 1815, from Eaton while the latter was in the prison where Torrey's father was an officer. Torrey was also greatly aided in his early endeavors through letters from and personal contacts with Schweinitz. Torrey published a *Catalogue of Plants . . . within Thirty Miles of the City of New York* (1819) and one volume of a *Flora of the Northern and Middle Sections of the United States* (1824), a work which began to assemble all of the known descriptive information on American floristic botany. To aid him in these floristic studies, in 1835 Torrey hired a young science teacher, Asa Gray (1810–1888), who had completed his medical education four years earlier but was not particularly interested in making a career of medicine. Gray, who was to become North America's foremost botanist of the latter half of the nineteenth century, joined in partnership with Torrey, and together they prepared a definitive and documented *Flora of North America* (1838–43). A monumental work of seven parts in two volumes, it served as an authoritative and enduring study of the North American flora, even though it left untreated the Monocotyledoneae and several families of Dicotyledoneae.

Rafinesque's Entry into American Botany

To this North American botanical fraternity came a nineteen-year-old immigrant naturalist, Constantine Samuel Rafinesque. During three years (1802–4), Rafinesque botanized as far west as the Allegheny Mountains of eastern Pennsylvania and as far south as Virginia. While traveling in America, he visited and became acquainted with "all the botanists," he said—including Barton, Darlington, Muhlenberg, and Pursh—by whom he was generally well received. Early in 1805, he returned to Europe with an herbarium of 2,400 species and 10,000 specimens that served as a basis for his later botanical publications. Headquartered in Palermo, Sicily, for the next ten years, he prepared botanical notes which appeared in Samuel L. Mitchill's *The Medical Repository* at New York (1806–11) in

Rafinesque and His Botanical Contemporaries

English, in Desvaux's *Journal de Botanique* at Paris (1808–14) in French, as well as other natural history materials published in Palermo both in Italian and in French.

After his return to North America in 1815, Rafinesque contributed to floristic and descriptive botany, naming and describing thousands of new taxa. He wrote local floras, published generic monographs, and prepared critical reviews of many of the floras written by his contemporaries. In his latter years, he published at his own expense floras of continental scope, among them his *Medical Flora* (1828, 1830), *New Flora and Botany of North America* (1836[–38]), *Flora Telluriana* ("1836" [1837–38]), *Alsographia Americana* (1838), *Sylva Telluriana* (1838), and *Autikon Botanikon* (1840). While North American botanists considered Rafinesque's early botanical publications credible, the contributions that appeared after his return to America, and especially those published after he returned to Philadelphia from Lexington in 1826, were severely questioned and criticized by his contemporaries.

Opinions Expressed by American Botanists in their Letters

Muhlenberg's and Baldwin's Opinions

Among the first to criticize Rafinesque's botanical work was the conservative and scholarly Henry Muhlenberg, who wrote to Stephen Elliott on June 16, 1809, concerning Rafinesque's "Prospectus . . . on North-American Botany . . . on the New Genera and Species of Plants Discovered by Himself . . ." (1808).

> Have you seen what Mr. Rafinesque Schmal[t]z had printed in the New York Medical Repository[1] and what he promises to publish hereafter? He makes a wonderful change and havoc amongst our plants and will do much harm if he keeps his promise. I know him personally and find a great number of my plants, which I gave him, superficially described without mentioning a word from whence he had them. Very often he makes a genus where hardly a species can be made, and where his specimen was quite imperfect. There is a medium in everything. In botany the *festina lente* is very necessary.

Muhlenberg wanted a cautious approach in the delimitation of taxa, and his comment sounded the warning that botanists should be ready to evaluate critically Rafinesque's proposals of new taxa. But Rafinesque never hesitated to express his views; he was a nonconformist, uncontrolled and uncontrollable.

The botanical correspondence between Muhlenberg and William Baldwin and between William Darlington and Baldwin was published by Darlington in

Ronald L. Stuckey

Fig. 8.1. Gotthilf Henry Muhlenberg. Courtesy of the Muhlenberg College Library.

Reliquiae Baldwinianae (1843) and contains a few comments about Rafinesque. On January 7, 1811, Muhlenberg expressed some concern about Rafinesque's new plants described from the state of Delaware: "Mr. Rafinesque was rather too quick in naming plants, and may have been mistaken in some names." A week later, on January 14, Baldwin replied: "I hope I shall be able to send some of the plants you have mentioned, as discovered and named by Mr. Rafinesque; and perhaps some others that escaped his notice." And he mildly observed that "This indefatigable Botanist has, perhaps, independent of his new discoveries, unnecessarily changed some of the *Linnaean names.*"

Following Muhlenberg's death in 1815, Baldwin's principal correspondent was his very close friend, Darlington, to whom on September 17, 1818, he made the following caustic observation concerning Rafinesque's *Florula Ludoviciana* (1817): "It is a shocking production, to come from one who has placed himself at the head of the botanical profession in our country,—and who finds fault with, and criticises all his predecessors and contemporaries." This last in allusion to the often scathing reviews Rafinesque had been publishing in *The American Monthly Magazine and Critical Review.*

Baldwin's comment might indicate that American botanists were in no mood to contend with the disruptions of this foreigner. Subsequently, Rafinesque

Rafinesque and His Botanical Contemporaries

began to have difficulty getting his manuscripts accepted for publication in America, as Baldwin related to Darlington on February 4, 1819:

> I had an opportunity, for the first time, of attending a meeting of the Academy of Natural Sciences. Two communications, on New Genera, from the redoubtable pen of Rafinesque, member of ten thousand learned societies, had been examined by a committee,—and rejected, as unworthy of publication![2] I am truly glad . . . that they have sufficient independence to reject the wild effusions of a *literary madman*. He is now in the city; but as he is *huffed* at the Academy, I had not the honor of seeing him there, and did not think it worth while to seek after him.

This event occurred in the same year that Benjamin Silliman (1779–1864) returned his manuscripts and would henceforth refuse to publish Rafinesque's papers in the *American Journal of Science*.[3] Because Rafinesque's manuscripts were being suppressed by the few North American scientific journals then in existence, he now published abroad, established his own journals as he had first done in Sicily, and published his books privately at his own expense. The first number of one of these self-published journals elicited this opinion from Lewis David von Schweinitz in a letter of April 12, 1832, to John Torrey: "I wonder whether you have seen the most extraordinary & impertinent publication which Rafinesque

Fig. 8.2. William Baldwin.

has just issued—on every possible subject, under the title of Atlantic Journal [Rafinesque (1832)]. He is doubtless a man of immense knowledge—as badly digested as may be & crack-brained I am sure. His short reviews of 23 recent works—among which your ed. of Lindley [Torrey (1831)] is likewise paraded—are truly comical."

Fig. 8.3. Lewis David von Schweinitz. Courtesy of the College of Physicians of Philadelphia.

Eaton's Opinions Expressed to Torrey

Amos Eaton enjoyed the friendship of many scientists, but no scientific confidant was closer than John Torrey. As described by McAllister (1941: 213), Eaton guided Torrey over many a pitfall of youth and inexperience, while Torrey was sympathetic, loyal, and always willing to assist Eaton in the search for scientific knowledge. Eaton, who did not meet Rafinesque until 1826, described some of his idiosyncrasies to Torrey on October 5, 1817: "I am glad Mr. Rafinesque has not set you all wild. Why can not he give up that foolish European foolery, which leads him to treat Americans like half-taught school boys? He may be assured, he will never succeed in this way. . . . His new names with which he is overwhelming the science will meet with universal contempt. Cannot some friend induce him to return to sober reason, and thus make himself highly useful and much esteemed?"

Rafinesque and His Botanical Contemporaries

On March 24 of the following year, Eaton wrote: "What is the matter with Rafinesque? In the March number of the New York Monthly Magazine[4] we are not amused with the usual quantity of the new names and wonderful acts of Mr. R. written by himself. Now do appoint some member of your society to write half a dozen pages every month in praise of Rafinesque. Then I presume he would be willing to devote part of his time to some other subject." Eaton added in sympathy and ultimate frustration: "I have defended him in New England, until I am ashamed to mention his name. His name is absolutely becoming a substitute for egotism. Even the ladies here, often adorn their witticisms with the name of Rafinesque, applied in the same. They talk of the science of Rafinesquism; meaning the most foolsome and disgusting manner of speaking in one's own praise."

Eaton's opinion of Rafinesque changed somewhat after they met eight years later, when Eaton was botanizing with his students along the Erie Canal, and Rafinesque was returning for the last time to Philadelphia from Lexington (Clinton 1910; Rezneck 1959). Of this encounter, Eaton wrote to his wife on April 3, 1826: "When we were at Rochester, the celebrated Rafinesque overtook us. He joined our party and is now with us, and is to continue on to Troy. I shall invite him to our house. He is a curious Frenchman. I am much pleased with him; though he has many queer notions."

Fig. 8.4. Amos Eaton. Courtesy of Ronald L. Stuckey.

Ronald L. Stuckey

After another decade of their acquaintance, when Rafinesque proposed giving lectures in Troy, Eaton publicly praised him in 1835 in a local newspaper, the *Troy Whig* (McAllister 1941: 257). "Even those who are disposed to pronounce Mr. R. an extravagant enthusiast, all agree, that he is a scholar of the first order; of vast reading and great classic learning. His nice discriminating talents have never been questioned. . . . 'Shades of variety' as evidence of specific differences comprise all his supposed scientific heresies."

Torrey's Opinions Expressed to Eaton, Darlington, and Short

John Torrey was perhaps Rafinesque's closest botanical friend, and their relationship lasted nearly twenty-five years. Torrey often praised Rafinesque, but overall his opinions appear inconsistent. Writing to Eaton in 1817, he singled out Rafinesque as a naturalist, but cautioned against his enthusiasm for novelty: "The circular address of Rafinesque [1816] which I send you will give you some idea of his character. He is the best *naturalist* I am acquainted with, but he is too fond of novelty. He finds too many new things. All is new! new! He has an opinion that there are no plants common to Europe and America."

Fig. 8.5. John Torrey.

Rafinesque and His Botanical Contemporaries

In reply to Eaton's letter of October 5, 1817, Torrey wrote on January 5, 1818: "Raf. certainly deserves to be ridiculed—his vanity is absolutely intolerable. . . . Is it not preposterous for a man to pretend to write a flora of a country he never even visited [Rafinesque (1817)]. If Raf. was to travel in Louisiana himself, & the very same ground that Robin did he could not find the plants he has described [in *Florula Ludoviciana*]— I shall say no more about it, but let you judge for yourself." Later, in a letter of March 21, 1818, to Eaton, Torrey returned to the *Florula Ludoviciana* (1817), writing sarcastically: "This work is the most curious medley I ever saw. The author without ever being in the country whose plants he describes, has discovered 50 or 60 new species. . . . I expect he will soon issue proposals for publishing the botany of the moon with figures of all the new species!"

On April 16, 1818, Torrey wrote further to Eaton about Rafinesque's mania for new species: "Rafinesque has just started on a three-months expedition. He goes through Philadelphia to Pittsburg[h], and intends to explore the Borders of the Lakes before he returns. You may imagine how many new discoveries he will make. He was almost Crazy with anticipation before he left here." In the same letter, Torrey also advised Eaton on the appropriateness of the name of Rafinesque's new genus *Clintonia*. "You ask me whether you shall adopt the Genus Clintonia. Although I think that the plant which has been dedicated to Mr. Clinton, should not be called either Dracenia or Convallaria, yet I do not think that [DeWitt] Clinton's attainments in Botany at all warrant its bearing his name. Nor would I adopt it if it did,—merely because it emanates from Mr. Rafinesque, for I should be afraid it would ruin my work."

A remark about Rafinesque's undeserved reputation appears in Torrey's letter of May 18, 1819, to Eaton: "The public is so much prejudiced against Mr. R. that whatever comes from his pen, let it be ever so good, is treated with greatest ridicule. It is very unfortunate for this laborious naturalist to have obtained such a character—a character which I am sure he does not deserve." Yet, Torrey himself sometimes became disgusted with Rafinesque's persistence in discussing "perpetual mutability" and other matters during overextended visits. To William Darlington, on August 12, 1831, he wrote: "Old Rafinesque is here—he talked me into a violent headache today—and threatens me with three or four more days of talk before he leaves town! He is very amusing and knows a great deal about plants and other matters—but will never advance science a single step. You doubtless know him well." Torrey described Rafinesque's scientific achievements in a similar fashion to the Swiss botanist Augustin Pyramus de Candolle (1778–1841) on August 30, 1838: "Some of Mr. Rafinesque's works possess considerable merit, but they are so full of errors, that they are seldom consulted or quoted by our naturalists. . . ."

In his correspondence with Charles Wilkins Short, Torrey commented about Rafinesque on several occasions. On October 24, 1831, he wrote concerning some Kentucky plants he had received from Short:

> I have ascertained that the leguminose plant "No. 14" is Virgilia lutea Mx. In the unexpanded state of its flowers I did not recognize it. Mr. Rafinesque has with more than his usual propriety separated this plant from the old species of Virgilia. It certainly forms a new genus. As so many mistakes have been made in our botanical books respecting it, and Mr. R. is the first to correct them, we cannot deny him the credit to which he is entitled, notwithstanding his numerous faults and blunders. His genus Cladrastis will no doubt be adopted.[5]

In a letter of October 23, 1834 asking Short about a *Hypericum*, Torrey had this adverse comment on Rafinesque's herbarium specimens: "Do let me have all the rare Hypericum of Kentucky. Do you know the Hyp. dolabriforme of Michx. I have specimens under this name, collected in Kentucky by Rafinesque—but they are miserable bits, and I am ashamed to keep them in my herbarium. Any rare Hypericums, especially the shrubby species, would be particularly acceptable at present."

C. W. Short's Opinions Expressed to J. C. Short, Torrey, and Darlington

Ordinarily, Charles Wilkins Short, a quiet, calm, kind, and most dignified gentleman, would not say anything to discredit another individual. When moved to anger, however, he must have dipped his pen in pure vitriol. He had very high standards of scientific workmanship, and could not tolerate Rafinesque's slipshod methods of collecting, preserving, and describing plants. Short (1833) had written a very creditable paper on these essential activities of the field botanist, and he followed faithfully the procedures he had outlined. From the opinions expressed in his letters, it appears evident that Short wanted to speak kindly of Rafinesque, but could not bring himself to do so because of his misgivings both about Rafinesque the man and the quality of his work.

Short had been warned of Rafinesque's excesses as early as November 15, 1818, in a letter from his closest Philadelphia friend, Dr. Edward Barton, whose words he repeated in a hurried letter to brother John Cleves Short on December 4:

> [S]ome intelligence which I have lately from the Scientific Athens of America [i.e., Philadelphia] induces me to question M[r]. Rafinesque's discriminating powers, and to alter in some measure the high opinion I had form[ed] of

Rafinesque and His Botanical Contemporaries

his proficiency in the Science of nat. history—My friend Dr. Edward Barton, of whom you have often heard me speak, and who since his return from Europe, still manifests that friendship and attention which sprung from the similarity of our dispositions & pursuits whilst students together, writes to me in this manner. "I am sorry that your new plants should fall into the hands of that maniac Rafinesque—he is so well known, that nobody pays any attention to his discoveries, and even should he hit upon something new, ten to one, whether it would be noticed. He is absolutely deranged—He makes Floras as [John] Melish does Maps—as they are called for—he will sit down and make you a Flora of a place he never saw with a fine parcel of new Genera, species &^{c}.— M^r. Correa [da Serra] says he is apprehensive Raf. will make a new genus of him[self]."

Short's later opinions were expressed in three letters to Torrey. The first, of December 17, 1834, was in reply to Torrey's request for information on the genus *Hypericum*.

I know nothing of the *Hypericum dolabriforme* Mx. nor do I see any such species mentioned by him. I suspect therefore it is one of Rafinesque's fabrications, and that the plant is not found in Ky. I have no faith in, nor patience with Raf. and they have been still lower (if possible) of late than ever. He desired me not long since to send him certain Ky. plants, expressing a great

Fig. 8.6. Charles Wilkins Short.

desire to possess them, and promising anything in return. Willing to accommodate him, and wishing at the same time to test the truth of his numberless new discoveries in this state, I told him he sh[oul]d have what he asked of me, provided he would send me some of the new genera and species which he had met with here. This he declined doing "because he could not exchange his new plants!—he could only sell them!" This confirms my suspicions that he really has them not, and that his pretended discoveries exist only in pretension. I am the more confirmed in this opinion from his having acknowledged that he never met with *Iresine Celossoides* in this state, than which a more common plant does not exist in certain localities which he must have visited. All this, however entre nous; I have no wish to injure the poor fellow, and if he can live by imposing on others I have no objections. I am sorry, however, that I have met with so few *Hypericum* this season, since they seem to be at present an especial object of your investigation.

The next summer, on August 11, 1835, Short lashed out to Torrey even more violently against Rafinesque:

Is not Rafinesque a madman! And have you honestly any confidence in him?— I sent to him this spring a considerable collection of plants, at his earnest desire, and beg'd him to let me have in return some of the many new genera and species which he had discovered in this country, that I might be aided in my search for and examination of them. Instead, however, of this his return package contained a miserable parcel of things, most of them as familiar to me as a "thrice told tale," together with some faded fragments of exotic affairs, render'd imposing and venerable by being labelled from "Etna," "Egypt," "Palestine" &ᶜ.&ᶜ.— He pronounces our common Dodecatheon a new species!, and does me the honor of offering me an Apotheosis among these twelve gods, by dedicating this species to me, which added to his twelve new species of that genus will make the full bakers dozen!—This blows up all my calculations, for I had set it down as one of my discoveries that you had gone too far in making two species, of D. Meadia & integrifolia.[6]

Can we not manage to silence this endless discoverer of new things, by raising among the botanical corps of the country an annuity for life? This if not a "divitial scheme" might at least afford him a decent competence, and prevent him from being under the necessity of gulling, or attempting to gull, the public with pulmels, fire proofs & pearl hatching,[7] together will [prevent] new genera & species without number in all the departments of nature—If I could lend a helping hand to effect this, and toward suppressing all his published

Rafinesque and His Botanical Contemporaries

nonsense I should be sure of having render'd a greater service to Science than I shall ever do in any other way.

Nevertheless, on October 25, 1840, a month after Rafinesque's death that probably was not yet known in Kentucky, Short suggested, without effect, that a goldenrod might be named in his honor:

> I have a goodly stock of the plant, gather'd in the best stage for examination, which I will send you; and will for the present conclude my remarks on it by observing that, should it prove sufficiently distinct both from *Solidago* and *Brachyris* [*Baccharis*] to form a new genus, another individual has a better claim to it than myself. Rafinesque, as you are aware, sent specimens of it to Decandolle in 1824; and although I have less confidence (entre nous) in that individual's pretensions than you seem to entertain, still as the original discoverer of the plant in question, he is entitled to any honor it may confer. I am not aware that any genus has yet been named after him;[8] and if his admits of being sufficiently euphorized, by all means crown him with this "golden wreath."

Short and Darlington were dedicated as botanists to exhaustive field work to assemble botanical specimens and knowledge, to building extensive herbaria through exchanges of specimens with other botanists, to the preparation of authoritative local floras, to the publication of their scientific work, and to the preservation of botanical history in a written form (Stuckey 1983). As early as January 15, 1829, Short wrote to Darlington about some confusion of names among some specimens that were improperly labeled. He explained:

> Some of my specimens you will very probably find improperly labelled, and some are not marked at all. This results in a great degree from their being collected at times when I had not books of reference at hand, and from my not taking time afterwards to do so; much confusion, moreover, may have arisen in their names, from the circumstance of my having shown them to Dr. or Mr. Rafinesque, who unhesitatingly called them by some name or other, which at the time I may have written down, although I have not the utmost confidence in their accuracy. No. 2 he professed alone to be unknown to him, and pronounced something new. I was unable under the circumstances in which it was found to distinguish its class, order, or genus.

When he was making botanical excursions within a fifty-mile range of Lexington and had found several plants he had not previously encountered, Short asked for Darlington's opinion of Rafinesque in a letter of July 1, 1832.

Ronald L. Stuckey

[A]lthough I cannot boast of the good fortune of my predecessor here, who in the same space, and with much less laborious search, would not have been satisfied with less than some ten or dozen new genera and thrice that N° of Species!— But what else, or what less, could have been expected of a "philosopher" in the back-woods of Kentucky who is even now, in the heart of Philadelphia, and in the ten thousand times trodden walks of Bartrams garden, making new discoveries daily!!— In sober seriousness what do you think of Rafinesque? Have you any faith whatever in his alleged discoveries? Will all the patient labours of American botanists, for the next half century, be able to extricate the science from that interminable chaos into which he will throw it, if any credence is given to him?—I had long since taken the liberty, on my own responsibility, of setting him down as an unblushing and shallow-pated pretender to everything; but this decision of mine has been somewhat shaken by seeing that Torrey is disposed to give him very considerable credit, and that the Editor[9] of the Phila. Jour. of Geology &[c] (how far competent to decide I know not) places him at the very head of American Naturalists—Nevertheless I do most heartily wish that the "philosopher" and "pulmist" could have been satisfied to restrict his investigations to the antediluvian history of China or the post diluvian chronicles of the thousand and one aboriginal nations of America.

Darlington concurred in this view of Rafinesque. On January 12, 1833, he wrote to Short:

> Your opinions of that singular genius Mr. Rafinesque, accord very much with my own; though I have but little personal knowledge of him—not having seen him for nearly thirty years. He makes sad confusion in natural science, by his itch for novelty; & never can be *authority* for any thing. I think Mr. De Candolle quotes him with too much confidence—though he often inserts marks of doubt, after his *new* genera and species. The great mischief, in natural science, is imposing *new names*, and starting new arrangements, & creating new genera, unnecessarily; and this is the begetting sin of Rafinesque.

Darlington summarized his conclusions about Rafinesque as follows, in his account of the "Progress of Botany in North America" (1849): "This eccentric and rambling naturalist—although he had, by long experience and observation, acquired considerable knowledge, and moreover issued several other botanical publications, in subsequent years,—yet he made statements so little reliable, held such peculiar views, and withal was so addicted to extravagant innovations in nomenclature, that his productions, generally, were rather injurious than beneficial to the science."

Rafinesque and His Botanical Contemporaries

Fig. 8.7. William Darlington.

Later, comparing the results of his efforts at rediscovering the genera that André Michaux named and described from Kentucky in his *Flora Boreali-Americana* (1803) with those supposedly discovered by Rafinesque, Short wrote to Darlington on July 13, 1838:

> It affords me much pleasure, indeed, to have the means of establishing so many of the discoveries of the excellent and enterprizing Frenchman [Michaux], whose reports, in relation to some of our western plants, were almost discarded. It has been my good fortune to find, in profuse abundance in many different localities, his *Bellis integrifolia* about which even Nuttall expressed a doubt; and his *Cunila glabella, Leptanthus ovalis, Pachysandra procumbens, Euphorbia dentata* &c.&c. have all occur'd to me within the last few years, and have all, I think been sent to you.
>
> Very different from this has been the result of my search after the many new genera and species of my old friend Rafinesque; many—by far the most of which, I fear, will ever repose in his own imagination, or within the Labor of that sealed volume to all human eyes—the *Autikon Botanikon* [Rafinesque (1840)]!

Ronald L. Stuckey

Gray's Published Opinion

Following Rafinesque's death, Asa Gray published a partial list of Rafinesque's botanical publications and wrote an evaluation of his work. Gray recognized that botany was Rafinesque's favorite pursuit in natural history, and that in his early days he showed considerable potential and was in some respects greatly in advance of other botanical writers, but that his labors had been disregarded or undervalued on account of his peculiarities. Of Rafinesque's earlier contributions, Gray (1841: 230–31) wrote:

> In these, he had certainly shown no little sagacity; and, considering his limited advantages, he must be deemed a botanist of unusual promise for that period, notwithstanding the defects which, increasing in after life, have obscured his real merits, and caused even his early writings to be in a great measure disregarded. The botany of the United States offered, at that time, a fine field to a botanist acquainted with natural affinities; and Rafinesque was the only person in this country, who had any pretensions to that kind of knowledge. All we can justly say is, that he possessed talents which, properly applied, would have raised him to a high rank in the science, and that he early apprehended the advantages of the natural classification, although he was by no means well grounded in structural and systematic botany. As early as 1814, indeed he sketched a general classification of organized beings [Rafinesque (1814)] to which he continued to attach great importance; but there is nothing new in it

Fig. 8.8. Asa Gray.

Rafinesque and His Botanical Contemporaries

except the names, the botanical portion being merely an anagram of Jussieu's leading divisions. His fuller developments of this system certainly contain much that is novel, but at the same time very absurd.

Of Rafinesque's later writings, Gray (1841: 234) remarked:

Many of Rafinesque's names should have been adopted; some as matter of courtesy, and others in accordance with strict rule. But it must be remembered, that the rule of priority in publication was not then universally recognized among botanists, at least as in present practice, (the prevalence of which is chiefly to be ascribed to the influence of De Candolle;) the older name being preferred *cæteris paribus,* but not otherwise. It is also true, that many of the scattered papers of Rafinesque were overlooked by those who would have been fully disposed to do justice to his labors, had they been acquainted with them; and a large portion of the genera proposed in his reviews of *Pursh, Nuttall, Bigelow* &c.,[10] were founded on their characters of plants which were doubtfully referred to the genera in which they were placed, or were stated to disagree in some particular from the other species. One who, like Rafinesque, followed the easy rule of founding new genera upon all these species, could not fail to make now and then an excellent hit; but as he very seldom knew the plants themselves, he was unable to characterize his proposed genera, or to advance our knowledge respecting them in the slightest degree. In his later publications, this practice is carried to so absurd an extent as entirely to defeat its object.

And Gray (1841: 237) summed up his evaluation of Rafinesque's work by writing that

A gradual deterioration will be observed in Rafinesque's botanical writings from 1819 to about 1830, when the passion for establishing new genera and species, appears to have become a complete *monomania.* This is the most charitable supposition we can entertain, and is confirmed by the opinions of those who knew him best. Hitherto we have been particular in the enumeration of his scattered productions, in order to facilitate the labors of those who may be disposed to search through bushels of chaff for the grain or two of wheat they perchance contain. What consideration they may deserve, let succeeding botanists determine; but we cannot hesitate to assert that none whatever is due to his subsequent works.

Gray went on to compare Rafinesque with Thomas Nuttall, even in a magazine of general circulation, the *North American Review* (1844): "Probably few naturalists have ever excelled [Nuttall] in aptitude for such observations, in quickness of eye, tact in discrimination, and tenacity of memory. In some of these respects, perhaps, he may have been equalled by Rafinesque,—and there are obvious points of resemblance between the later writings of the two, which might tempt us to continue the parallel;—but in scientific knowledge and judgment, [Nuttall] was always greatly superior to that eccentric individual [Rafinesque]."

Gray's criticism of Rafinesque's botanical contributions was the harshest that had been printed, and undoubtedly it set guidelines followed by subsequent botanists, who essentially ignored Rafinesque's works. Gray's influence persisted for nearly a century before a new generation of botanists began the task of sorting through Rafinesque's proposed botanical names, descriptions of new taxa, and surviving herbarium specimens.

Short's Opinions Following Rafinesque's Death

Short continued to remark about Rafinesque's work after his death. His final comment to Darlington was written on May 10, 1841: "So, poor old Rafinesque is dead; and all his labours 'in the deep bosom of the ocean buried!'—What has become of his collection?[11]—It must have been a queer medley; and yet I should like to have the picking of it over, for I long to see some of the many new genera and species which he found in all parts of Kentucky, which I have look'd in vain for—I have in my Herbm. a few specimens from him, so miserably bad that they have been eye-sores to me whenever I have met them; but I suppose I must keep them now as mementos of the eccentric giver."

Short also conducted an extensive botanical correspondence with Asa Gray during the late 1840s and throughout the 1850s, and the subject of Rafinesque comes up in several of these letters. Short maintained that Gray, in his writings, had been unfair in his appraisal of Nuttall by linking and comparing him with Rafinesque. On two occasions, October 3, 1856, and October 19, 1859, Short remarked:

> I have a greater respect and regard for this Pioneer of Western Botany than you seem to have had; for I remember on some occasion you compared him to Rafinesque! Now every body knows that poor Raffy was a most bare-faced liar, not to say a rogue; and the only possible way of apologizing for his gross frauds and deceptions is by Mr. Durand's charitable supposition that he was deranged.[12] On the contrary, have you not found Nuttall truthful and reliable on all occasions?
>
> Have you not done [Nuttall] injustice in some of your notices regarding him, in comparing him to Rafinesque? The latter I always regarded as an

Rafinesque and His Botanical Contemporaries

unprincipled charlatan, whilst in Mr. Nuttall I never saw anything but straight forward truth and honesty. I hope that we shall be favoured with a full and just biographical memoir of him.

If Gray answered on this subject, his replies are not known.[13]

Typical of those botanists who followed in the wake of Rafinesque was George Engelmann, who was interested in receiving some books by Rafinesque that had been offered by Short. On June 12, 1855, he wrote to Short: "Your offer of Rafinesque's writings I thankfully accept; though I believe with you that the trouble of sifting the little good wheat from the mass of chaff is vastly greater than to go to work anew without paying any attention to him." After examining the books, Engelmann wrote again, on October 13, 1855: "I hope you will not think too hard of me for my neglect in acknowledging the receipt of those books of Rafinesque's. They have been in my hands for some time and I was a good deal interested in looking through them; not so much I must confess at the amount of research and knowledge displayed in them as at the barefaced impudence with which he urges his decisions and assails better men. The criticism on Nuttall's Genera [*The Genera of North American Plants . . .* (1818)] is particularly rich!"

Fig. 8.9. George Engelmann. Courtesy of Ronald L. Stuckey.

Ronald L. Stuckey

Conclusions

The private opinions of Rafinesque expressed by eastern North American botanists during the first half of the nineteenth century were numerous and for the most part uncomplimentary. Although Eaton, Torrey, and Gray praised his scientific efforts in certain respects, they also did not hesitate to condemn his personal and scientific shortcomings. It is clear from Short's comments that he had nothing constructive to say about Rafinesque's personality and scientific work. Since most of the uncomplimentary comments were expressed in private correspondence, they generally were not available to the botanical public for evaluation; but Gray's opinion, expressed in his published memorial of Rafinesque, served as a summation of the opinions that had been generated among botanists. Gray's publication certainly influenced American botanists during the latter half of the nineteenth and early part of the twentieth centuries, during which time Rafinesque's botanical work was largely ignored. Because of the subsequent clarification of the rules of priority and homonymy in botanical nomenclature, Rafinesque's published scientific botanical names must now be taken into consideration in taxonomic studies.

Notes

1. Rafinesque's "Prospectus . . ." appeared there in the journal's "second hexade," 5 (Mar.–Apr. 1808): 350–56, and a translation of it was printed the next year in Desvaux's *Journal de Botanique*. Other Rafinesque contributions had been published in *The Medical Repository* in 1804 and 1806.

2. [Baldwin forgot to mention that one paper was on a fossil jellyfish that, though later published by Silliman, probably struck the Academicians as an essay about an improbable creature. He also must have been unaware of the animosity of Thomas Say, chairman of the academy's Publication Committee. See "Fall from Grace," p. 206.) **Ed.**]

3. [For details, see "Fall from Grace," chapter 11. **Ed.**]

4. Rafinesque's review of Jacob Bigelow's *Florula Bostoniensis* (1814) appeared in the *American Monthly Magazine and Critical Review* 2 (Mar. 1818): 342–44.

5. Rafinesque's new genus *Cladrastis* was accepted, with *C. lutea* Michx. as the type species, although some recent authors have concluded that *C. kentuckea* (Dum.-Cours.) Rudd is the older name for this species.

6. No species of *Dodecatheon* dedicated to or named for Dr. Short is known to have been published. *D. meadia* was named by Linnaeus and *D. integrifolia* was named by André Michaux.

7. [What Short calls Rafinesque's "divitial scheme" was a plan in response to the Panic of 1819 to make shares of stock and other things of value act as a circulating

174

Rafinesque and His Botanical Contemporaries

medium in place of money. Pulmel was Rafinesque's secret nostrum for the cure of tuberculosis. He advocated fireproof buildings, especially to house libraries, and had advertised a paint claimed to fireproof wood, while "pearl hatching" refers to his observation that pearls could be made to grow in Ohio River bivalve mollusks. **Ed.**]

8. [This neglect was always a sore point for Rafinesque. Having named dozens of genera for contemporaneous botanists whom he thought worthy of the honor without ever having the distinction reciprocated, he even tried to establish a Rafinesquia genus himself. This act only earned him more ridicule. **Ed.**]

9. George W. Featherstonhaugh, editor of the *Monthly American Journal of Geology and Natural Science* (Philadelphia), called Rafinesque "a perfect master of the subject" when he announced the publication of the naturalist's forthcoming paper "Visit to Big-Bone Lick," Kentucky, which appeared 1 (Feb. 1832): 355–58, and dealt with the osteological remains found at Big-Bone Lick. [But the editor changed his tune after Rafinesque exposed Harland and Featherstonhaugh's "fossil jaw bone of *Rhinoceroides alleghaniensis*" as nothing but a curious piece of stone. Of Rafinesque's own magazine, the *Atlantic Journal*, Featherstonhaugh now wrote that "nothing but a rash presumption of a general ignorance, that would dishonour us all, and of the public inability to discover the worthlessness of such a farrago as he has now let loose upon us, could have encouraged him to produce what is entirely beneath the dignity of criticism" (1 [May 1832]: 509). **Ed.**]

10. Rafinesque's reviews of the floras of Pursh, Nuttall, Bigelow, etc., were published in the *American Monthly Magazine and Critical Review* during 1818 and 1819.

11. The fate of Rafinesque's herbarium, most of which was destroyed, has been described in considerable detail by Merrill (1949), Pennell (1942, 1945), and Stuckey (1971a,b).

12. Short's reference is to Elias Durand's letter of June 4, 1856, that describes his attitude toward Rafinesque and tells how he obtained and then proceeded to discard most of Rafinesque's herbarium (see Pennell, p. 53, and Croizat, pp. 180–81).

13. No letters from Gray after 1856 are in the Short collection in the library of the Filson Club, Louisville (Davies 1945).

Letters from which Portions Are Quoted

Baldwin, William. 1811. Letter to Henry Muhlenberg, Jan. 14, from Wilmington, Del. From Darlington, 1843: 17–19.

———. 1818–19. Letters to William Darlington, Sept. 17, 1818, Feb. 4, 1819, from Wilmington, Del. From Darlington, 1843: 281–82, 299–302.

Barton, Edward. 1818. Letter to Charles W. Short, Nov. 15, from Philadelphia. *In* Charles W. Short Manuscript Collection, Filson Club Library, Louisville, Ky.

Ronald L. Stuckey

Darlington, Wm. 1833. Letter to Charles W. Short, Jan. 12, from West Chester, Pa. *In* Charles W. Short Manuscript Collection, Filson Club Library, Louisville, Ky.

Eaton, Amos. 1817–18. Letters to John Torrey, Oct. 5, 1817, from Northampton, Mass.; Mar. 24, 1818, from Brimfield, Mass. From McAllister 1941: 255–56.

———.1826. Letter to Anna B. Eaton, Apr. 3, from Rochester, N.Y. From McAllister 1941: 256.

Engelmann, George. 1855. Letters to Charles Wilkins Short, June 12, Oct. 13, from St. Louis. From Coker 1941: 144–45, 148–49.

Muhlenberg, Henry. 1809. Letter to Stephen Elliott, June 16, from Lancaster, Pa. From Merrill 1948: 6; Merrill and Hu 1949: 26.

———. 1811. Letter to William Baldwin, Jan. 7, from Lancaster, Pa. From Darlington 1843: 15–16.

Schweinitz, Lewis David von. 1832. Letter to John Torrey, Apr. 12, from Bethlehem, Pa. From Shear and Stevens 1921: 269–71.

Short, C. W. 1818. Letter to John Cleves Short, Dec. 4, from Hopkinsville, Ky. *In* Short, Harrison, and Symmes Family Papers, Library of Congress, Washington, D.C.

———. 1829–41. Letters to William Darlington, Jan. 15, 1829, July 1, 1832, July 13, 1838, from Lexington, Ky.; May 10, 1841, from Louisville, Ky. *In* William Darlington Papers, The New-York Historical Society, New York City.

———. 1834–40. Letters to John Torrey, Dec. 17, 1834, Aug. 11, 1835, Nov. 22, 1836, Mar. 20, 1837, from Lexington, Ky.; Oct. 25, 1840, from Louisville, Ky. *In* John Torrey Letters, The New York Botanical Garden Library, Bronx, N.Y.

———. 1856–59. Letters to Asa Gray, Oct. 3, 1856, Oct. 19, 1859, from Louisville, Ky. *In* Asa Gray Collection, Gray Herbarium Library, Harvard Univ., Cambridge, Mass.

Torrey, John. 1817. Letter to Amos Eaton, exact date unknown, from New York City. From James 1883.

———. 1818. Letters to Amos Eaton, Jan. 5, Mar. 21, from New York City. From McAllister 1941: 255–56.

———. 1818. Letter to Amos Eaton, Apr. 16, from New York City. From Rodgers 1942: 31.

———. 1819. Letter to Amos Eaton, May 18, from New York City. From Robbins 1968: 539.

———. 1831. Letter to William Darlington, Aug. 12, from New York City. *In* William Darlington Papers, The New-York Historical Society, New York City.

———. 1831–34. Letters to Charles W. Short, Oct. 24, 1831, Oct. 23, 1834, from New York City. *In* Charles W. Short Manuscript Collection, The Filson Club Library, Louisville, Ky.

———. 1838. Letter to A. P. DeCandolle, Aug. 30, from New York City, Archives, Conservatoire de Botanique de Genève, Switzerland.

Rafinesque and His Botanical Contemporaries

References

Clinton, George W. 1910. Journal of a Tour from Albany to Lake Erie in 1826. Ed. Frank H. Severance. *Buffalo Historical Society Publications* 14: 227–305.

Coker, W. C., ed. 1941. Letters from the Collection of Dr. Charles Wilkins Short. *Journal of the Elisha Mitchell Scientific Society* 57: 98–168.

Darlington, William. 1843. Reliquiae Baldwinianae: Selections from the Correspondence of the Late William Baldwin, M.D., Surgeon in the U.S. Navy, with Occasional Notes, and a Short Biographical Memoir. Kimber & Sharpless, Philadelphia.

――――. 1849. Progress of Botany in North America, pp. 17–33. *In* Memorials of John Bartram and Humphrey Marshall, with Notices of Their Botanical Contemporaries. Lindsay & Blakiston, Philadelphia.

Davies, P. Albert. 1945. Charles Wilkins Short, 1794–1863, Botanist and Physician. *Filson Club History Quarterly* 19: 131–55, 208–49.

Featherstonhaugh, G. W., ed. 1831–32. *Monthly American Journal of Geology and Natural Science.*

G[ray], A[sa]. 1841. Notice of the Botanical Writings of the Late C. S. Rafinesque. *American Journal of Science* 40: 221–41.

[Gray, Asa.] 1844. The Longevity of Trees. *North American Review* 59: 189–238.

James, Jos. T. 1883. A letter from Dr. Torrey to Amos Eaton. *Botanical Gazette* 8: 289–91.

McAllister, Ethel M. 1941. Amos Eaton, Scientist and Educator. Univ. of Pennsylvania Press, Philadelphia.

Merrill, Elmer D. 1948. C. S. Rafinesque, with Notes on His Publications in the Harvard Libraries. *Harvard Library Bulletin* 2: 5–21.

Merrill, E. D., and Shiu-ying Hu. 1949. Work and Publications of Henry Muhlenberg, with Special Attention to Unrecorded or Incorrectly Recorded Binomials. *Bartonia* 25: 1–66.

Pennell, Francis W. 1942. The Life and Work of Rafinesque. *Transylvania College Bulletin* 15 (7): 10–70.

――――. 1945. How Durand Acquired Rafinesque's Herbarium. *Bartonia* 23: 43–46.

Rafinesque, C. S. 1814. Principes Fondamentaux de Soiologie ou les Loix de la Nomenclature et de la Classification de l'Empire Organique ou des Animaux et des Végétaux. Aux dépens de l'Auteur, Palermo.

――――. Circular Address on Botany and Zoology. Printed for the author, Philadelphia.

――――. 1817. Florula Ludoviciana; or, a Flora of the State of Louisiana. C. Wiley & Co., New York.

――――. 1840. Autikon Botanikon. Icones Plantarum Select. Nov. vel Rariorum, plerumque Americana, interdum African. Europ. Asiat. Oceanic. &c. Philadelphia.

――――, ed. 1832–33. Atlantic Journal, and Friend of Knowledge; a Cyclopedic Journal and Review of Universal Knowledge.

Rezneck, Samuel. 1959. A Traveling School of Science on the Erie Canal in 1826. *New York History* 40: 255–69.

Robbins, Christine Chapman. 1968. John Torrey (1796–1873). His Life and Times. *Bulletin of the Torrey Botanical Club* 95: 515–645.

Rodgers, Andrew Denny, III. 1942. John Torrey: A Story of North American Botany. Princeton Univ. Press, Princeton.

Shear, C. L. and Neil E. Stevens, eds. 1921. The Correspondence of Schweinitz and Torrey. *Memoirs of the Torrey Botanical Club* 16: 119–300.

Short, Charles W. 1833. Instructions for the Gathering and Preservation of Plants for Herbaria; in a Letter to a Young Botanist. *Transylvania Journal of Medicine and Associate Sciences* 6: 60–74.

Stuckey, Ronald L. 1971a. C. S. Rafinesque's North American Vascular Plants at the Academy of Natural Sciences of Philadelphia. *Brittonia* 23: 191–208.

———. 1971b. The First Public Auction of an American Herbarium Including an Account of the Fate of the Baldwin, Collins, and Rafinesque Herbaria. *Taxon* 20: 443–59.

———. 1983. Dr. William Darlington's Botanical Contacts on the Western American Frontier. *Transactions & Studies of the College of Physicians of Philadelphia*, ser. 5, 5: 213–43.

Torrey, John, ed. 1831. An Introduction to the Natural System of Botany: or, a Systematic View of the Organization, Natural Affinities, and Geographical Distribution of the Whole Vegetable Kingdom; Together with the Uses of the Most Important Species in Medicine, the Arts, and Rural or Domestic Economy. By John Lindley. G. & C. & H. Carvill, New York.

Rafinesque: A Concrete Case

"HENRICUS QUATRE" [LEON CROIZAT]
(1894–1982)

WHERE OPINIONS ARE sharply divided, prudence dictates that time be allowed to do its work. This writer would abide by the dictates of prudence as a matter of course, and refrain from taking sides, if he were not cognizant of facts seemingly unknown to most. To all appearances, no indexer is taking note of the botanical names proposed by Rafinesque for groups higher than the genus. It would be an outright error to assume that this body of nomenclature is unimportant. It consists, indeed, of at least six hundred names scattered over the pages of many publications, but centering in the main around five Rafinesquian systems of classification published in 1814, 1815, 1820, 1837, and 1840.

Impelled by faith in his mission as the true founder of nomenclature above the genus,[1] Rafinesque ruled[2] that all names of this nature must be substantives—*Rosacia*, for example, instead of *Rosaceae*. In furtherance of this fiat that Rafinesque asserted to be required for consistency's sake,[3] he christened all groups of his making above the generic rank in a manner which, under less personal standards, has been and still is reserved for genera and subgenera. This can only mean, unfortunately, that the Rafinesquian names above the genus do sometimes match those published by other authors in different categories, witness Subfamily *Cnestidia* Raf., a close match of Planchon's genus *Cnestidium;* Family *Pyrenaria* Raf. and Family *Celosia* Raf., which duplicate outright Blume's and Linnaeus's genera of the same names. Other instances to the same effect could be adduced.

Leon Croizat

From the standpoint of technical nomenclature, a family and a genus cannot be homonymous on account of their respective ranks being different. However, a cardinal principle of name giving in science is that confusion must be avoided, such as will arise when identically the same, or very similar names, are circulated at cross-purposes. In this light, the names published by Rafinesque for groups above the genus cannot be ignored. They either must be indexed, or appropriate action must be taken to "sterilize" them forever.

A large body of opinion holds that the Rafinesquian nomenclature ought to be disposed of without going to the trouble of indexing it in detail. Possibly as large a body of opinion holds precisely the opposite, deeming that publication in print is sacred, and that no one may ignore a word once it is cloaked in the black robes of the press. It should be understood that while opinion in North America is about evenly divided, in Europe there are very few who believe that Rafinesque ought be paid attention to at all.

This writer is concerned only with the *nature* of the Rafinesquian output. With hundreds of Rafinesquian names as yet unindexed, will a hero be found to catalog them? Above all, why must anybody toil upon this, if this mess of nomenclature is worth less than zero?

There is on record a clean-cut statement by C. A. Weatherby to the effect that the Rafinesquian publication "Represents a kind of pseudo-scientific work, the nomenclatural results of which may well be legislated out of existence."[4] This judgment is far from incompatible with another, this time by Asa Gray. Gray, unlike Weatherby, had firsthand knowledge of Rafinesque and his doings, and what Asa Gray states cannot be ignored lightly. He declared[5] that certain of the Rafinesque names might be received, but that no attention whatever should be paid to others, in particular those published after 1830. The reason was that after 1830 Rafinesque no longer could be held sane, which, Asa Gray added, was the belief also of those who had known Rafinesque most intimately.

Stark in its wording is the comment of a Frenchman, Elias Durand, who was among those who had precise understanding of Rafinesque and his doings. In a letter written to Dr. Charles Wilkins Short, dated June 4, 1856, in which he accounts for his purchase of Rafinesque's herbarium, he says:

> Rafinesque's own specimens were trash, incomplete, good for nothing; He was the worst collector I ever knew; But I found in Collins' herbarium, in your own and Dr. Buckley's plants, enough to repay my forty five Dollars, my soap & cologne water.
>
> There I saw the wonders he boasted of—The series of what he called his very rare specimens at 25 cents apiece, and a very small piece too; a

Rafinesque: A Concrete Case

disfigured flower and a few leaves, not spreading but more or less reflexed and rumpled; as for instance, his 8 or 9 species of Dodecatheon, *all made up from the scapes more or less furnished with flowers of the common species, with leaves belonging to other genera;* his new genera and species from the Apalach-Mountains (a locality he loved above all) which were simply exotic plants well known, but disfigured and with addition of parts that did not belong to them. . . . But enough about that poor insane—Requiescat in pace! (italics added)[6]

These statements from the pen of one who knew Rafinesque best, and was indeed willing to pay hard cash for Rafinesque's botanical belongings, are as illuminating as anything could be, and perfectly tally with Asa Gray's verdict that Rafinesque lived long years in insanity, if not the whole of his life. They show us Rafinesque playing with sundry scraps of material, most of it poorly pressed, and compounding out of this rubbish high-sounding "Rare specimens" to fetch fancy prices in imagination. These "Rare specimens" were fakes, and since they did not sell (had they sold, the fake would have been readily detected), they can only be the work of a delusional insane person. Such people as a rule credit with imaginary value manufactures of their own doing, and live by fictitious assumptions throughout, which is well known to those who have adequate knowledge of lunacy, in the precise sense of the term. Durand and Asa Gray knew perfectly well what they spoke of when they dismissed Rafinesque as insane.

This is not all, because other statements to the same general effect occur here and there in *Rafinesquiana*. Remarking upon plates illustrating certain Rafinesquian fungi, and drawn up by Rafinesque in person, a mycologist comments, "[Some of these fungi] remind me very strongly of certain quaint objects which I have seen in collections of Japanese ceramics. I venture to say that no so absurd fungi as these ever were, or will be found on the face of the earth."[7] Asa Gray spoke of the Rafinesquian genus *Geanthus* or *Geanthia* in these words, "There is little risk in asserting that no such plant exists in the United States."[8]

Other bells clang the same tune. David Starr Jordan, who was in no sense averse to Rafinesque, tells[9] that in the very midst of botanical notes in French written in one of Rafinesque's books he spotted the following in good English *"The girls at Fort George eat clay!"* which certainly has no relation whatever to a subject dear to flora, but well may interest a psychiatrist. As it is well known, Rafinesque received— *as a consolation award only*—a gold medal from the Société de Géographie of Paris. In an elaborate testament, Rafinesque disposed that this medal should be kept forever "[i]n the family of Rafinesque as a honorable record of a reward of merit." Pennell, who relates the fact and reports that the gold medal was sold for $16.55, has this to say of the doings of Dr. James Mease, Rafinesque's executor, "It is hard to charge such a faithful friend as Dr. Mease with breach of trust; *rather I suppose that he considered the will fantastic and inconceivable of execution*" (italics added).[10] This is a shrewd remark; in fact, it is the only remark fit to account for Dr. Mease's "breach of trust." Quite as much as Durand had found the specimens of Rafinesque fantastic, so the good Dr. Mease had found the will of Rafinesque impossible. Everything tallies to perfection in this tale, whether the participants in it are articulate, like Durand and Asa Gray, or silent, like Dr. Mease. Everybody knew that Rafinesque was insane, and everybody acted accordingly. Here was the "naturalist" who had nonetheless invented other *mirabilia* such as (Pennell, p. 24) "*A New Kind of Artillery*, a single discharge of which will destroy *One thousand Men in Arms*, one mile off, or sink a large Ship of War. This awful Invention will be communicated Secretly to all such governments who will grant me [*Rafinesque*] a Patent or Privilege for my *Divitial Invention*." The numeric precision apparent in these declarations will, no doubt, also interest the psychiatrist.

To close this sketchy *collectanea*, Pennell may once more be asked to speak to us (p. 54) this time as follows, "[Rafinesque] rushed from one group of plants or animals to another in an ecstasy of discovery. He could scarcely spare the precious time for thorough work, and the tedium of minute dissections always irked him. Hence it is that so many of his presentations remain unsatisfactory. Then, and all his life long, his work was too hastily done for his volumes to become masterpieces."

This is plain, though sugarcoated. Rafinesque's "descriptions" are of the same cloth as his "rare specimens." In his mental condition, no one needs to do what most men in their wits understand as honest work. His brain eaten by delusions of grandeur, a man in this sad state grabs a common daisy, lifts it up to the heavens, and enters into an "ecstasy of discovery." It no longer is a daisy but a rare and new fungus, a *Geanthia*, a girl with mouth full of clay, etc., etc. So inspired, Rafinesque is ready to function as Linnaeus; that is, to "reform" science. Accordingly, he pens the following, for example, "*Petalonia*. Gen. Nov. Arb. Corolla 5-mera, sep. magnis, styl. 3, seminis plur. A perfectly new good genus overlooked by all Linneists and discovered by me."[11] This, to Rafinesque, is adequate, for it suits his needs. Does this scrap of paper stand for anything else, in Rafinesque's mind, if not supreme proof that, he Rafinesque, somiologist, systematist, doctor of medicine, pulmist, inventor, etc., etc., is "reforming" science and achieving thereby "colossal" results with "immense" toil? Lunatics of the kind are well known in insane asylums, though they generally prefer to be recipients of revelations from Gabriel the archangel, rather than from Linnaeus the naturalist.

Rafinesque: A Concrete Case

In the present writer's matured judgment, *Rafinesque was an arrant lunatic,* and his output cannot be viewed in any other light. There remains the question to settle—and this is a capital one—how and in what measure Rafinesque was insane.

This question is capital because, in judging the effects of insanity, we cannot hope to proceed in the abstract. The core of the question is the following: a man may be afflicted with persecution mania of the very deepest hue and remain a painter of great talent throughout. This man may secrete himself in a garret, live under padlock and key out of his own volition, shake in his boots at the sound of steps coming upstairs, and be out of his wits nearly the whole day and night long. This very same man may also become another being when at the easel. He may be quite as talented with the brush as he is disturbed mentally otherwise. In sum, though an insane *whole,* this man *as part of himself*—that is, as a painter—may exhibit genius.

As a matter of fact, disturbances in one sector of the mind may well call for compensation in other sectors. The gift of exalted imagination that is characteristic of the poet, and the unerring, fast-working intuition that dictates to a musician what is to come after certain notes stand on paper already, is often purchased at the price of deviations from the norm in sexual behavior or a similar field of supposedly normal activity. He who sees things of beauty that no other man can see by himself cannot be like everybody else. Inspiration is dearly bought, and genius may indeed be next to door to insanity.

However, we need not fear being unable to discriminate as between the gift of divine vision to bless Leonardo da Vinci and the distorted perspectives begetting *Geanthia* of Rafinesque. *The work bears witness to the man.* The genius leaves behind something which is orderly, precise, logical, and because of its being this way, is also beautiful to everybody who understands that science and art are, in their loftiest expression, facets of the same prism. Whether Dante sings, Bach composes, or Leonardo da Vinci dreams, something of great power stands before us, and we are well aware of it. The crystal-clear thought of a scientist of high order who masters the elements, and with sure hand steers through untold details to hidden, vital conclusions, quivers before us, too, as does the harp of a bard. The work speaks for the man, and steady piffle never came out of a mind of high caliber, though out of this mind things may not always flow unalloyed.

Judged by his work, Rafinesque is by no means a genius, for his work is precisely piffle and rubbish. Not a word need be spent to argue the contrary with those who, for example, believe that Rafinesque was a "forerunner of Darwin," for they know no better, and ought to learn before speaking.

Leon Croizat

If Rafinesque is not a genius, nor is he a normal man, he is likely to be a lunatic, which is precisely what this writer believes, as stated. The question remains, then, was he a lunatic when writing of natural science? If he was—and this writer has already intimated this much—the *Rafinesquian output has no status in science*. This output cannot have status in science, because by definition science is good faith active in quest for truth. There is no good faith, and even less truth, in a lunatic. A lunatic lives in a world of his own, works in certain ways because he has to assuage certain of his own cravings, sees what nobody else can see for he has visions of his own, and, in so many words, is dead before his age. To admit him among scientists is to perpetrate an error in judgment—whether this error can be supported with certain scraps of paper or not hardly matters. Those in the scientific world who, satisfied with the candor of a putterer, are willing to accept fifth-rate output from of his pen and to make allowance for his lack of proper terminology and perspective, cannot accord the same measure of tolerance to the lunatic who tries to steal into the precincts of science via the printing press. Let those who read these lines fancy what they would themselves advise in regard to a man who, smitten with delusion of being called upon to "reform" botany, proceeds to "recast" standard manuals of taxonomy, erecting in the process untold synonyms together with a few "good" names caught by chance.

It seems to this writer unfortunate that American authors who most sedulously labored in *Rafinesquiana* never possessed, perhaps, the linguistic accomplishments needed to peruse Rafinesque's work in French and Italian, which is a key to the understanding of this man in the first place. This omission is certainly none of Rafinesque's fault, for he took good care to emphasize that work, and in forceful English. Bemoaning what he took to be the sorry plight of the natural sciences, he stated, "If my suggestions in 1814 in Principles of Somiology, and in 1815 in Analysis of Nature had been attended to, it might have been otherwise," to which he appended the following: "as early as 1802 I began to perceive the necessity of rectifying the presumed Orders of Jussieu, and after many observations in both hemispheres, I published my Natural Classes in 1814, and my 66 Natural Orders in 1815."[12]

It is well to observe that the stress laid by Rafinesque on his output of 1814 and 1815 is fully borne out by the record. These are the two crucial years in the Rafinesquian life and miracles, and no one may understand the man and his toils who is not fit to understand this. In these years, Rafinesque labored under the weight of an onset of delusion that in more or less fitful degrees remained with him to the last. This onset was foretold by faint rumblings of megalomania in years before, but these rumblings were as nothing compared with the occurrences in those

Rafinesque: A Concrete Case

twenty-four months. Asa Gray, by the way, is also borne out by the record when he states that the Rafinesquian output between 1830 and 1840 (the year of the death of this lunatic) ought to be wholly neglected. The record shows that in this decade Rafinesque sank mentally to the end, when he stood indeed in mental eclipse.

In the two fateful years, 1814–15, Rafinesque published at least three works, namely, (a) *Précis des Découvertes Somiologiques ou Zoologiques et Botaniques,* which came out of the press shortly after June 3, 1814; (b) *Principes Fondamentaux de Somiologie ou les loix* [*sic*] *de la Nomenclature et de la Classification de l'Empire Organique,* issued in the second half of the same year; (c) *Analyse de la Nature ou Tableau de l'Univers et des Corps Organisés,* published in the early months of 1815. While all this was being quickly dished out, the same kitchen labored at other undertakings. Rafinesque was by then also editing a private periodical, the *Specchio delle Scienze o Giornale Enciclopedico di Sicilia.* These titles bespeak the importance of the subjects involved, as Rafinesque saw them, for they are of cosmic proportions. During all this time, it should be added, Rafinesque was active in Sicily.

Critical readers of this pabulum can easily learn that they face a man of deranged mind. The second work mentioned, *Principes Fondamentaux de Somiologie,* etc., consists of 186 dicta and rules having as their purpose to warn the world at large that creation must henceforth be classified and named in the Rafinesquian manner. In the pages of this pamphlet Rafinesque decides that *Calamagrostis* DC., for example, is objectionable, and must be renamed *Amagris* Raf.; that Linnaeus should not have called a certain plant *Hippocrepis* L., leaving it to Rafinesque to identify it under the name *Hippocris* Raf., etc., etc. All this is trifling compared with the manner in which our hero "classifies" *Homo sapiens* L. This featherless biped can be stewed in every sauce, and proper categories are ready to tag him out as Sicilian, red-haired, dwarf, virile, male, mulatto, hunchback, or born with two instead of one head. The difference between the last two attributes is that the former falls under the subdivision "Difformité," the latter, on the contrary, under "Monstruosité," always according to Rafinesquian *somiology.* It should be remarked—for this is of interest to psychiatry when not to botany—that Rafinesque pays great attention to "monstrosity" whenever he meets it.

Somiology is a Rafinesquian term connoting joint study of all living beings. As regards his toils of lawgiver in *somiology,* Rafinesque claims nothing less than this much:

> This writing is in imitation of those of Linnaeus directed only to
> instructed Scholars and to those who desire to become so, or have the noble
> desire of Instructing themselves methodically and profoundly. Linnaeus

reformed Botany and Zoology in the depths of Sweden, and the neighborhood of the pole, I undertake a like labor on the southern shores of the Mediterranean, in Sicily, and close to barbarous Africa, in a place in short where almost no one will appreciate my labors, but enlightened Europe will recognize their importance and time will crown my efforts, this deep-seated conviction will sustain and animate me: the partisans of ancient opinions will nevertheless struggle against me, as they struggled against Linnaeus and all the reformers of the Sciences; but the clear facts will always triumph over them.[13]

This text is imbued with a strange hue of messianism, not at all unknown in writings which reach the wastebasket of institutions for the mentally deranged, and are composed in their wards by ladies or gentlemen oftentimes of mild disposition but scarcely clear about their worth to the world. Rafinesque visualizes himself as a new Linnaeus, and the comment he pens, to the effect that he labors near barbarous Africa, while Linnaeus toiled in "the neighborhood of the pole," is typical of writings either of great significance or none at all, except in this case to psychiatrists.

The *Précis des Découvertes Somiologiques* is an open letter addressed to the well-known botanist C. H. Persoon. It seems this pamphlet of less than sixty pages in tiny octavo is the keystone of *Rafinesquiana*. In this opuscule Rafinesque speaks aloud for himself, not merely milling out numerous genera and species, but also telling the world why he does so, and what he intends to gain by it. Quoth Rafinesque (addressing Persoon):

I have undertaken since the beginning of this year [1814], the editing of a literary journal (the only one of its kind here), of which I publish an issue every month, under the title *Specchio delle Scienze o Giornale Enciclopedico di Sicilia*. In writing the Journal I have had the aim of making myself useful, of spreading abroad in Sicily the too limited taste for the Sciences and arts, and finally of getting myself read by the generality of readers by putting myself at their level; but the present state of literature and the sciences in this country, makes me fear that it may be entirely out of place. All the Sciences, but especially the physical sciences are within my scope in it and natural history occupies there a distinguished place. One half is devoted to original Essays and the other to selections and miscellanies. Six numbers of it have already appeared, which make up the first volume. I have begun to work out some of my discoveries for this periodical compilation, and I have announced in it the principal [one], that which has for its object a reform in the study of organized bodies.

Rafinesque: A Concrete Case

I consider this reform to be the great *Desideratum* of natural history, indeed in the present state of Zoology and Botany I observe three capital defects in their study: 1. an unsteady Nomenclature, 2. imperfect distributions, 3. changeable or illogical definitions; thus each of these objects clamors loudly for the reform which I have undertaken, the happy result of which will be directed to fix invariably the Nomenclature, the Classification, and the Definitions of Organized Bodies. You will readily conceive what incalculable advantage such a plan portends; but you may object with a doubt of its success, you may ask me how I dare flatter myself with success while the most celebrated modern Botanists and Zoologists have renounced [it] or miscarried [in it]. I reply that the Scholars who could have undertaken this fine work have been repelled by the difficulties of the task or disgusted by the painful and unpleasing labors it demands: you yourself, who could have effected it in part for plants in your *Sinopsis plantarum,* have preferred to follow the tracks of your predecessors and to walk somewhat slavishly in the steps of the great Linnaeus. Finally having seen that this labor remained to be done in order to perfect the Sciences of Animals and Vegetables, I have had the boldness to undertake it and the good fortune to carry it out in a few years of assiduous work and profound meditation, not without having groped about for a long time and often failed in my first efforts. The idea that Linnaeus had had no more means than I, apart from his genius, when he succeeded in wholly reforming Botany and Zoology, has encouraged me and sustained my zeal: I said to myself why doubt of success? why may I not imitate this great man while my ardor is alike and my means similar?[14]

These are megalomaniac ravings, for, as we shall presently learn in the most precise manner, the boasts we here read bear no relation whatever to accomplishment. The mind of the lunatic who visualizes himself a reborn Linnaeus in these words is the very same one which in later years delighted with concocting strange "specimens," or arranging farcical testamentary dispositions. Such a mind well could spawn names of biological entities by the thousands in the delusion that this meant to "reform" science. If anything sane ever came out of this polluted well, out it came by chance, which Asa Gray lucidly perceived when summing up the services of Rafinesque to botany.

Having fed Persoon the generalities quoted above, Rafinesque proceeded to acquaint him[15] with certain facts material to the execution of the Rafinesquian plans. Accordingly, Rafinesque stated that he had already prepared a work of large scope, the *Critique des Genres,* in the pages of which Linnaeus had his match, and generic nomenclature would forever be settled. He averred further that he had written a second colossal opus to be known to posterity as *Ordre des Genres*, which contained not less than twenty-five hundred new genera of plants and cost him twelve years of toil. In this monument, all genera, even the newest and most obscure, would find their resting place in classification forever. Rafinesque characteristically added[16] that people as gross as the Sicilians could have no feeling for the sublimity of this science, and the very few among them who, perhaps, knew anything at all of the subject were eaten up by the direst pangs of jealousy and envy.

This writer could quote without difficulty other claims and boasts of the same nature, for such abound in the Rafinesquian writings of 1814–15, shortly before Rafinesque's flight to the United States from Sicily. These claims cannot be dismissed as the byproduct of generous Sicilian wine, because Persoon was by no means their sole recipient in the French language. Rafinesque had them in print also in good Italian,[17] affirming most categorically that his immense work had been finished down to the very last genus ("Senza l'eccezione di un solo genere . . . questo immenso lavoro è già compiuto"). Be it noticed that this was printed about a month earlier than the open letter to Persoon.

Beyond doubt, Rafinesque believed these claims himself, for, had it been otherwise, he would hardly dare set them down in print *in Sicily*. Persoon, with whom he had only corresponded for the last four or five years, might not know Rafinesque well, but the Sicilians among whom he lived—the botanist Bivona, for example—undoubtedly knew better and could not be fooled by loose talk.

This being in the nature of the evidence, a critical reader of the Rafinesquian *opera omnia* must be shocked beyond words when he opens the pages of the *Analyse de la Nature*, written almost contemporaneously with the texts quoted and published but a few months later. The Rafinesquian classification in this work covers ten classes, sixty orders, or thereabouts, and some three hundred families. After dealing sketchily with but three of these orders in the first class, certainly less than one-tenth of the entire projected opus, Rafinesque goes dead and drops in the body of the text the following:

> N.B. I had the intention of pursuing the enumeration of all the Genera of Vegetables on the same plan as the three preceding orders, just as I have executed it in regard to Animals; but having maturely considered the details of my vast enterprise, I perceived that they merit still further elaborating and perfecting; several new genera of Humboldt, Bonpland, Tussac, Brown, Labillardière . . . are besides known to me only by name, and it could be that the majority should be the types of new families. I am following here a wholly new road in which I am almost devoid of guides, while for Animals I

have found a beaten track and sure guides. These considerations have led me to suspend the enumeration of Families and Genera; but I am going to continue to trace the plan of my natural distribution of Vegetables, and to give the table of Classes and Orders, with examples chosen among the Families and Genera which belong to them: such a work will complete the sketch of my vegetable natural method, since the table of the three first Orders has already offered an example of its practical application, and the families which I shall have to omit are very few in number. Moreover the details which I omit with the sole view of perfecting them, will soon see the [light of] day in a special work entitled DISTRIBUTION OF VEGETABLES.[18]

Now, then, the author of this amazing confession is the very same writer of colossal boasts penned only a few months before, as has been demonstrated. He lied to himself and others at least once, and managed the whole in the spirit of complete irresponsibility. While admitting that the *Critique des Genres* and the *Ordre des Genres* must be fabulous, at least by implication, he assuaged his feelings by promising himself, and the world, a new *Distribution des Végétaux*. This is not all, for caught on the short side of the field in one respect, he found himself competent to work at a fast clip elsewhere. He broke down the Kingdom of the Elements, which he proclaimed as the subject of a new "science," *Sochology*, into two subkingdom, six classes, and twelve orders. The whole rings as follows, *Socaplogy, Rytology, Leplogy, Gasaplogy, Sereology, Phlegology, Metallogy, Socadology, Gasology, Anopatology, Atmisology, Ychrology, Sycreology, Thermiology, Phlogology, Eleiology, Sphaltology, Coniology, Oxidology, Aiology.*[19]

This writer feels that abundant material is contained in the few passages just quoted to label Rafinesque an arrant lunatic in the sense which medicine and law give this term. We have before us a man who utters fantastic and colossal boasts and indeed believes to be a reborn Linnaeus, a great reformer and lawgiver. This man, smitten by a feverish urge to go into print, pours out page after page to inform us that he has already written works of immense scope, which he identifies by name. At the flip of an eyelash, and next to uttering these boasts, this man changes his tune, acknowledging himself short of everything he should have according to claims made a few months before. However, while recognizing this, this very same man utters other boasts to replace the former ones, and coins *Gasaplogy, Socadology*, etc., etc., in order to "reform" the study of *Sochology*. In his last effort[20] this very same man remains true to his being of a quarter century earlier, for he publishes a new "Classification of the Natural Sciences and Objects," in which we detect a certain *Stocology* perhaps reminiscent of the *Sochology* of 1815, but also a new swarm of "sciences," such as *Abarialogy* to deal with the "Abarials" or "Imponderables"; *Gazomy* to investigate no longer gasses but the "Ablepsoms" or "Invisibles"; *Stiontosy* to handle "Stony beings"; *Telorontosy* to study "Immense beings"; *Peloronyosy* to probe "Monstrous beings"; etc., etc.[21] Peculiarly, the pamphlet which contains these amenities also works out a new a "classification" of the Cyperaceae.

This writer could without the slightest difficulty go on quoting abstracts from *Rafinesquiana* that prove their author to have been insane, and, which is of the essence of this review, to have handled the natural sciences because of his being insane, and believing himself a reborn Linnaeus. It stands to reason that the Rafinesquian output is replete with things of mystery, be they fungi which are fantastic, or claims that sober reason never is adequate to explain. Lost in this maelstrom of lunacy are certain "good" genera and species, most of which have long ago entered current usage. This outcome is not because of Rafinesque's work, but despite it, for lunacy impelled this man to consume an immense amount of botanical fodder to live up to the delusion of being *Linnaeus redivivus*. In the digestion of this, something came to the mill which, by the law of probabilities, had worth. *Geanthia* Raf. was thus born with *Cladrastis* Raf., and it meant little to the author of both these names that the former was spurious, the latter "good." Indeed, he had neither eyes nor ears for such details, for all he cared for was to play with symbols and fictions answering an inner voice that urged, "Rafinesque, you are Linnaeus, a great reformer, a great genius. The world hates you and misunderstands you because of your immense greatness and colossal services to mankind. Do take pen in hand, and write something down about *Gazomy, Stiontosy, Carex, Geanthia, Cladrastis* which you—and you alone—can understand"!

In the light of the evidence, this writer is of the opinion that the names published by Rafinesque above generic rank should not be indexed at all. The writer believes that sanity is of the essence of science, and that under no circumstance can a lunatic be treated as the equal of a person who is sane. Expediency, the rule of homonyms, etc., will never be so strong as to force the writer to entertain a single one of the Rafinesquian synonyms as worth the paper upon which it is printed. Obviously, to the extent that certain Rafinesquian names have received currency through work of taxonomic significance (that is, not at all special indexes [such as E. D. Merrill had proposed and was preparing when Croizat wrote— **Ed.**]), this writer will use them.

Finally, utterly refusing value to the Rafinesquian names above the rank of genus, the writer of this review suggests that they be finally outlawed by inserting in the International Rules of Botanical Nomenclature the following: "Art. 21 . . . Note 5. Names published by C. S. Rafinesque which were not received in the Index Kewensis, Suppl. viii, 1930, are not considered to be legitimate nomenclature."

This proposal is direct and not at all arbitrary. It cuts short a flood of polluted nomenclature contributed by a lunatic, *who wrote botany because he was of unsound mind.* This, as far as this writer knows, is a sad and probably unique case in the history of botany.

I leave it to the wisdom of the competent Committee to determine whether *all names* of Rafinesque not found in the Index Kewensis, Suppl. viii, 1930, ought to be outlawed, or whether only the names published or proposed by this lunatic *above the rank of genus* must be meted out this fate. Being chiefly interested in the latter, I choose to be silent as regards the former.

Notes

1. *Principes fondamentaux de somiologie* (Palermo, 1814), 33, no. 84. "Neither Linnaeus nor any other somiologist has ever established the rules of Nomenclature of these various groups [above the rank of genus], thus I shall be the true and first founder of their Nomenclature; since I am going to propose to subject them to fixed and stable laws, like those of Genera, and I shall execute in respect of them a work like that which Linnaeus had the genius to conceive formerly for Genera and Species. The absolute necessity of such a work has become obvious; indeed it should seem very absurd and illogical to neglect thus the Classes, Orders and Families etc., while giving such care to designating correctly the generic and specific denominations." [Trans. Arthur J. Cain. Croizat's internal documentation has been transferred to endnotes so that the entirety of the texts he alludes to can be quoted as here and following. **Ed.**]

2. Ibid., 34; no. 86. "With regard to the nomenclature of Orders and Families as well as their divisions, of which there is a considerable number, it is necessary to assimilate it entirely to that of Genera; the classical rule given already will be the principal foundation for them also, and its capital defect being at present to admit almost exclusively plural and often adjectival names, one must keep oneself from imitating this example: for these names cannot become plural except when they are to express collectively the objects which they designate, and one can admit adjectival names only among the secondary denominations etc." [Trans. Arthur J. Cain]

3. "Tableau analytique des ordres natureles, familles naturelles et genres, de la classe endogynie, sous classe corisantherie," *Annales Générales des Sciences Physiques* 6 (1820): 79.

4. [Kew] *Bulletin of Miscellaneous Information* Nos. 6–9 (1935): 409.

5. *American Journal of Science and Arts* 40 (Jan.–Mar. 1841): 237. Quoted here by Stuckey, pp. 170–71.

6. Francis W. Pennell, "How Durand Acquired Rafinesque's Herbarium," *Bartonia* No. 23 (1944–45): 42–46. [Croizat's quotation, into which some errors had been introduced, is here corrected from the text as given by Pennell, including the erratic punctuation. **Ed.**]

7. W. R. Gerard, "Reliquiae Rafinesquianae," *Bulletin of the Torrey Botanical Club* 12 (Apr. 1885): 37–38. [Gerard's "certain quaint objects" is a Victorian euphemism for *phallus*, which, of course, some fungi do resemble. **Ed.**]

8. *American Journal of Science and Arts* 40 (Jan.–Mar. 1841): 226. [Croizat conveniently overlooks the fact that on the next page Gray speaks of "Mr. Rafinesque's sagacity at that early period." **Ed.**]

9. [This is from Jordan's biographical sketch of Rafinesque in *Popular Science Monthly* 29 (June 1886): 212–21, an account marred by even more egregious errors than this one. The note about clay-eating girls—at Fort Edwards, says Jordan, not Fort George—occurs in one of Rafinesque's field notebooks now in the library of the Smithsonian Institution. Jordan did not appreciate the miscellaneous nature of these notes. They are merely memoranda of bits of information Rafinesque wanted to remember, and include most of his interests, as well as snatches of doggerel poetry. Here he had apparently come across a reference to the eating disorder now called Pica. No doubt Jordan found clay eating an exotic proclivity, but any of us who have ever swallowed Kaopectate also have indulged in a bit of geophagy. **Ed.**]

10. [Pennell's account is given here, p. 51. His misunderstanding of Mease's handling of the Rafinesque estate is cleared up on p. 94. **Ed.**]

11. [Croizat gives no reference for this citation, and—strange to behold—the generic name *Petalonia* is not indexed by Merrill nor has it been discovered in Rafinesque's published writings by anyone else. The diagnostic characters are so general that Croizat probably devised them as a joke. **Ed.**]

12. *Flora Telluriana* (Philadelphia, 1836), I: 9, 45.

13. *Principes fondamentaux de somiologie* (Palermo, 1814), 10. [Here, and for the following translations, Croizat's quoted text has been supplanted by Arthur J. Cain's. Although Croizat's is a workmanlike version, English was not his mother tongue nor that of either his editor or compositor. Not only is the British Cain's translation superior, its nuances even cast Rafinesque in a more adverse light—if that is possible. I have, however, changed Cain's British spelling conventions, and have normalized punctuation wherever necessary to make meaning clear, for Rafinesque's text suffered from shoddy typesetting in Palermo. **Ed.**]

14. *Précis des decouvertes et travaux somiologiques* (Palermo, 1814), 4–6. [Cain trans. **Ed.**]

15. Ibid., 7–10.

16. Ibid., 53–54.

17. ["Esempio del Metodo Sinottico di Botanica, Illustrato nel Primo Ordine della Prima Classe dal Sig. C. S. Rafinesque,"] *Specchio delle Scienze* 1 (Apr. 1814): 154–56.

Rafinesque: A Concrete Case

18. *Analyse de la Nature* (Palermo, 1815), 177–178. [Cain trans. The ellipsis sign appears in Rafinesque's original text. **Ed.**]

19. Ibid., 23–24. [It is not surprising that either Croizat or his compositor (or both) were unable to put these strange neologisms into print without making numerous typographical errors. Consequently, as listed here they have been taken from Cain's translation, where original terms such as Rafinesque's French "Séréologie" becomes Sereology and "Gazologie" becomes Gasology. See Cain, pp. 135–36, for analogues of such ungainly terms in the literature that helped to furnish Rafinesque's mind. **Ed.**]

20. ["Classification of the Natural Sciences and Objects,"] *The Good Book and Amenities of Nature* (Philadelphia, 1840), 5–12. [It is hard to understand why Croizat, in his effort to demonstrate Rafinesque's folly, failed to mention that here we also will find "Zooscopy" should we need a name for the "science" of predicting the future by means of animals, such as by reading signs from their entrails. **Ed.**]

21. [Since Croizat managed to misspell five of these "new sciences" of 1840, Rafinesque's spelling from his own text has been inserted here. Such a correction may be misplaced historical priggishness, yet it is worth noting that these bizarre terms were once taken so seriously that the New York Lyceum of Natural History—of which Rafinesque was a founding member—used several of them to designate the specialties of some of its members. But only one of Rafinesque's names for new sciences has gained general acceptance. This is "malacology" (1815), which he himself repudiated in favor of "apalogy." **Ed.**]

Rebuttal

ELMER D. MERRILL
(1876–1956)

IT IS DOUBTED IF, in the entire history of descriptive biology, there is any other author who has suffered more from the weight of authority than Rafinesque. The leading biologists of his time, both in Europe and in America, ignored his numerous nomenclatural proposals to an extraordinary degree, whether he was correct in his conclusions or not. The lesser lights of the times could only follow their leaders. There was at that time no strict priority rule, and individual authors were free to exercise considerable latitude in the selection of plant names. About the best that we can do, a century or more after the Rafinesque proposals were published, when it is discovered that one of his correctly published generic names has priority over a more recent and currently accepted one proposed by some other author, is promptly to add the now offending earlier Rafinesque name to the already over-long list of officially rejected ones. The reason, of course, is that this procedure obviates many undesirable changes in currently accepted and widely used plant names. It is perhaps worthy of note that within the present decade, at least two new generic names have been proposed which were unnecessary, for *Ipheion* Raf., 1837, was based on the same type as *Beauverdia* Herter, 1941, and *Stethoma* Raf., 1838, has more than a century of priority over *Psacadocaymma* Bremekamp, 1949.

Whatever else may be said regarding Rafinesque, one must admit that he was very adept in selecting short euphonious generic names for the new entities

Rebuttal

that he proposed.[1] For those who are intrigued with the meanings of generic names, most of which are derived from classical sources, Rafinesque's work presents many puzzles. Because he firmly believed in short names and strenuously objected to sesquipedalian ones, he did not hesitate to eliminate entire syllables in order to attain his objective. He rarely explained the origin of a name proposed by him, but one striking case illustrates his method. In publishing the name *Diodeilis* he states "abridged from *Diodontocheilis.*" The end result, *Diodeilis*, would scarcely indicate to the most skilled classical scholar its source as from the Greek, meaning a plant the flowers of which have a two-toothed lip.

His new generic names total about twenty-seven hundred, a vastly greater number than those proposed by any other botanist, and nearly twice as many as those recognized by Linnaeus in the middle decades of the eighteenth century when the binomial system was established. Yet only about thirty of these Rafinesque names have been generally accepted, although on the basis of strict priority at least 160 would qualify.

Rafinesque's Critics

I have been mildly amused by statements repeated to me, said to have been made by this or that contemporary. I suspect that in some cases the criticisms were due to an uneasy feeling on the part of John Doe or Richard Roe, on learning of my contemplated excursions into the Rafinesque field, that perhaps they may have slighted their own bibliographic obligations by ignoring his work. They may insist that their own conclusions are correct and that these should be accepted by others, even as Rafinesque maintained this attitude persistently in reference to his own work. Yet they blandly disregard the fact that in ignoring the validly published names of Rafinesque, their contemporaries and successors may perhaps find as much justification in ignoring or modifying the conclusions of those who object to Rafinesque and all that he proposed. I have been accused of wasting my time by one[2] who naively admits that he knows all about these Rafinesque names, which is really taking in considerable territory.

His Own Worst Enemy

How disastrous Rafinesque's erratic work was to his future reputation is indicated by the fact that of all the generic names he originated in the plant kingdom, only an infinitesimally small percentage are universally accepted and used by all botanists. This low score may be somewhat increased by adding the somewhat more than seventy cases that have been demonstrated to have clear priority over later names of other authors now currently used. It is a foregone conclusion that while some of the valid early Rafinesque names will be accepted, most of them will eventually be included in the official list of *nomina generica rejicienda*. In that list the unflattering total of ninety of Rafinesque's names have already found their places.

However, it is a forlorn hope that most contemporaneous and future systematic botanists will accept the philosophy of those who argue that the large and more or less polymorphous genera must be divided into smaller ones. They will argue that the interrelationships of groups can be indicated just as clearly by retaining subgenera and sections, and that it is not necessary to raise these minor groups to generic status or to make the inordinate number of transfers of specific names the latter course requires. On the basis of fairness, however, this modern school of "splitters" was eminently correct in accepting Rafinesque names for the entities that both he and they recognized and for which the former originated the generic name. Too often in the history of botany those who would recognize minor categories as genera have selected names *de nova*, regardless of those which others may have proposed at earlier dates. To a large degree this failure on their part resulted in the rather extraordinary number of about 160 cases where, on the basis of strict priority, Rafinesque's generic names should have been accepted.

That about ninety of Rafinesque's generic names have been officially discarded in spite of their having priority is largely a reflection on Rafinesque and the utterly careless type of work he sponsored. Judged by any reasonable standards anyone who will critically examine the record can only conclude that in his highly erratic work, Rafinesque was utterly irresponsible. No matter if his conclusions were correct, as they were on occasion, he failed to make himself clear, failed in general to provide even reasonably good descriptions, failed too often even to mention the really striking differential characters by which this or that entity could safely be recognized, failed in his feeble attempts at documentation, and failed in his associations with contemporaneous botanists.

Rafinesque's contemporaries seriously objected to his erratic work, considered it valueless, and further that it could and should be ignored. Yet universal acceptance has been accorded to many of his genera, including such well known ones as *Distichlis, Sitanion, Peltandra, Clintonia, Hexalectris, Isotria, Ofaiston, Enemion, Adlumia, Polanisia, Cladrastis, Nemopanthus, Pachistima, Cymopterus, Lomatium, Musineon, Oreoxis, Osmorhiza, Oxypolis, Ptilimnium, Spermolepis, Blephilia, Agalinis, Aureolaria, Pagesia, Agoseris, Erechtites, Othake, Ratibida,* and *Serinia.* This is an impressive list even if we are forced to admit that only a very small percentage of Rafinesque total of twenty-seven hundred new generic names is included.

Rebuttal

Rafinesque's Character

Versatile, positive, in a sense aggressive, with a phenomenal memory, an unusually wide knowledge of languages, uncontrolled and uncontrollable, Rafinesque proceeded on his way apparently convinced that his mission was to correct the errors of all of his contemporaries and predecessors in the wide realm of biology and many other fields of human endeavor. He scorned the opinions of his contemporaries and in some cases—in no uncertain manner—pointed out their weaknesses and errors. Apparently when his contributions were refused by the editors of this or that periodical, as was not infrequently the case, he expanded his private publishing activities to the limits of his available income. This private publishing activity was initiated in Palermo, Sicily, in 1809, or perhaps in 1808, was continued there until 1815, reinitiated on a small scale first in Philadelphia, in 1816, slightly expanded during the period that he lived in Lexington, Kentucky, 1819–25, and reached its apogee after he returned to Philadelphia, and particularly in the decade between 1830 and 1840.

This intense publishing activity of Rafinesque was an indication of his lack of good judgment, there being very little demand for the technical works that he insisted on preparing and publishing. In this last hectic decade of Rafinesque's life, he became even more Rafinesquian than he had been in the earlier years, his publishing activities continuing up to the very end. It was apparently to a very considerable degree the dissipation of his always limited financial resources to cover the printing costs of his numerous privately published works, from the sales of which the financial returns must have been very small, that left Rafinesque a veritable pauper at the end.[3] It may be worthy of note that eighteen privately published volumes were issued between 1830 and 1840 that contained from 100 to 265 pages, and another long series from 75 to 100 pages each, the two series containing a total of somewhat in excess of 3,700 printed pages; and these figures do not include the numerous smaller pamphlets and take no account of his extensive contributions to periodical literature that appeared in the *Saturday Evening Post* and the *Casket*.

We may admit that in some respects Rafinesque was ahead of the times. He realized the significance of the natural system of classification long before his colleagues in the United States even considered it. He recognized and largely based his work on the type principle. He properly subdivided certain large genera, anticipating our modern concepts of generic limits, although in many cases he extended the subdivisions entirely beyond reason. He outlined briefly his ideas on organic evolution. He was impatient of the limitations and shortcomings of his colleagues. However, the very fact that he published voluminously, and on an enormous range of subject matter, does not make him great; the fact that he proposed more new generic names than any botanist who ever lived and a plethora of ill-founded new species again does not add to his stature. It is the lasting nature of one's contributions to a science that qualifies this or that individual for his place of fame in the history of a subject, and here Rafinesque definitely failed to qualify. He had the opportunity of actually becoming one of the great botanists of his time, but this opportunity he thoroughly missed.

There are those who rather brutally say it is too bad that Rafinesque did not lose his life in the shipwreck in which he lost his herbarium, library, and manuscripts in 1815, for later generations of biologists would then have been spared the endless task of trying to unravel his thousands of nomenclatural enigmas. And there are doubtless those moderns who will insist that anyone who concerns himself with such an erratic individual as Rafinesque a century after the latter's death deserves just as little consideration as does that prince of erratic botanists. After several years of effort, devoted in part to a consideration of the unending series of problems raised by Rafinesque's work in botany alone, my frank conclusion is that in taxonomy and nomenclature we would have been infinitely better off today had Rafinesque never written or published anything appertaining to the subject.

The Issue of Rafinesque's Sanity

Rafinesque has been accused by many writers of being insane, and if one accepts this verdict, it offers a very simple way of eliminating all Rafinesque nomenclatural problems by merely ignoring them on the basis that he was mentally incompetent. This was what most of his contemporaries did, and is doubtless what some modern botanists would prefer to do. But who shall say?

Rafinesque was unstable, he was neurotic, perhaps at times he was on the verge of insanity, but far be it from me even to attempt to prove that he was actually insane. There are individuals who insist that this was the case and that they can prove it, but insane or not, as Durand and others have claimed, he was at times remarkably normal, even if we must admit that at other times he was often more illogical than logical. He lacked balance, perspective, judgment, being markedly egotistical and erratic. He was always planning vast projects that were quite beyond the scope of his means to consummate, but these traits do not make him insane.

After all, what is insanity, and just when does one go beyond the pale? This is a matter for the alienist and the psychiatrist and not for the ordinary botanist or zoologist to decide, no matter what the degree of exasperation may be when one attempts to work out a Rafinesque enigma in nomenclature. Being somewhat

Rebuttal

curious as to what an experienced psychiatrist's reactions might be from reading various papers by and about Rafinesque, I asked the Boston psychiatrist Dr. J. M. Woodall to scan certain articles published by Rafinesque, including his autobiography, *A Life of Travels,* and certain papers by other authors about Rafinesque. This he very kindly did, and has courteously supplied the following report. It is interesting to note that his conclusion is that Rafinesque suffered from psychic disorder, but this was neurotic rather than psychotic. Doctor Woodall's report follows:

Psychiatric Evaluation

In an attempt to make an objective delineation of the personality of Rafinesque from his own account of his life, one is forced to contend with the temptation his writings invite to effusive speculation and value opinion. However, a careful study of his utterances, behavior, habits, and performances as revealed in his autobiographical material is productive of an abundance of characterological data, much of which justifies a conclusion as to his personality and character structure.

The composite impression one experiences of Rafinesque is that of a person in the grasp of an inexhaustible energy, a restless and tireless wanderer, an indefatigable examiner and reporter of natural phenomena, a consummate egoist of consuming intellectual appetite and versatility, whose prodigious productivity seems to have been the result of compulsive forces rather than a volitional direction. One gains the impression that he was inhabited by a driving force that made of him but a vehicle for its intellectual and physical expression and gave to his activities an automatic and indiscriminate character usually lacking where conscious choice and seasoned judgment are operative. There is a compulsion and inevitability to the outpouring of the grist of his intellectual mill that imparts to the process a quality that is detached, unbridled, and undisciplined.

Evidence adequately warrants the conclusion that Rafinesque was a genius. He possessed a versatility of kaleidoscopic range. His work frequently was said to be careless, incorrect, and unreliable. His ceaseless, compulsive mental activity, rapidity of execution, and magnitude of output—representations of his character organization—adequately account for whatever lack of integrity his work possessed. I can find no evidence for dishonesty or that he was motivated by a desire to deceive. He was not a charlatan inasmuch as there was no conscious desire to misrepresent.[4] His errors went uncorrected because he could not afford to let himself know they existed. Unreliability is a justifiable accusation. It is the natural expectation in a person almost completely lacking in self-criticism. He was guilty of failure to remove his errors. His judgment was obviously impaired but this should not be confused with dishonesty. Of what quality is that portion of his work found to be correct?

Aside from his extraordinary intellectual and energy endowments, qualities to be accounted for by heredity, his bizarre habits and performances are to be understood only when considered in the light of his personality reactivity and general character organization. Inasmuch as there is but meager knowledge of his formative years it would be sheer guesswork to attempt to state why his character structure took the form that it did. Suffice it to say that character organization is the product, in general terms, of the interplay of environmental influences upon an inherited constitution; and we know but little about either of these elements in the early life of Rafinesque.

Character trends bear a reciprocal relationship to the pattern of the behavior responses. The latter are but an expression of the former. The peculiar performances of Rafinesque were an expression of the demands of his character components. His behavior, otherwise meaningless, becomes intelligible and rational when viewed as an expression of his adaptive struggle. In such a struggle attitudes and patterns of behavior are formulated for dealing with the environment that have as their objective the preservation of the respective character organization, whether normal or abnormal. The patterns assumed by his behavior responses therefore subserved a protective function and it is this protective function that gave to them their compulsiveness, rigidity, and indiscriminate application. These patterns were expressive of his individuality. Furthermore, his security was based upon the maintenance of these established patterns of response insofar as they represented his methods of dealing with the world according to his lights.

Within the actual character edifice of Rafinesque the ego or self is of chief concern insofar as within its framework are to be found his ideas and attitudes of personal worth—values motivating his adaptive strivings. His ego was enlarged and hypertrophied to an abnormal degree. Its expansiveness was of such proportions as to constitute megalomania. About this inflated image of self as a central core his entire character organization revolved and crystallized. His life's energies flowed into behavior expressions that best allowed him to experience reality in accordance with his inflated ego values. His adaptive struggle became one of finding confirmation of his self-inflation in his relationship to life.

His actions therefore were entirely consistent with his concept of self. His attitudes were in conformity likewise with his ego values. No element of doubt as to the high quality of his work was allowed to enter his mind. He felt that he had a mission in life: to instruct, to improve, and to reform. An overbearing manner reflected feelings of superiority.

Rebuttal

In his contact with life whatever tended to conflict with his inflated image of self whether person or situation—had to be depreciated. His colleagues were experienced as inferiors or peers but never as superiors. Some who disagreed with him were considered to be incapable of appreciating the high quality of his efforts and were therefore regarded as inferiors. Others who failed to confirm his narcissistic attitude were experienced as enemies and appeared in the light of persecutors. Admiration and prestige were obviously essential to his sense of well being.

If a name must be given to the type of personality possessed by Rafinesque that of paranoid is the one of choice. His inflated image of self, his tendency to projection and his ineffectuality are all characteristic of a paranoid type of character structure. He was destined throughout his life to feel that others were slighting him. He suffered much from their failure to recognize his worth. "My works, researches, travels, collections, etc., will remain as proof of uncommon zeal, although unrequited and unrewarded."

From the evidence at hand I do not consider Rafinesque to have been insane. His reality was not qualitatively different from that of other people, although unquestionably its value was quantitatively altered. While in constant conflict with reality he nevertheless met its demands. His distorted character structure forced him to cope with life under conditions which were difficult. His capacity to enjoy life was markedly impaired due to the limited means of attaining satisfaction and security imposed upon him by his distorted character structure, but nevertheless he experienced life in essentially the same terms as those around him. His character trends caused him to stand in his own way, to live a life of relative frustration and to accomplish much less than was consistent with his extraordinary intellectual capacities.

Rafinesque suffered from a psychic disorder despite the fact that he met environmental demands. This disturbance was neurotic in character rather than psychotic. When the character traits of a person are of such nature that collectively they interfere with an efficient and appropriate adaptation and therefore with the effectuality of that person, albeit such character traits represent individualized patterns of adaptive response, that person may be considered neurotic. In this sense we may consider Rafinesque to have been neurotic and his life to have been a representation of such neuroticism.[5]

Notes

1. [But see Arthur J. Cain, p. 133, for a very different opinion about Rafinesque's euphonious names. As Alexander Pope counsels, "Who shall decide when doctors disagree?" **Ed.**]

2. [This is as close as Merrill came to naming Croizat as the vilifier of his work. Croizat himself wrote that he was responding to a note published in 1948 in Merrill's own *Journal of the Arnold Arboretum,* but Croizat must have been aware of the project for some time since Merrill—working with handwritten slips—could not have indexed all the published Rafinesque generic names overnight. When the final product was printed in 1949 it required 178 two-column quarto pages to give the whole "Index Rafinesquianus" and 50 three-column pages of fine print just to index the "Index." **Ed.**]

3. [In 1949, Merrill was, of course, unaware of the details of "The Last Days of Rafinesque" narrated here. **Ed.**]

4. [Cf. Oestreicher's contradictory view, pp. 237–40. **Ed.**]

5. [As with expert opinion on most psychological matters, other views have been expressed. In a little-known note published in the *Newsletter of the Connecticut Botanical Society* 20 (Winter 1992): 3, the late Joe D. Pratt argued with noteworthy persuasiveness that Rafinesque suffered from "bipolar affective disorder"—i.e., he was manic-depressive. Though not a psychiatrist, Pratt spoke on the subject with considerable authority; he himself had suffered from the condition all his life. **Ed.**]

The Fall from Grace of that
"Base Wretch" Rafinesque

CHARLES BOEWE

WHEN HE WAS defending Rafinesque's sanity in 1949, Merrill could not have known that the very next year a pamphlet recently published by a former associate would in fact spark the unprecedented endeavor to solve the "Rafinesque nomenclatural problems"—as Merrill himself had called them—by ruling that Rafinesque was "mentally incompetent." In 1950 that is what happened when participants of the Seventh International Botanical Congress in Stockholm were exhorted to proclaim that most of Rafinesque's published botanical discoveries never legally existed and that the work of their insane discoverer should be expunged from the record.

This unusual international conspiracy began to germinate as early as 1935, when another Harvard botanist, C. A. Weatherby, agreeing with Asa Gray's earlier opinion, wrote that the plant genera established in all of Rafinesque's later books represented "a kind of pseudo-scientific work, the nomenclatural results of which may well be legislated out of existence"[1] by other botanists. As indeed it lay within their power to do. Over the years since Rafinesque flourished, the world's botanists, meeting in solemn conclave, had legislated for themselves an elaborate International Code of Botanical Nomenclature that requires that the first-used Latin name for a plant stand forever, if it was validly published as defined by the code, though exceptions also occur—one being the change that follows if the plant itself is reclassified. But just as the world's botanists had devised this rule of perpetual priority, so, it was argued, they had the power to make another exception in the uncommon case of botanical binomials published by a madman.

The International Code(s) of Botanical Nomenclature were needed to formalize the principles of priority that had been advocated since Linnaeus himself and had been developed in the practice of succeeding naturalists, especially as upheld by such leaders as Rafinesque's friend A. P. de Candolle. International congresses of botanists had begun with one in Paris in 1867, followed by one in Vienna in 1905, then Cambridge in 1930, and so on—each resulting in a published code bearing its name.

When Rafinesque was publishing his discoveries, and splitting established genera into new ones requiring additional names, priority was much more a personal matter of what one could persuade—or even coerce—his colleagues to accept. Rafinesque demanded—virtually challenged—other botanists to search out and adopt his own published plant names, however obscure the source, and taunted them when they failed to succeed. It took eleven packed pages for him to review Frederick Pursh's *Flora Americae Septentrionalis* (1814), where he cited chapter and verse of all the plant names followed by "Raf." that Pursh had overlooked, including those in *Florula Missourica*, which nobody then, or since, has ever seen.[2] At the same time it must be admitted that Rafinesque was a formidable bibliographer himself, for with primitive resources at his disposal he cited and used publications that are hard to come by even today.[3] His detractors have found his erudition no more endearing than his edgy temperament and overweening egotism.

Any of us can appreciate the natural human desire to receive credit for one's own contributions, but by the middle of the twentieth century, when Croizat took up his crusade to expunge Rafinesque's names from botany, the principle of priority also had come to have an even more important function in the life sciences, particularly in botanical nomenclature. Knowledge in the physical sciences is said to cumulate, but knowledge in the life sciences—especially the naming of new plants—*ac*cumulates, with the result that chaos would ensue if the same plant were known by two or more scientific names. Priority of valid publication seemed to be an objective, impersonal, automatic device to purge the record of needless redundancy.[4] Hence, a ruling on the admissibility of the contribution of one who had added as much to the published record as Rafinesque had done was no light matter.

As we have seen, the attempt to follow up on Weatherby's brusque suggestion that Rafinesque's writings deserved to be outlawed was spearheaded by Leon Croizat. This peripatetic Italian, who had been dismissed from his job at the Arnold Arboretum, where Merrill was compiling his *Index Rafinesquianus,* wrote his Rafinesque "exposé" while teaching in Venezuela, where he had settled. He published his attack in Italy and then circulated it internationally in preparation for the

The Fall from Grace of that "Base Wretch" Rafinesque

Seventh International Botanical Congress in Sweden. It is unclear why one already much accustomed to controversy weaseled a bit by publishing under the pseudonym "Henricus Quatre." Probably what he had in mind was not the first Bourbon but, rather, Henry IV of England, who neatly solved the problem of heretics by burning them, *pococurante*—for Croizat's intemperate conclusion (p. 191) was that Rafinesque's plant names had been a "flood of polluted nomenclature contributed by a lunatic, who wrote botany because he was of unsound mind."

Stuckey shows that most American naturalists had decided by the time of Rafinesque's death that his publications were not worth being sought out and consulted. They, too, had concluded that these often obscure publications—many of them in languages few Americans could read—were the product of an unsound mind. But there is more here than meets the eye. Left unmentioned in the denunciations of their colleague was the extreme annoyance the other American naturalists all felt with his strident advocacy of the French "natural method" of classification that he had carried on in private correspondence for some time, had insisted on in his reviews of their books, and had demonstrated in his own *Florula Ludoviciana* (1817), for to a man the other naturalists in the United States were Linnaeanists. As with any other professionals, the naturalists welcomed additional data; but also—like any other professionals, then or now—they fiercely resisted having to change the way they thought about that data.

It must be remembered that in "taking Linnaeus as my master," Rafinesque meant only that he was going to follow—and "improve" on—Linnaeus as lawgiver for the classification of plants and animals. The actual system Linnaeus developed in his own taxonomic work, the so-called "sexual system" in botany that classified according to the number and arrangement of stamens in flowering plants, not only was "indelicate" in Rafinesque's opinion but, more important, was less relevant than the French so-called "natural systems" he preferred, systems that classified according to various perceived "affinities" among plants. A rough analogy of the difference might be expressed by saying that if Linnaeanists were to classify furniture, they would group beds, tables, and chairs in the same genus because all have four legs, leaving a milking stool to fall into a different genus because it has three legs. However, advocates of natural systems would put both chairs and milking stools into the same genus because both are made to sit on; and they would put beds and hammocks into the same genus because both support sleepers, even though the latter has no legs at all. Linnaeanists defended their system because the way it placed plants caused them to be easily remembered, and hence it was considered an aid to students. Though the controversy has only historical interest since the Darwinian revolution in biology, the advocates of "natural" systems were moving in the direction of the phylogenetic classification of the

Charles Boewe

future. In this matter, there is no question that Rafinesque was indeed in advance of his time, as he often claimed; and before his death, first John Torrey, then most of the others, came around to his persuasion.

Despite his advanced views, Rafinesque also brought trouble on himself by his slipshod work methods, his haughty egotism, his paranoia, as Merrill shows, and—one ought to add—his irascible temperament. Among the earliest in America to write reviews of natural history books, Rafinesque flayed his colleagues for faults both real and imagined with never a thought that in so small a community they would have a chance, sooner or later, to pay him back in the same coin. To take a single example, Thomas Say never forgave him for the mildly adverse review Rafinesque published in 1817 about the first number of Say's *American Entomology*.[5] This perceived slight, added to the embarrassment occasioned when Say published Rafinesque's description of a fish that also had appeared earlier in the *American Journal of Science*, was enough for him to see to it that Rafinesque contributions were no longer welcome in the *Journal* of the Academy of Natural Sciences of Philadelphia, whose publication committee he headed.

And there is even more here than meets the eye. John Torrey would have been the logical person to memorialize Rafinesque, but in 1841 Asa Gray, who had not yet attained his professorship at Harvard, at age thirty was very much on the make and needed to validate his authority as a leading American botanist that he did, in fact, become later on. One might suppose he took on the sad duty of writing an obituary article as an obligation to his profession and as a tribute to the old man, whom he had known personally in his youth. In a letter to W. J. Hooker in Britain, January 15, 1841, he said, "Poor Rafinesque, you know, perhaps, is dead; and I have attempted the somewhat ungracious task of giving some account of his botanical writings. . . ."[6] More ungracious, however, was the concealed fact that Gray had volunteered to write the article about "poor Rafinesque" and, as soon as he finished it, wrote again to editor Silliman urging him to make it the lead article of the next issue of the *American Journal of Science*.[7] Why? Gray intended this article to be the coup de grâce for Rafinesque's writings—once and for all to outlaw the man and most of his works. Even though he says he did not have a copy of *A Life of Travels* before him, Gray knew perfectly well that Rafinesque was not a Sicilian; and he shrewdly concocted the ad-hominem anecdote about Rafinesque having sent a paper to "a well known scientific journal, describing and characterizing, in natural history style, *twelve new species of thunder and lightning!*"—a canard that has been repeated over and over since, both in print and in lecture rooms, by snickering pedants.[8] Gray achieved his objective: to make it seem absurd even to think of consulting the published ravings of this madman in the grip of a "monomania" for making new species.

206

The Fall from Grace of that "Base Wretch" Rafinesque

Croizat did not achieve his. Perhaps because irascible, paranoid egotists are as common among botanists as in other professions, the botanical legislators at Stockholm chose not to include sanity of the author as a condition for valid publication. Calmer heads prevailed, so the question of whether a crazy scientist can produce sane science was not settled in 1950. As a result, the name Rafinesque remains a thorn in the side for many taxonomists yet, including zoologists as well as botanists.

And so it has gone ever since Rafinesque returned to these shores in 1815. Because the insane seldom are aware of their insanity, perhaps the best corroboration of Rafinesque's mental clarity is his own acknowledgment that he suffered himself to be "laughed at as a mad Botanist" in his rambles around Kentucky, in order "to be a pioneer of science."[9] But whatever the validity of psychoanalyzing the dead, Merrill's psychiatrist surely had hit on Rafinesque's salient flaw—his swollen ego—which did cause others to regard him as crazy. Then, too, the psychiatrist had been acute enough to detect his patient's persistent paranoia, probably because Rafinesque complained bitterly about the "foes of science" at Transylvania, among whom he numbered Transylvania's president Holley as the chief offender.

Yale-educated Horace Holley, formerly a Unitarian minister, was one of the early republic's truly great educators. Before his appointment, a visitor to Lexington described the slovenly appearance of Transylvania's single classroom building with the students lounging around on the steps, chewing tobacco and spitting in all directions except into the wind. Holley disciplined the students, reformed the organization of the institution, and tried to recruit a distinguished faculty. He introduced the freshman, sophomore, junior, and senior class division we recognize today. Coming to an enterprise that was little more than a glorified secondary school, he took justified pride in raising the university's standards to the point where students soon were able to write both Latin and Greek by their junior year, by which time they also had read Cicero, Ovid, Horace, and Juvenal. His curriculum included natural philosophy—roughly what we mean by physics today—which fitted into the classical curriculum by tradition, as well as mathematics, which included such practical specialties as trigonometry and surveying.

Among the faculty he inherited, some were talented, others were misfits, all were prima donnas. Among the misfits was the conceited Charles Caldwell, who never quite reconciled himself to what he considered the raw society of Lexington after having known the sophisticated salons of Philadelphia. But Holley himself, impressed by Caldwell's prestige among Philadelphia's physicians, had recruited Caldwell to run the university's medical school, which already had some outstanding —if unruly—professors, three of whom (Daniel Drake, Benjamin Dudley,

Charles Boewe

and William Richardson) on one occasion indulged in a three-way shoot-out. Holley also inherited the Latin teacher, John Roche, who so imbibed the wisdom of *in vino veritas* that he lay drunk much of the time. And, what must have been a real trial, he inherited a professor of chemistry who seldom gave a lecture. Even the trustees from time to time adjured James Blythe to lecture. Blythe had been Transylvania's acting president for fourteen years before Holley took over. He surely was sulking in his tent, but it is likely also that he preferred not to reveal his ignorance about the subject he was supposed to be teaching. With only an honorary doctorate of divinity as an academic credential, he probably did not know an acid from a hole in the ground, though he was deeply experienced about the latter. With leisure to spare during his acting presidency, he had dug with his own hands the pit for the university's privy.

To Holley, Rafinesque was the "Constantinopolitan" who had been thrust upon him by trustee John D. Clifford, Rafinesque's friend since 1802 when they first met in Italy. Although Rafinesque had accepted his professorship without a salary (depending, like the medical faculty, on selling tickets to prospective auditors), when Holley tried to discharge him the trustees rallied to his support, made him secretary of the faculty, librarian, and keeper of the university museum—and agreed to pay him three hundred dollars a year. And Holley himself was under attack by others during much of his tenure. His conservative critics condemned his theology—he had declared from the pulpit that, while God's grace was a sure route to salvation, good works also might be pleasing in the sight of the Lord—and his critics were aghast at reports that he favored dancing, smoked cigars, enjoyed fine wines, and had an eye for a pretty ankle. Eventually they drove him out of town.

Both personally and professionally, Holley labored under difficult conditions to shape Transylvania University into a worthy parallel of such existing universities as Harvard, Yale, and Princeton. He not only succeeded, but he brought Transylvania to the forefront in some respects, one being enrollment—this despite the fact that the eccentric cosmographer John Cleves Symmes tried to recruit "100 brave lads" from his student body to help explore the interior of the earth, which Symmes declared was hollow in his public lectures, one of which was well attended in Lexington.

Interrupting Holley's Herculean task was Rafinesque's repeated demand that natural history also have a hearing. Though we can see now that the explosion of knowledge which took place in natural history in the early decades of the nineteenth century put botany and zoology on the cutting edge of science—much as particle physics or molecular biology are for our own time—such a revolution

208

The Fall from Grace of that "Base Wretch" Rafinesque

had little relevance to Holley's struggle to plant the Greek and Latin classics firmly on the frontier. The proper role for natural science must have seemed to him the one adopted without a murmur by Charles Wilkins Short, who pursued botanical research as an adjunct to his yeoman service teaching materia medica to the medical students.

At most tolerating Rafinesque, Holley tried to persuade Benjamin Silliman to take the place of the reluctant chemistry professor—which shows that, far from being a foe of science, as Rafinesque declared, Holley wished to strengthen his faculty in those branches he considered valid by recruiting the best scientific talent to be had. But Silliman preferred to remain at Yale, where he taught chemistry and geology, eventually secured the establishment of the Sheffield Scientific School, and founded *The American Journal of Science,* which became the most prestigious scientific journal in America. There Silliman published eleven short papers by Rafinesque in the first volume (1818–19) of his journal, when something happened that has been seen as a turning point in Rafinesque's career and, in wider context, the symbol of a watershed in the development of science in America: the replacement of the broad-gauge field naturalists by laboratory-based, narrow specialists.[10]

When he published Asa Gray's obituary notice on Rafinesque's botanical work, Silliman saw fit to append his own footnote to the article where he explained that in 1819 "I became alarmed by a flood of communications, announcing new discoveries by C. S. Rafinesque, and being warned, both at home and abroad, against his claims, I returned him a large bundle of memoirs. . . . This will account for the early disappearance of his communications from this Journal. The step was painful, but necessary; for, if there had been no other difficulty, he alone would have filled the Journal, had he been permitted to proceed."[11]

Silliman's comment, often cited since, deserves further explication. Taking its points from last to first, it may well be that Rafinesque could have filled the journal single-handedly, for the flood of communications in 1819 resulted from discoveries he made during the previous year in his trip from Pittsburgh down the Ohio River as far as Shawneetown. The Kentucky years gave him a rich harvest in new flora and fauna, and when he left Kentucky in the spring of 1826 he shipped to Philadelphia forty crates of collections and continued to exploit these materials the rest of his life. However, Silliman failed to mention that he published a final Rafinesque contribution as late as 1821—the description of a fossil jellyfish, a *lusus naturae* that surely ranks in the annals of pseudoscience with Rafinesque's Audubon-inspired Devil-Jack Diamond-Fish, even though authentic fossilized siphonophores have been reported since from the Flinders Range in Australia.

And Silliman's journal continued to carry third-person notes about Rafinesque's activities, even though it would no longer accept articles from his hand. In 1836 the journal devoted the best part of two pages to announcements of Rafinesque's current book publications and to his offer to buy or exchange plant specimens. So he was not entirely ignored by the *American Journal of Science,* as has been thought.

Far more significant in Silliman's decision must have been the warnings he said he received. Who could have issued warnings so forceful? As we have seen already, many contemporary botanists were ready to bad-mouth Rafinesque, and zoologists such as Richard Harlan, Thomas Say, and James E. DeKay also could be added to their number.[12] What has not been understood is that the person who gave the first shove of Rafinesque's endangered reputation down slippery slopes was not a naturalist at all. This person's poison-pen letters began to have their effect while Rafinesque was a member of the Transylvania faculty, but this enemy was far removed from the university itself, and even farther removed from the coterie of American naturalists. As we shall see, what he shared with Rafinesque was an interest in the prehistoric "antiquities" of the Ohio Valley.

Already deprived of the Journal of the Academy of Natural Sciences of Philadelphia as a medium for his publications, in Lexington Rafinesque had turned to various publication media, including newspapers, pamphlets published at his own expense, and Lexington's *Western Review and Miscellaneous Magazine,* a monthly which struggled to survive during the period 1819–21, under the editorship of William Gibbes Hunt. Most of the magazine's contributors were associated with Transylvania and included faculty, trustees, the president, and his wife, Mary Holley. In Lexington, too, Rafinesque had developed a new enthusiasm so far removed from the later concerns of natural history that his earliest bibliographers passed over his contributions in this field as hastily as possible, missing several as a consequence. His new interest was one he shared with his patron, trustee John D. Clifford, on "circumvallations"—Indian "forts," as several near Lexington still are called—or, as Rafinesque persuaded himself, the Ancient Monuments of America. One of his earliest publications in this field appeared in Hunt's magazine, and was respectfully addressed to Postmaster "Caleb Atwater, of Circleville," Ohio.[13]

Atwater is generally credited with being the first to produce a book-length survey of the Ohio Valley prehistoric mounds based on attentive examination of them by himself and others. His book included woodcuts of artifacts found in and near some of the mounds, descriptions of the earthen structures themselves, spec-ulation about their builders—too much speculation, most of it secondhand any-way—and handsome, if stylized, engraved maps of many of the sites. Other people, including Rafinesque, also had drawn manuscript maps of prehistoric

The Fall from Grace of that "Base Wretch" Rafinesque

sites, and no doubt Rafinesque felt some proprietary interest, since he had been publishing his observations for several months even though his printed contributions appeared too late for Atwater to make use of them had he wanted to. However, Rafinesque had been in direct correspondence with Atwater for an even longer time and, taking advantage of Postmaster Atwater's franking privilege, had used him as a conduit for archaeological reports he was transmitting to the American Antiquarian Society in Massachusetts.

Despite his peevishness and proneness to take offense, there is no question about Rafinesque's willingness to share his knowledge with others—whether plant specimens, fossils, vocabularies of Indian languages, or, in the case of Atwater, maps and descriptions of prehistoric sites. He only expected equal measure in return and acknowledgment of his contributions. Yet, when Atwater's book appeared as a major part of *Archaeologia Americana*[14] he thanked everyone imaginable, including John D. Clifford, but remained obdurately silent about Rafinesque.

Rafinesque's anonymous review of the book, in Hunt's magazine, was unknown until the 1982 revision of Fitzpatrick's bibliography of writings by and about Rafinesque.[15] For the work of an author scorned, the review was remarkably impartial, which gives some credence to Atwater's own belief that it was the joint production of Hunt and Rafinesque. According to Atwater, Hunt left the manuscript for Rafinesque to see through the press, and "this base wretch" then "inserted in every part of the review, the basest insinuations against me and inserted more than one hundred as base falsehoods as were ever uttered by man!"[16] At any rate, it was Rafinesque who roused Atwater's ire and got the blame, not Hunt. In a copy of the magazine now preserved at the Cincinnati Historical Society, Atwater scrawled such angry comments as "Only R's say so who has not seen them" beside a list of sixteen prehistoric sites the review said Atwater had overlooked. About all one can find today likely to give offense in the review is a supercilious comment about Atwater's style, which "though animated, is diffuse, and not always correct. He is not even exempt from grammatical errors, nor is he uniformly accurate in his orthography." If, as Buffon maintained, the style is the man himself, perhaps this affront to his amour-propre was enough to throw Atwater into a sputtering rage.

Rafinesque soon became aware that Atwater blamed him for the review and, taking full responsibility for it, he wrote with great sang-froid four years later that having "corrected some inaccuracies of his in a Review of his labours . . . I have incurred his *displeasure*"—and Rafinesque may have become aware of the consequences, for he went on to remark that the displeasure "has shown itself in a manner rather singular and unwarrantable." As indeed it had.

Rafinesque's review appeared in September, and the following January Atwater wrote to Samuel Latham Mitchill, a fellow contributor to *Archaeologia Americana,* asking whether he had received the volume in New York and snarling that "as to Prof. Raf. as he *now* calls himself, or Smaltdz as he *was* called, until the sea washed away his *actual* name," he "injures us considerably in Kentucky. But he cannot last long anywhere. I shall take care, that his true name, real character and private history shall be well understood there, very soon."[17]

Using his postmaster's franking privilege, he fired off letters in all directions. He asked Prof. Parker Cleaveland in Maine to return an essay of his that included information on conchology, courtesy of Rafinesque, "to correct it by striking out every word depending on the veracity of a person, who ought to be ranked among the worst of impostors, in literature and science, now living in the world."[18] No letter of his has been found among the Silliman papers at Yale, but language so similar to Silliman's—"being warned, both at home and abroad"—turns up in Atwater's correspondence elsewhere that surely he was the one who put the bee in Silliman's bonnet. To the American Antiquarian Society in Massachusetts, Atwater wrote that "Prof. Raf. is writing a great deal for the 2nd Vol. [of *Archaeologia Americana*] but before you publish any thing of his, where facts are wanted, I would advise you to ask the opinion of Professor Silliman and Pres. Holley"— both of whom he must have tried to set up—"and to consult any periodical work, published in London or Paris. In the meantime, I can inform you, that in Europe, his statements are not believed in any case whatever. These things I knew not, úntil since his review, when letters from all quarters poured in upon me."[19]

Certainly Atwater tried to prejudice Holley, as he probably had Silliman, but to his credit Holley asked for proof. We do not know what evidence, if any, was sent him, but Holley remarked to his brother, "I have received a letter from Caleb Atwater with many severe remarks upon Rafinesque, in consequence of one that I wrote to him not long ago, asking the names of the Journals in Europe, in which the public were cautioned against believing our Professor. Atwater is petulant, and evidently a little nettled by a review of his work in Hunt's magazine, written by Rafinesque."[20]

Finally Atwater's wrath was spent, for when Rafinesque asked later that year for the return of essays and maps he had tried to transmit through Atwater to the American Antiquarian Society, its president, Isaiah Thomas, replied that he would attempt to oblige; but he added that he had received "a very strange and unhandsome letter from C. Atwater," saying "that he shall withdraw himself from all Societies—that he has quitted all Antiquarian and Geological researches—and intends to drop all correspondence on those subjects."[21] Later Atwater relented and proposed a second volume, which never appeared.

The Fall from Grace of that "Base Wretch" Rafinesque

The bad blood between Atwater and Rafinesque must have been common knowledge at one time in southern Ohio; in its review of Rafinesque's *Ancient History or Annals of Kentucky* (1824), the *Cincinnati Literary Gazette* expressed mock surprise that he had located the Garden of Eden in Asia, whereas his friend, Dr. Samuel Latham Mitchill, had more patriotically surmised it probably was in North America. "We can only account for his dissent . . . from the fact that the doctor's theory . . . is published in *Archaeologia Americana*, a work that must for ever be of doubtful authority, while its pages contain the name of *Caleb Atwater*."[22]

Though long since forgotten, the petulance of a provincial postmaster had done its work, starting Rafinesque's reputation down a long decline from which it has never wholly recovered. As Stuckey has shown, his botanical colleagues had other bones to pick with him, but Atwater was the first to try to annihilate him. Rafinesque's exclusion from the pages of the *American Journal of Science* may have coincided with a watershed in the history of American science, but he owed the start of his troubles to an immediate cause no greater than another ill-received book review, and at that a book that had nothing to do with natural science as we now understand it. Whatever else the episode may show, it also reminds us that even paranoids do have enemies.

Notes

1. [Kew] *Bulletin of Miscellaneous Information* Nos. 6–9 (1935): 409. Among the international fraternity of botanists, however, there has been disagreement on this as on most other issues. A decade later the Czech botanist J. Paclt declared (*Taxon* 9 [1960]: 47–49) that "both the first and most ingenious system of descriptive biology is undoubtedly that of C. S. Rafinesque," which "corresponds, in fact, to that of the more recent plant and animal taxonomy."

2. *American Monthly Magazine and Critical Review* 2 (Jan. 1818): 170–76; (Feb. 1818): 265–69. Unable to find a publisher in this country for *Florula Missourica*, Rafinesque sent the manuscript (which he expected would make a pamphlet of forty pages) to William Swainson in England. When it was not published there, he translated it and sent it to another friend, Baron Bory de Saint-Vincent, who was unable to publish it either in Brussels or Paris. The booklet described and named plants discovered by the English botanical traveler John Bradbury, who had settled near Louisville. Since Rafinesque agreed with Bradbury that Pursh had stolen Bradbury's discoveries from a duplicate set of specimens in England, perhaps he concluded that Pursh should have stolen his own descriptions and names as well. Neither manuscript has ever been found. I have elaborated on Bradbury's career and Rafinesque's connection with him in "John Bradbury (1768–1823), Kentucky's Forgotten Naturalist," *Filson Club History Quarterly* 74 (Summer 2000): 221–49.

3. A good example is Rafinesque's recovery of an obscure Danish work on cottons of the West Indies, *Anmerkungen über den Cattunbau* (1791–93), by J. P. B. von Rohr, which Rafinesque probably knew through an equally obscure anonymous French translation (1807). Several species of the genus *Gossypium* that Rafinesque named from this literary source, without having seen the plants themselves, are generally accepted today. This despite the brouhaha in his own time and since over the impropriety of his having also conferred plant names in *Florula Ludoviciana* (1817) without having seen the plants they pertain to.

4. A. Hunter Dupree, "The Measuring Behavior of Americans," in *Nineteenth-Century American Science,* ed. George H. Daniels (Evanston, Ill., 1972), 32–33.

5. Rafinesque's chief complaint was that to include all of the American insects, Say's book would cost too much when completed. The two also were hopelessly at odds over what natural history books should be. Rafinesque wanted them cheap (and his look like it, despite their prices on the rare-book market today). Say wanted them beautiful (and illustrated by hand-colored drawings by his wife, his are). Say's mistake, however, was his forlorn hope that Americans would be as interested in books with "beautiful" bug pictures as they were in Audubon's beautiful bird pictures.

6. Jane Loring Gray, ed., *Letters of Asa Gray* (2 vols., Boston, 1894), I: 278.

7. Asa Gray to Benjamin Silliman, Oct. 5, 1840, and Dec. 31, 1840, both letters at the American Philosophical Society. The Gray Herbarium Library has Asa Gray's own copy of Rafinesque's *Medical Flora.* Its pages have not been cut.

8. Further details about the article are given in n. 25 of Pennell's essay. Silliman actually commissioned S. S. Haldeman to write a "Notice of the Zoological Writings of the Late C. S. Rafinesque," which appeared the next year in the *American Journal of Science* 42 (1842): 280–91. More temperate in his comments than Gray, Haldeman was critical of the brevity of Rafinesque's scientific descriptions and of his credulity in accepting the observations of others, but it appears that he went out of his way to straighten out the facts of Rafinesque's nativity, even giving lengthy quotations from *A Life of Travels,* which Gray denied having seen.

9. Rafinesque, *New Flora of North America* (4 parts, Philadelphia, 1836), I: 11–12.

10. The way in which field naturalists like Rafinesque were made obsolete by advancing specialization in all the sciences is discussed in John C. Greene, *American Science in the Age of Jefferson* (Ames, Iowa, 1984), and Charlotte M. Porter, *The Eagle's Nest: Natural History and American Ideas* (University, Ala., 1986).

11. [Silliman], *American Journal of Science and Arts* 40 (1841): 237n. Announcements of Rafinesque's recent publications and his offer to purchase plants occurred in vol. 29 (1836): 393–94.

12. Harlan and Rafinesque had nothing but contempt for each other, Harlan speaking of Rafinesque's "utter ignorance" because of his "insulated situation" in Kentucky, and

The Fall from Grace of that "Base Wretch" Rafinesque

Rafinesque "exposing" a Harlan fossil as an ordinary piece of stone. As developed earlier, Say was miffed by Rafinesque's review of his book and wrote to DeKay that "I cannot understand the writings of Mr Raffinesque & therefore do not consult them"—a remark DeKay was happy to pass on in a letter (Jan. 29, 1825) to Charles Lucian Bonaparte. And after further reflection, DeKay wrote again to Bonaparte (Mar. 5, 1825) that "Nothing[,] I repeat[,] should induce us to countenance such quackery" as Rafinesque's, and "the ensuing generation will certainly follow our example." Harlan's sentiments are found in his book *Fauna Americana* (Philadelphia, 1825), vii, and DeKay's in the Bonaparte correspondence at the Bibliothèque of the Muséum Nationale d'Histoire Naturelle in Paris.

13. Rafinesque, "Description of the Ancient Town Near Lexington," *Western Review and Miscellaneous Magazine* 2 (May 1820): 242–44, a follow-up on an earlier brief article about the same site in the December 1819 issue of the same magazine. An even more substantial article titled "On the Upper Alleghawian Monuments of North Elkhorn Creek" also was written as an open letter addressed to Atwater and was published in the *Western Review and Miscellaneous Magazine* 3 (Aug. 1820): 53–57. All of these have been reprinted in Charles Boewe, ed., *John D. Clifford's Indian Antiquities: Related Material by C. S. Rafinesque* (Knoxville, 2000), 73–82.

14. *Archaeologia Americana* was the general title for early volumes of the Transactions and Collections of the American Antiquarian Society. Atwater's contribution, titled "Description of the Antiquities Discovered in the State of Ohio and Other Western States," appeared in the first volume (1820), pp. 105–267, which also contained a number of related articles of lesser length.

15. The unsigned review appeared under the title of the book being reviewed. Reasons for the attribution are given in my *Fitzpatrick's Rafinesque: A Sketch of His Life with Bibliography* (Weston: M&S Press, 1982), 241. Most of the review is reprinted in *John D. Clifford's Indian Antiquities,* 99–109; Atwater's marginalia on the Cincinnati Historical Society's copy are given on p. 147; and Rafinesque's subsequent published rejoinder to Atwater appears on pp. 110–16.

16. Caleb Atwater to Isaiah Thomas, Oct. 12, 1820, American Antiquarian Society. Horace Holley may have been the source of Atwater's belief that Hunt originally wrote the review; Holley said as much in a letter to his brother Orville, Oct. 14, 1820, Crosby Papers, Univ. of Louisville.

17. Caleb Atwater to Samuel Latham Mitchill, Jan. 7, 1821, Cincinnati Historical Society. "Smaltdz" is Atwater's error for Schmaltz, the matronym Rafinesque hyphenated to his surname in Sicily to avoid being considered a Frenchman when it looked as though the island, then controlled by the English, might be invaded by the French. The reference to the sea shows that Atwater knew Rafinesque was shipwrecked when he returned to the United States in 1815.

18. Caleb Atwater to Parker Cleaveland, Nov. 4, 1820, Cincinnati Historical Society.

19. Caleb Atwater to Isaiah Thomas, Nov. 22, 1820, American Antiquarian Society. Atwater's letter must have had some effect on Thomas, for in truth the American Antiquarian Society, of which Rafinesque was a member, never published any of the several articles he sent there.

20. Horace Holley to Orville L. Holley, Feb. 22, 1821, Crosby Papers, Univ. of Louisville. Though European journals were critical of Rafinesque later in his career, no such reference has been found this early by his bibliographers. In fact, Kurt Sprengel listed all of Rafinesque's botanical discoveries appearing in Silliman's journal in his *Neue Entdeckungen im Ganzen Umfang der Pflanzenkunde* 1 (1820): 142–46; 2 (1821): 206–8—which was more European recognition than most American botanists received at the time. Holley went on in the same letter to say that Rafinesque's "correspondents in Europe compliment him, and increase. He has shown me letters from Cuvier, and from some of the distinguished naturalists in Germany and England."

21. Isaiah Thomas to C. S. Rafinesque (letterbook copy), Sept. 3, 1821, American Antiquarian Society.

22. *Cincinnati Literary Gazette* 2 (Dec. 25, 1824), 202–4. Other Cincinnati wits also were amused by the antiquarians' dispute over the location of the Garden of Eden. In an erudite bilingual pun on Samuel Latham Mitchill's middle name, Thomas Peirce, *The Odes of Horace in Cincinnati* (1822), attributed the discovery of Eden on the banks of the Ohio to "Professor Brickibus, M.D."

Part III

The Philologist

Rafinesque's Linguistic Activity

VILEN V. BELYI

Introduction

THE EARLIST ATTEMPTS, in the seventeenth and eighteenth centuries, to understand indigenous American languages were those of Christian missionaries. The history of the post-missionary period of American linguistics had its beginnings in the language-directed activities of the so-called nonprofessional linguists. Among them, the names of Peter Stephen Du Ponceau, John Pickering, James Hammond Trumbull, Albert Gallatin, and John Wesley Powell loom large in the early scientific study of American Indian languages. Still, this list of noble names, I believe, suffers a serious omission without that of Constantine Samuel Rafinesque, who contributed much to this branch of language science, but is rarely, if at all, mentioned among the precursors of American linguistics. I am wholly aware that, from the standpoint of modern linguistics, much in Rafinesque's linguistic writings is farfetched and wrong. But his mistakes and false etymologies were, as Brinton aptly said of his *Ancient Annals of Kentucky* (Rafinesque 1824a), "not a whit more absurd than the laborious card houses of many a subsequent antiquary of renown" (1885: 150–51). My position regarding Rafinesque's linguistic writings is that it would be erroneous to require of him the standards of the linguistic studies of today. Nor is my respect for Rafinesque's linguistic accomplishments diminished by the recent demonstration of the fraudulence of the *Walam Olum* (Oestreicher 1994), the alleged creation myth and migration legend of the Delaware Indians published by Rafinesque (1836a, I: 121–61). Whatever mistakes and blunders our linguistic forefathers have made, we nevertheless stand on their shoulders.

Vilen V. Belyi

Discussion

A precocious boy, Constantine Samuel Rafinesque-Schmaltz read widely, and, possessing a great talent for foreign languages, he learned many of them.[1] He came to the United States in the spring of 1802, landing at the port of Philadelphia. It was there that he first saw Indians. Still, during this period he did not show any special interest in American Indian languages. From 1806 to 1815, he lived in Sicily, and after his return to America in 1815, he lived for three years in New York, and then moved to Lexington, Kentucky. It was in Kentucky, as he stated, that he "began to study earnestly American history and Arch[a]eology, with the Ethnography and Philology of the American nations" (Rafinesque 1836b: 63). This study led him much farther than he expected since "it became needful to review the whole of comparative philology and primitive arch[a]eology, in order to obtain satisfactory results" (1836b: 63).

Although at this time awareness of European scholarship was rare in the United States, and news of German discoveries such as those detailed in Jacob Grimm's *Deutsche Grammatik,* published in its second edition in 1822 (Grimm 1822), was not always welcome, Rafinesque was one of the few who were well versed in European scientific accomplishments and especially in the development of comparative linguistic methodology.[2] As his manuscripts reveal, he was at home with the studies of such European scholars as Leibniz, Charlevoix, Adelung and Vater, Alexander and Wilhelm von Humboldt, Reeland, Friedrich von Schlegel, Court de Gébelin, and Klaproth. Moreover, he knew the works of Heckewelder, Barton, Hervás, Edwards, Molina, and Clavigero dealing with American Indian languages. A great champion of comparative linguistic studies, Rafinesque considered the comparative method as the means to an illustrious end. In his unpublished "Introduction to American Ethnology," he states that "Comparative Philology has lately become a *Science* of great importance: being cultivated by the most learned linguists, divested of useless etymologies and reduced to regular principles, it has assumed a new form and given rise to a new science, Ethnology[,] which enquires into and ascertains the History of Nations by means of their compared languages and Dialects" (Rafinesque 1824b: 1; emphasis mine). In a letter, also of 1824, to Thomas Jefferson seeking vocabularies, he observed that "Comparative Philology is now becoming in Europe the base of History" and said that he had begun "to explore with some attention the wide and fruitful field of American Ethnology," adding that he was "collecting materials for a more extended Work or Works on the Ancient History, Antiquities, Languages and Ethnology of North America" (quoted in Betts 1944: 379–80).

Rafinesque's Linguistic Activity

Many problems with the ancient history of America, he insisted, can be solved through the aid of comparative philology when it gets rid of "useless etymologies" and is "reduced to regular principles" (Rafinesque 1824b: 1). This demand for the regularity of principles, a cornerstone of the modern study of language, emerges, no doubt, from the natural-history approach so characteristic of Rafinesque's ideas regarding the objectivity and logical consistency of any science. His appeal for the scientific reshaping of the study of language foreshadows the later views of Powell, Boas, and Bloomfield in their demands to make the science of language objective, exact, and, in the end, axiomatic. Because of this aspect of Rafinesque's methodology, his practice may be seen as a forerunner of structural linguistics.

Now, the question is, what are these regular principles of the comparative method? Rafinesque expressed them as follows:

> 1. Words are more important than Grammars in the comparison of Languages. 2. The most important Words are the Names of material objects, primary Nouns and adjectives, besides the cardinal Numbers. 3. By selecting or comparing a certain number of such words in two Dialects, the number of them evidently analogous will be the amount of analogy between the Dialects. 4. This is susceptible of mathematical demonstration, and the number of words in two Dialects being known or inferred, the total number of analogies may be calculated by progressive induction. (Rafinesque 1824b: 2)

The decisive factor in determining language affinity is the analogy of vocabularies, of which Rafinesque distinguished seven degrees: "1. Identical words, as *Nih* in Otoh and *Nih* in Nipegon. 2. Similar words, as *Ijah* in Wocon and *Iyah* in Catahba. 3. Consimilar words, as *Lan* in Aztec and *An* in Mandan. 4. Analogous words, as *Kig* in Penobscot and *Okih* in Santikani. 5. Affiliated words, as *Pa* Shoshonih: *Paavi* Tuhih. 6. Derivative words, as *Noni* Tamanac: *Peni* Maipuri. 7. Remote words, as *Ondesha* Huron: *Ishoini* Atnah" (Rafinesque 1824b: 2–3). The implication of these quotations is that the only sound (and really scientific) criterion of language affinity is lexicon. He expressed this conviction as axioms of comparative philology (see Boewe, p. 247). And in suggesting such a model for the classification of American Indian languages, Rafinesque anticipated the ideas of Gallatin,[3] of Powell,[4] and, partly, of Hale.[5]

It was Rafinesque who laid the cornerstone of the American model of determining language relationships through similarities of words or stems, or material content of a language, contrary to the generally accepted dictum of European philologists that it was grammar, or internal structure or form, that should be of more weight. In her lengthy article "Grammar or Lexicon? The American Indian Side of the Question from Duponceau to Powell," Mary R. Haas (1969) does not mention Rafinesque's contribution to the shaping of this specifically American model for ascertaining language relationships, though he was one of the precursors of such prominent classifiers as Gallatin, Powell, and Hale.[6] The history of the vocabulary-oriented approach in American linguistics is not complete unless Rafinesque's practice of collecting and comparing vocabularies is taken into account.

While he was living in Lexington (1819–26), Rafinesque began collecting materials for his *Tellus,* a projected monograph devoted especially to aboriginal America history and ethnology (Rafinesque 1836b: 73). To achieve this aim, he concentrated on the systematic collection, analysis, and treatment of Indian vocabularies, asking Thomas Jefferson to contribute the vocabularies collected on the Lewis and Clark expedition (Betts 1944: 380). Vocabularies once in Jefferson's possession had been lost, but later in Philadelphia, Rafinesque succeeded in borrowing Du Ponceau's large manuscript collection devoted to aboriginal philology and ethnology (1836b: 84), and despite his poor health Rafinesque accomplished a good deal of work on the comparative analysis of thirty-five Indian languages and two Eastern languages in the manner of Adelung and Klaproth (Betts 1944: 380).

As a part of his work for *Tellus,* Rafinesque began compiling the *Synglosson,* a five-volume set of vocabularies of languages of Asia, Africa, Europe, and Polynesia. Neither *Tellus* nor the *Synglosson* was ever published, but in practicing the comparative analysis of vocabularies of such wide range, Rafinesque was indeed a pioneer. Moreover, he was the first in America to insist on systematic fieldwork. In 1825 he persuaded Maj. T. L. McKenney, the superintendent of Indian Affairs, to print circulars for the collection of one-hundred-word vocabularies in "all the Languages spoken by the Indians of the United States" (Rafinesque 1836b: 76).[7]

Rafinesque nowhere explained why he thought it preferable to choose a one-hundred-word vocabulary in studying language affinities. The only reason, it seems, was pragmatic convenience and the simplicity of calculating relational indices for the languages concerned. Neither Gallatin nor Powell insisted on specific numbers of vocabulary items in elaborating their classifications.[8] In his understanding of the necessity of uniting efforts in laying down the fundamentals of aboriginal linguistics as a scientific institution, and in his understanding of the necessity of eliminating dilettantism, Rafinesque was a forerunner of Gallatin, as well as of Gibbs and of Powell.

However, Rafinesque's linguistic activity always was subordinated to the solution of the problem of aboriginal ethnogenesis, since the genetic classification of languages in his time was a means to a quite definite end: that of showing the kinship of American Indian languages to languages of the Old World—sometimes with Hebrew, sometimes with other languages, especially those of Asia. Rafinesque's linguistics was a linguistics for ethnology's sake. But if Gallatin was the father of American ethnology, it would not be too much to call Rafinesque the father of American historical ethnolinguistics. Rafinesque seems to have been the first to attempt a mathematical approach to the establishment of the probability of prehistoric ethnic contacts, being thus the forerunner of both Pliny Earle Chase (1865) and Morris Swadesh (1952).

Rafinesque's lexicostatistics was based, he writes, on a "formula rigoreuse et mathematique" (1831: 17). He further writes that he had discovered this so-called "Synoremic formula" in 1822, had announced it in 1824, and had published it in 1828. Such a publication has not yet been found, however. Nevertheless, the principles of Rafinesque's lexicostatistics may be deduced to a certain extent from his articles "On the Panis [i.e., Pawnee] Language and Dialects" (1824c), where he applies the formula to several related Indian languages; his "Introduction to American Ethnology" (1824b), discussed above; and "The Fundamental Base of the Philosophy of Human Speech, or Philology and Ethnology" (1832c), where he discusses the Synoremic formula and applies it to several languages as examples, including English and French. The epistemological postulates of the Synoremic formula presuppose the following:

> 1. That a small number of . . . words taken almost at random in two languages or dialects, are sufficient to indicate their degree of analogy. . . .
> 2. That the degree of similarity, analogy or affinities between 2 or more languages ought to be expressed numerically. 3. That when needful to pursue the enquiry still further or very minutely, the deviations or variations of sounds in the compound words might be divided into 5 or 10 series of successive or combined changes, additions or elision of sounds and letters; whose numbers should express the analogy, and by a division of the total by 5 or 10, the whole numerical and strict amount of identity is ascertained. (Rafinesque 1832c: 49)

As far as Rafinesque's linguistic practice is concerned, his Synoremic formula is not very impressive. Still, it was one of the first quantitative approaches to the problem of language affinity, an issue that continued to be important in Europe. Rafinesque's contemporary, the celebrated French explorer Capt. Jules Dumont d'Urville (1790–1842), came across it in Rafinesque's (1831) memoir on the origin of Asiatic Negroes.[9] Although he thought Rafinesque was sometimes too facile in certain disputed questions of language affinity, he concluded that "in savage languages, where analysis, that sort of Ariadne's ball of thread, is entirely lacking, it is necessary to have recourse to a more empirical means; that which the author of the Memoir has proposed and used appears at once ingenious, simple, and convenient" (Hymes 1983: 77; translation by Hymes).[10] And in his lexicostatistics of languages of Oceania, Dumont d'Urville made wide use of Rafinesque's method of calculating language affinity according to the six degrees of relationship found between two forms expressing the same meaning in two different languages (Hymes 1983: 83). On the other hand, Dumont d'Urville also remarked that Rafinesque's methodology "would of course fail in some circumstances, such as with the well known languages of Europe . . . [where] the procedure would show the French *jour* entirely separate from Latin *dies*" (Hymes 1983: 77; translation by Hymes).[11] Nevertheless, despite the deficiencies of the Synoremic formula and his hasty generalizations, Rafinesque has to be considered a pioneer in lexicostatistics and glottochronology.

For Rafinesque, the disputed issue of aboriginal ethnogenesis could be solved only through the recognition of two routes of migration: one via the Bering Strait, the other via the Atlantic. Trying to demonstrate his hypothesis about the Atlantic route for Central and South America, Rafinesque studied vocabularies of such aboriginal languages as Tarascan, Cherokee, and "Shona," comparing them with Greek, Latin, and Italian. Of course, from the standpoint of modern linguistic science, much of his comparative analysis would be beneath criticism. The arbitrariness of words he compared is often self-evident, and his etymologies are often naive and amateurish. Still, Brinton was right; Rafinesque's work was no worse than that of many of his contemporaries.

To his credit, Rafinesque flatly denied the myth of "the ten lost tribes of Israel," which was common currency among his contemporaries (Priest 1838: 58–82).[12] From his data, Rafinesque (1832a) concluded that, long before Columbus, the inhabitants of the Old World had reached the American continent, and that, in the antiquity of their civilization and the level of its development, the prehistoric Americans were equal to the nations of Africa and Europe. This conclusion was directed against the widespread views that denigrated aboriginal history, culture, and languages.

Along with collecting and studying aboriginal vocabularies, Rafinesque (1832b: 4–5) was second to none as a classifier; he made a thorough study of aboriginal systems of writing and was among the first to suggest how they might be classified. Thus, considering different elements of the systems used and their relations to the content and form of spoken language, he identified twelve kinds of aboriginal writing (including pictography, syllabic writing, wampum, quipu, runic glyphs, etc.). And what was more progressive, he pointed out that with the aid of these graphic systems it was possible "to express ideas; all of which find equivalents in the east continent" (1832b: 4). This view was in sharp contrast with the dominant one of his time, the core of which was the postulate that abstract ideas could not be expressed through aboriginal languages.

In addition, Rafinesque was one of the first to attempt to decipher the Maya glyphs. His analysis of these glyphs led him to conclude that they lacked any similarity with Chinese, Persian, or Egyptian glyphs, but rather showed some similarities with the Old Libyan alphabet. His studies of these glyphs resulted in his suggestion (Rafinesque 1827) that the Mayan system of writing was partly syllabic, thus anticipating modern Mayanists.

Critically observing the state of the genetic classification of aboriginal languages in his time, Rafinesque concluded that "one of the most glaring errors of speculative philosophers on the subject of America, is to be found in their assertion that American languages and nations are multiplied beyond conception, and cannot be reduced to order" (Rafinesque 1832f: 6). This linguistic agnosticism, he said, was due to "a superficial knowledge of the matter, and a wish to assert extraordinary things. If the same wish had been evinced respecting Europe, they could have found 60 languages and nations in France, and 100 in Italy, by considering the various provincial French and Italian Dialects as so many languages" (1832f: 6). He insisted upon accuracy in distinguishing languages and dialects as the necessary basis for a scientific approach to the genetic classification of languages in general and the American Indian languages in particular. The practical development of these principles gave rise to Rafinesque's first (1824b) and later improved (1832f) classifications of aboriginal American languages.

Thanks to his realistic, nonspeculative insight, he came to the conclusion that it was possible to reduce an estimated eighteen hundred aboriginal dialects to twenty-five genetic languages (fourteen in North and eleven in South America) (1832f: 6–7). Moreover, he thought it quite possible to reduce these twenty-five genetic languages further. "Even these 25 Languages and Original Nations," he writes, "may perhaps be reduced to 18 by more accurate investigation" (1832f: 8). "And in the whole of North and South America hardly 25 original languages and nations are met with, although actually divided in[to] 1500 tribes and dialects; as the actual European languages, only 6 in number originally, are now divided into 600 dialects . . ." (1832d: 56).

And thus, seven years before Gallatin (1836) and sixty-two years before Brinton (1891), Rafinesque suggested his synthetic classification scheme aimed at reducing the seemingly enormous number of aboriginal languages.[13] It is worth recalling that Brinton, who considered the number of linguistic stocks in both North and South America to be forty-five, pointed out that thirteen of these belonged to North America. In Rafinesque's final classification scheme the number is fourteen (1832f: 6–7), and it is a mark of his insight and linguistic intuition.

Of course, neither Rafinesque's classification scheme nor the procedures he used can be accepted without serious reservations. Alfred Kroeber (1913: 390–94) showed that without considering geographical ties between nations, the method of vocabulary comparison used as the principal heuristic device cannot unmistakably determine ethnogenic affinity. Edward Sapir, whose classification scheme ascertains sixteen macrofamilies for North America and has frequently been adopted by linguists (e.g., Voegelin, Rivet, Hoijer, and Kroeber), points out that "the recognition of 50 to 60 genetically independent 'stocks' north of Mexico alone is tantamount to a historical absurdity" (1921: 408). Sapir's dictum is a historically distant echo of Rafinesque's view of the problem of the plurality of linguistic stocks of the New World.

In addition to searching for vestiges of affinity among New World languages, Rafinesque also hoped to establish a universal phonology and orthography as necessary prerequisites for the scientific study of American Indian languages. The great epistemological and methodological importance of phonological issues were quite clear to him. After critically examining Du Ponceau's English phonology project (Du Ponceau 1818), he attempted to create his own system of universal phonology (Rafinesque 1821c: 60). His manuscript dealing with the subject has not been found, but from his correspondence with Thomas Jefferson, it is evident that Rafinesque considered a universal phonology to be a necessary heuristic device for the science of language (Betts 1944: 379).[14] Whatever Rafinesque's understanding of phonology may have been, he saw the need for elaborating a universal system of sound description as the basis for scientific linguistics.

Rafinesque's linguistic activity was not confined to practical research in the field of aboriginal linguistics. The inconsistency between Latin-based orthographies used to record aboriginal languages in written form and their real phonetic appearance suggested to him the lack of identity between graphical and auditory forms of language in general and raised the questions both of the interrelation of these forms and of their functioning. To his credit, Rafinesque was one of the first to recognize the lack of identity between written and spoken language forms.

"Our actual English," he says, "is a natural deviation or dialect of [Old English], begun between 1475 and 1525, and gradually *improved* and polished under two different forms, the written English and the spoken English, which are as *different* from each other as the English from the French" (1832e: 44; emphasis mine).[15]

Rafinesque's Linguistic Activity

Although Rafinesque failed to develop this idea in his extant writings, still one sees that he anticipated here a complex of problems that emerged by the second half of this century and gave rise to the so-called graphical linguistics (Bradley 1928; McIntosh 1966). But despite his clear understanding of the difference between spoken language and written language, Rafinesque did not consider the latter to be alien to proper linguistic inquiry, contrary to Bloomfield's well-known opinion (cf. 1939: 6–8). It is interesting to note in this connection that his contemporary, Wilhelm von Humboldt, did not pay attention to this difference, which is methodologically essential for linguistic science (Humboldt 1907 [1827–29]).

As far as extant sources reveal, Rafinesque nowhere discussed the methodological and epistemological problems of science per se. Nor can one find in his writings explicitly stated views on the metaphysics of science. Still, a careful examination of his linguistic works, as well as remarks scattered throughout his writings, enable us to draw some conclusions about the implicit "metaphysics" of the ontological structure of his conception of science and his theory of method. Though he was far ahead of his contemporaries in solving linguistic problems, he was also a product of his age and wholly shared the then existing epistemological paradigm. He was averse to metaphysics as such and sometimes identified it with "idle vapid talk" (Dupre 1945: 74). His conceptions of science, of scientific truth, and of the duties of scientists were in sympathy with the ideas of Benjamin Franklin and Thomas Cooper, representing the American variety of the inductive-empirical method. Rafinesque believed in the power of science without reserve, and declared that "the boldest and the most incredible predictions in natural sciences, have shortly after been proved and demonstrated" (1821a: 22). At the same time, he viewed with sarcasm attempts to limit the scope of science to "working ideas" only. "I had often been attacked," he wrote, "in the Cincinnati papers, and some one said once that it would be more useful to introduce a new article of industry, (meaning probably some new patent frying pan) than to discover one hundred new fishes fit to fry" (182lb: 75).

Rafinesque's opposition to conceptualism in constructing scientific knowledge is of great epistemological and methodological importance. He understood very well the necessity of constructing scientific knowledge adequate to American ethnoscience based on the empirical specificity of American facts uninfluenced by preconceived schemes. "America," he said, "has been the land of false systems; all those made in Europe on it are more or less vain and erroneous" (1832b: 4). And he insisted that this principle guided him in his inquiries in aboriginal linguistics. This basic thesis of Rafinesque is of great epistemological power since it unequivocally calls for the immanency of knowledge and American ethnolinguistic reality per se and calls for approaching it from within and not from without. Methodologically, Rafinesque is a forerunner of that trend in American linguistics and ethnology that later was developed by Boas, Bloomfield, and their associates.

Conclusion

The linguistic heritage of Rafinesque has never been the subject of serious study. The present article is a modest attempt merely to draw attention to this side of a man to whom honor is overdue. It is only regrettable that scholars have not had at their disposal all of his linguistic legacy, published and unpublished, even though much of the latter has been lost. The lack of it greatly handicaps the analysis and appreciation of Rafinesque's contribution to American linguistics and of his place in the development of linguistic thought in the United States.

Notes

1. "I never was in a regular College, nor lost my time on dead languages; but I spent it in learning alone and by mere reading ten times more than is taught in Schools. I have undertaken to learn the Latin and Greek, as well as the Hebrew, Sanscrit, Chinese and fifty other languages, as I felt the need or inclination to study them" (Rafinesque 1836b: 8–9).

2. Even such a great figure in American philology as Noah Webster had no idea of Jacob Grimm's monumental work (Grimm 1822) and believed all languages to derive from biblical Aramaic (Finegan 1980: 63).

3. "The only object I had in view . . . was to ascertain, by their vocabularies alone, the different languages of the Indians within the United States; and, amongst these, to discover the affinities sufficient to distinguish those belonging to the same family" (Gallatin 1848: xcviii).

4. "The evidence of cognation is derived exclusively from the vocabulary. . . . It must be remembered that extreme peculiarities of grammar, like the vowel mutations of the Hebrew or the monosyllabic separation of Chinese, have not been discovered among Indian tongues. It therefore becomes necessary in the classification of Indian languages into families to neglect grammatic structure, and to consider lexical elements only" (Powell 1891: 11).

5. As in his Polynesian work, Hale compared vocabulary and aimed at preparing for Iroquoian "a pretty extensive comparative vocabulary . . . of all the existing languages of this stock, viz. the Huron (or Wyandot) and the Mohawk, Oneida, Onondaga, Cayuga, Seneca, and Tuskarora" (Gruber 1967: 35).

6. "In America, without any theories being formulated on logical grounds, it has on the contrary been the customary practice to disregard grammar, and to unite as related, languages whose words or stems were in considerable measure similar; but to class as distinct in origin, all idioms whose vocables did not appreciably resemble one another" (Kroeber 1913: 389).

7. Rafinesque's letter of Aug. 1, 1825, to Major McKenney, specifying the information needed, was printed and sent out to fifty-six recipients under the cover of another printed letter signed by Thomas L. McKenney. Gallatin's request for the same kind of information was acted on later that year by the Office of Indian Affairs.

8. Gallatin's vocabulary, adopted as the standard for the discovery of affinities between Indian languages, consists of 211 words. Gibbs (1867) thought that this quantity was sufficient for the task. It was only two years before his death that Gibbs finally broke with the Hale-Gallatin tradition by proposing a flexible list of at least 1,500 words and phrases, rather than the 211 words of his earlier circulars.

9. Rafinesque's 105-page essay titled "Mémoires sur l'Origine des Nations Nègres" (1831) is known only from his own file copy preserved at the American Philosophical Society in Philadelphia. Dumont d'Urville was one of a five-member committee appointed by the Société de Géographie to judge Rafinesque's memoir when it was entered in the society's prize contest. Rafinesque was awarded a small gold medal as a consolation prize.

10. "Dans les langues sauvages, où l'analyse, cette espèce de fil d'Ariane, manque entièrement, il faut avoir recours à un moyen plus empyrique; celui qu'a proposé et employé l'auteur du Mémoire nous paraît à-la-foils ingénieux, simple et commode" (Dumont d'Urville 1832: 179).

11. "Il faut avouer qu'il a des circonstances où cette méthode serait en défaut, particulièrement à l'égard des langues bien connues de notre Europe . . . procédé indiqué, le mot français *jour* serait regardé comme tout-à-fait disparte avec le latin *dies*" (Dumont d'Urville 1832: 179).

12. Rafinesque "is decidedly, and we may say severely, opposed to this doctrine, and alleges that the Ten Tribes were *never* lost and are *still* in the countries of the east about the region of ancient Syria, in Asia" (Priest 1838: 76). Compare the following: "[T]he pre-monarchial period of tribal dissensions and intertribal feuds had reduced many of the tribes to a state of weakness which resulted in their absorption by their stronger and more numerous neighbors. This process of tribal disintegration was accelerated by the Syrian and Assyrian wars leading up to deportation and exile, the 'ten' tribes constituting the Northern Kingdom being 'lost' through natural decimation in consequence of war and famine at home and through absorption by 'the people of the land,' the Syrians north of them . . ." (Hirsch 1905: 254).

Vilen V. Belyi

13. These dates assume that Rafinesque did first publish his "Tabular View of the American Generic Languages" (1832f) in 1829, just as he said he did, even if bibliographers have not been able to find it. Calling this article the "key to American Ethnology, Philology and History" (1832f: 8), he writes that it was first printed in the *Saturday Evening Post*. The 1832 reprint bears the date "July 4, 1829" (1832f).

14. In 1824 in a letter to Jefferson, Rafinesque listed "Elements of Universal Phonology" as one of his unpublished manuscripts (Betts 1944: 379). This manuscript has not been found.

15. Rafinesque writes that the article on the "Primitive Origin of the English Language" has been "extracted from my manuscript philosophy of the English, French and Italian languages compared with all the other languages or dialects of the whole world" (1832e: 44), a manuscript not known to exist.

References

Betts, Edwin M. 1944. The Correspondence between Constantine Samuel Rafinesque and Thomas Jefferson. *Proceedings of the American Philosophical Society* 87: 368–80.

Bloomfield, Leonard. 1939. Linguistic Aspects of Science. In *International Encyclopedia of Unified Science* 1: 4. Chicago: Univ. of Chicago Press.

Bradley, Henry. 1928. Spoken and Written English (Read at the International Historical Congress, April 1913). In *The Collected Papers of Henry Bradley*, 168–93. Oxford: Clarendon Press.

Brinton, Daniel Garrison. 1885. *The Lenâpé and Their Legends; with the Complete Text and Symbols of the Walam Olum, a New Translation, and an Inquiry into Its Authenticity.* Brinton's Library of Aboriginal Literature 5. Philadelphia: D. G. Brinton.

———. 1891. *The American Race: A Linguistic Classification and Ethnographic Description of the Native Tribes of North and South America.* New York: N. D. C. Hodges.

Chase, Pliny Earle. 1865. On the Mathematical Probability of Accidental Linguistic Resemblances. *Transactions of the American Philosophical Society* n.s. 13: 25–33.

Dumont d'Urville, Jules. 1832. Rapport à la société de géographie, sur le concours relatif à l'origine des nègres asiatiques. *Bulletin de la Société de Géographie* 17: 175–86.

Du Ponceau, Peter Stephen. 1818. English Phonology; or an Essay towards an Analysis and Description of the Component Sounds of the English Language. *Transactions of the American Philosophical Society* n.s. 1: 228–64.

Dupre, Huntley. 1945. *Rafinesque in Lexington, 1819–1826.* Lexington: Bur Press.

Finegan, Edward. 1980. *Attitudes toward English Usage: The History of a War of Words.* New York: Teachers College Press.

Gallatin, Albert. 1836. A Synopsis of the Indian Tribes within the United States East of the Rocky Mountains, and in the British and Russian Possessions in North America.

Archaeologia Americana: Transactions and Collections of the American Antiquarian Society 2: 1–422. Cambridge, Mass.

———. 1848. Hale's Indians of North-West America, and Vocabularies of North America, with an Introduction by Albert Gallatin. *Transactions of the American Ethnological Society* 2: xxi–clxxxviii, 1–130.

Gibbs, George. 1867. Instructions for Research Relative to the Ethnology and Philology of America [1863]. *Smithsonian Miscellaneous Collections* 7(11): 1–51.

Goode, George Brown. 1895. Review of *The Life and Writings of Rafinesque,* by Richard Ellsworth Call. *Science* n.s. 1: 384–87.

Grimm, Jacob. 1822. *Deutsche Grammatik,* 2nd ed. Göttingen: Dieterichsche Buchhandlung.

Gruber, Jacob W. 1967. Horatio Hale and the Development of American Anthropology. *Proceedings of the American Philosophical Society* 111: 5–37.

Haas, Mary R. 1969. Grammar or Lexicon? The American Indian Side of the Question from Duponceau to Powell. *International Journal of American Linguistics* (Native American Text Series) 35: 239–55.

Hirsch, Emil G. 1905. The Twelve Tribes. In *The Jewish Encyclopedia* 12: 253–54. New York: Funk and Wagnalls.

Humboldt, Wilhelm von. 1907. Ueber die Verschiedenheit des Menschlichen Sprachbaues. In [1827–29] *Gesammelte Schriften* 6: 111–303. Berlin: B. Behr's Verlag.

Hymes, Dell H. 1983. Lexicostatistics and Glottochronology in the Nineteenth Century (with Notes toward a General History). In *Essays in the History of Linguistic Anthropology,* 59–113. Amsterdam: John Benjamins.

Kroeber, Alfred L. 1913. The Determination of Linguistic Relationship. *Anthropos* 8: 389–401.

McIntosh, Angus. 1966. "Graphology" and Meaning [1961]. In *Patterns of Language: Papers in General, Descriptive, and Applied Linguistics,* ed. Angus McIntosh and M. A. K. Halliday, 98–110. Bloomington: Indiana Univ. Press.

Oestreicher, David M. 1994. Unmasking the Walam Olum: A 19th-Century Hoax. *Bulletin of the Archaeological Society of New Jersey* 49: 1–43.

Powell, John Wesley. 1891. Indian Linguistic Families of America North of Mexico. In *Seventh Annual Report of the Bureau of Ethnology for the Years 1885–86,* 1–142. Washington, D.C.: Government Printing Office.

Priest, Josiah. 1838. *American Antiquities and Discoveries in the West: Being an Exhibition of the Evidence That an Ancient Population of Partially Civilized Nations Differing Entirely from Those of the Present Indians Peopled America Many Centuries before Its Discovery by Columbus, and Inquiries into Their Origin, with a Copious Description of Many of Their Stupendous Works Now in Ruins, with Conjectures Concerning What May Have Become of Them.* Albany: Hoffman and White.

Vilen V. Belyi

Rafinesque, Constantine Samuel. 1821a. Enquiries on the Sidereal, or Upper Spheres. *Western Minerva, or American Annals of Knowledge and Literature*, 22–26.

———. 1821b. Fragments of a Letter to Mr. Bory St. Vincent at Paris. *Western Minerva, or American Annals of Knowledge and Literature*, 72–76.

———. 1821c. Philology. *Western Minerva, or American Annals of Knowledge and Literature*, 60.

———. 1824a. *Ancient Annals of Kentucky.* Frankfort, Ky.: The Author.

———. 1824b. An Introduction to American Ethnology. Ms., American Philosophical Society, Philadelphia.

———. 1824c. On the Panis Language and Dialects. *Cincinnati Literary Gazette* 2: 50–51.

———. 1827. Important Historical and Philological Discovery. *Saturday Evening Post* 6: 2.

———. 1831. Mémoires sur l'Origine des Nations Nègres ou Introduction à l'Histoire des Nègres Indigènes d'Asie, d'Afrique, Polynèsie, Amérique et Europe par C. S. Rafinesque, Professeur des Sciences Historiques et Naturelles à Philadelphie 1831. Ms., American Philosophical Society, Philadelphia.

———. 1832a. The Atlantic Nations of America. *Atlantic Journal* 1: 8–10.

———. 1832b. First Letter to Mr. Champollion, on the Graphic Systems of America, and the Glyphs of Otolum or Palenque, in Central America. *Atlantic Journal* 1: 4–6.

———. 1832c. The Fundamental Base of the Philosophy of Human Speech, or Philology and Ethnology. *Atlantic Journal* 1: 48–51.

———. 1832d. On the Zapotecas and Other Tribes of the State of Oaxaca. *Atlantic Journal* 1: 51–56.

———. 1832e. Primitive Origin of the English Language. *Atlantic Journal* 1: 44–48.

———. 1832f. Tabular View of the American Generic Languages and Original Nations. *Atlantic Journal* 1: 6–8.

———. 1836a. *The American Nations; or Outlines of a National History of the Ancient and Modern Nations of North and South America,* 2 vols. Philadelphia: C. S. Rafinesque.

———. 1836b. *A Life of Travels and Researches in North America and South Europe, or Outlines of the Life, Travels and Researches of C. S. Rafinesque, A.M., Ph.D.* Philadelphia: The Author.

Sapir, Edward. 1921. A Bird's-Eye View of American Languages North of Mexico. *Science* 54: 408.

Swadesh, Morris. 1952. Lexico-Statistic Dating of Prehistoric Ethnic Contacts with Special Reference to North American Indians and Eskimos. *Proceedings of the American Philosophical Society* 96: 452–63.

Zimmermann, Günter. 1964. La escritura jeroglifica y el calendario como indicadores de tendencias de la historia cultural de los Mayas. In *Desarrollo Cultural de los Mayas,* ed. Evon Z. Vogt and Alberto Ruz L., 243–56. México: Universidad Nacional Autónoma de México.

Unraveling the Walam Olum

DAVID M. OESTREICHER

IN 1836, CONSTANTINE SAMUEL RAFINESQUE announced in his book *The American Nations* (I: 121–61) that he had deciphered an ancient pictographic record whose story revealed the long-lost history of North America. Painted and engraved upon wooden tablets, it was an account of the peopling of the continent by the Lenape (Delaware) Indians that presumably had been passed down in the tribe for thousands of years. According to Rafinesque, he had obtained the tablets from "the late Dr. Ward," who had originally received them from a grateful Lenape patient he had cured. The "original" tablets were inexplicably lost; Rafinesque's notebook "copy" is the sole record of the hieroglyphs, or "synoglyphes," as he referred to them in French. He called the document *Walam Olum*, which he said meant "painted record" in the Delaware language. Its 183 hieroglyphs contain a creation myth, a flood myth, and an origin legend of the Delaware people. It allegedly documents how the Lenape crossed the Bering Strait from Asia and migrated southeast across the North American continent. It describes the Lenape conquest of a mound-building people who had already settled in the Midwest; it narrates the fracture of the Lenape into the numerous tribes of the Algonquian language family; finally, it tells of the Lenape's settlement along the Mid-Atlantic coast. The epic concludes with the arrival of Europeans in the Lenape homeland during the 1600s.

Rafinesque claimed to have acquired not only the original tablets but also a separate transcription of forgotten songs in the Delaware language that explained them. He declared that by 1833—after more than a decade of study—he was able to render a complete English translation of the epic history, which he published in his 1836 book along with a portion of the Lenape text. Included there also was a sort of epilogue to the *Walam Olum:* a "fragment on the subsequent period," "translated" by a mysterious John Burns, for which no Lenape original is known. Conveniently, the Burns "fragment" begins where the *Walam Olum* concludes—with the arrival of Europeans in North America—and it continues until about 1820, with the Delaware removal from Indiana into Missouri.

For more than a century after Rafinesque's death, the *Walam Olum* assumed increasing importance in the literature about American Indian origins. Some of America's most prominent historians, ethnologists, and linguists—including Daniel Brinton, Cyrus Thomas, and Frank Speck—believed it contained crucial evidence for prehistoric Amerindian migrations and the identity of the mysterious Mid-western Mound Builders. But there also were undercurrents of doubt from the first; the anthropologist Henry Rowe Schoolcraft, an expert on Eastern Woodland Indians, questioned the document's authenticity. Even Brinton, who made a new translation in 1884, acknowledged "the possibility that a more searching criticism will demonstrate it to have been a fabrication [and] may condemn as labor lost the pains that I have bestowed upon it" (*The Lenâpé and Their Legends,* v).

Meanwhile, growing archaeological and ethnological evidence persuaded most scholars that migrations from Asia across the Bering Strait land bridge accounted for the early peopling of the New World. Expeditions to Siberia and the northwest coast of America were demonstrating continuities between Asian and Amerindian peoples. The *Walam Olum,* apparent corroboration from Amerindian folklore, was welcomed by many scholars as support for this scenario.

In the 1930s the pharmaceutical tycoon Eli Lilly charged a team of linguists, archeologists, ethnologists, and historians with analyzing the *Walam Olum.* This mammoth project took more than twenty years. The team's report (published by the Indiana Historical Society in 1954) proclaimed "all the confidence in the historical value of the *Walam Olum* that [German archeologist Heinrich] Schliemann had in the accuracy of the Homeric epics" (xiv).

With the development of radiocarbon dating during the 1950s, however, sup-porters of the *Walam Olum* faced a dilemma. Some archaeologists, such as Herbert C. Kraft (*The Lenape: Archaeology, History and Ethnography,* 1986) soon found that fore-bears of the Lenape had inhabited the northeast coast for some twelve thousand years, thereby contradicting Rafinesque's much more recent chronology. As a result, many scholars began to think that while the *Walam Olum* might be authentic folklore, it was historically unreliable.

A few came to believe that the document was an outright fabrication. Among them were historian William A. Hunter (as reported by his friend the late

Unraveling the Walam Olum

Herbert C. Kraft); archaeologist James B. Griffin, who defected from the Lilly team (*Indiana Magazine of History* 51 [1955]: 59–65); and Pennsylvania's state archeologist, John Witthoft (in personal correspondence), who found grammatical inconsistencies in the text and a suspicious resemblance to linguistic lists and dictionaries compiled by eighteenth-century missionaries. No solid refutation was ever published, however, and the *Walam Olum* remained an enigma. The majority of scholars continued to endorse the document's authenticity. As recently as 1987, historian C. A. Weslager reaffirmed in a personal letter his belief that the *Walam Olum* was genuine, and he reeled off a list of distinguished colleagues who agreed with him.

During the late 1970s, when I first began conducting research among Delaware Indians in Oklahoma, I was surprised to learn that elderly Lenape speakers did not consider the *Walam Olum* part of their culture at all. Stating that they had learned of it recently from anthropologists, they found its text puzzling and often incomprehensible. On the other hand, many young Delaware, lacking knowledge of their native tongue, eagerly seized upon it as a glorious remnant of their ancient culture. In May of 1993, while researching the tribe's mythology for my doctoral dissertation, I decided to take a new look at the document.

Before long, it became clear that preposterous grammatical constructions filled the text. In it, animate and inanimate genders are consistently confused; there is no use of the obviative (a marker integral to Algonquian languages that distinguishes the object of a phrase from the subject); prefixes are often used as suffixes and vice versa, as though the speaker had no notion of their proper usage; and often the words consist merely of uninflected verb stems (i.e., lacking any conjugation), precisely as one would find in a dictionary. The inflections are provided only in Rafinesque's English translation, with the result of making sentences appear to be conjugated, but the inflections are entirely absent in the Delaware text. Those verbs in the *Walam Olum* that are inflected bear no relation to the context of the sentences they are a part of, as though each verb has been lifted from a grammar and stuck indiscriminately into a sentence. In short, it began to appear that Rafinesque was translating from English to Lenape, rather than the other way around; his manuscript was the rough draft of a forgery.

To confirm my suspicion, I called upon an old friend, the late Lucy Parks Blalock, *Oxeapanexkwe* "Early Dawn Woman," of Quapaw, Oklahoma. Then eighty-seven years of age, Mrs. Blalock was among the last fluent speakers of Unami, a Lenape dialect. We spent four months reviewing every word in the *Walam Olum* during an intensive series of taped interviews. As our work proceeded, it soon became evident that the text is replete with fractured constructions, and includes Lenape versions of English idioms unknown in the Delaware language.

In *Walam Olum* IV: 52, for example, the Delaware word *Talegawil* is translated as "Talega head, or Emperor." The Delaware word *wil* does indeed mean "head," but it can only signify an *anatomical* head, and is never used figuratively for a leader, as it is in English. Only someone unfamiliar with the Delaware language could have used the word in such a manner. Similarly, the translation of *Walam Olum* III: 4 relates that "the comers [i.e., the Amerindian immigrants to America] divided into tillers and hunters." The Delaware word in the manuscript for "divided" is *pokwihil*, which Rafinesque specifically notes means "divided or broken." However, the verb *poquihilleu*, "he/it is broken," cannot designate the breaking up of a group of people. In the Delaware language, this verb stem, and indeed the very root *pok-*, connotes something being physically broken into pieces or a person bent with age. To indicate people separating or dividing into groups, an entirely different verb, *chpihileyok*, "they are separating," is required.

These and other English idioms in the *Walam Olum* could not have been uttered by a native speaker thousands of years ago, 170 years ago, or this morning. Even today, the examples provided are entirely foreign to Delaware speech, despite the fact that all the remaining speakers are bilingual and use such idioms when speaking English. The misuse of such terms in the *Walam Olum* provides solid evidence that the text was composed by a non-Delaware speaker who located Delaware words in specific published sources but lacked the understanding necessary to apply them in any meaningful way.

The Lenape words employed to fashion the supposed original Delaware text of the *Walam Olum* were taken mainly from the works of two Moravian missionaries: David Zeisberger's *Grammar of the Language of the Lenni-Lenape* (1827) and John Heckewelder's list of Lenape personal and place names published in the *Transactions of the American Philosophical Society* in 1834. These were the major published sources on Delaware available when Rafinesque's English version of the *Walam Olum* was published in 1836.

His alleged Delaware epic names eighty-six chiefs. And twenty-six of these supposedly ancient chiefs bear exactly the same names as the seventeenth- and eighteenth-century chiefs mentioned in Heckewelder's list. This despite the Lenape custom that a name may only be bestowed once and cannot be shared by any two individuals, living or dead. Most of the remaining chiefs' names are ungrammatical expansions, contractions, or amalgamations of personal or place names listed by Heckewelder, or are "substantives," "adjectives," or "adjective verbs" from Zeisberger's *Grammar*, or fragments of words from both Heckewelder and Zeisberger.

In addition, Rafinesque frequently did not know how to break apart into individual units the Delaware compound words he found in his sources, thereby leaving a trail of broken stems in his text. For example, in *Walam Olum* IV: 2 we find the chief's name *Wapallanewa*, "White Eagle." In Heckewelder's list,

Wapallanewachschiechey is given for "Bald eagle's nest," a place name written as one word. However, Rafinesque did not know how to separate "bald eagle" from "nest," with the result that in his name for Chief White Eagle, part of the word for nest is still attached!

The pictographs of the *Walam Olum* also were created from previously published materials. They are hybrids concocted from ancient Egyptian hieroglyphs, ancient Chinese Ku-Wen script, the less ancient Ojibwa Midewiwin pictographs, and even some Mayan symbols. In searching through Rafinesque's unpublished drawings and notes in the archives of the American Philosophical Society in Philadelphia, I was able to pin down nearly all of his original sources for the pictographs.

But why did Rafinesque create this elaborate hoax? And why would he knowingly incorporate into a supposedly Amerindian epic words and hieroglyphs derived from a variety of Old World cultures? The answer, I believe, lies not only in his desire for wealth and fame but also in his passionately held view of how Indians came to North America, and in the discoveries and dilemmas of his time.

During the early nineteenth century, ancient Persian cuneiform and Egyptian hieroglyphics were being deciphered, Herculaneum and Pompeii were being unearthed, and exotic civilizations around the globe were captivating the Western imagination. Before Charles Darwin's *On the Origin of Species* appeared in 1859, the Bible was still the scholar's mainstay, to which newly discovered continents, cultures, and conflicting mythologies presented a growing challenge. On both sides of the Atlantic, scholars turned to the infant sciences of comparative ethnology and philology to corroborate the Biblical account of human origins.

In America the great mystery was the American Indians: Where did they come from? Was there one creation of humanity, as depicted in Genesis, or were races created separately on each continent? If there was a single creation, where was the cradle of mankind? How and where did American Indians survive the biblical flood? Was there once a primordial language understood by all humanity? Did an advanced civilization in North America predate the American Indians and erect the mysterious earthen mounds in the Midwest? These and many other questions fascinated the public, as well as scholars.

Rafinesque first advanced his theories while teaching at Transylvania University in Lexington, Kentucky, from 1819 to 1826, but his views received little scholarly acceptance—partly because he made so many wild claims of all kinds. He marketed a cure for tuberculosis, devised an investment program "never to be liable to losses," and invented a design for "sink-proof" boats. He also offered the public an "incombustible varnish and paint" that would "save the lives of 100,000 persons doomed to be burnt alive."

As early as 1819, publishing houses began to shun him, forcing Rafinesque to publish his own books. After settling in Philadelphia in 1826, he failed to gain

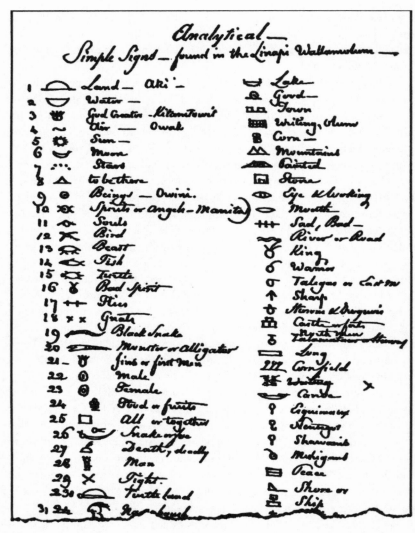

Fig. 13.1. A Walam Olum *worksheet. Were the strikeouts carefully planted by a diabolical forger hoping to trick posterity, or were they corrections made by an honest scholar with a difficult problem to solve? Courtesy of the American Philosophical Society.*

membership in the prestigious American Philosophical Society. Nevertheless, he continued to use the society's library not only to do basic research but also to create his counterfeit masterpiece. Frustrated and feeling unappreciated, Rafinesque apparently became convinced that no one would ever accept his views unless he manufactured some dramatic "evidence."

In 1830, six years before the publication of the *Walam Olum*, Joseph Smith, founder of the Church of Jesus Christ of Latter-day Saints, announced that he had found a set of golden tablets buried in upstate New York, written in the "language of the Egyptians" (*The Book of Mormon*, Nephi 1:2; Mosiah 1:4) and proving that the Indians and the Mound Builders were descendants of ancient Jews. The Mormon Bible purported to be an ancient epic recorded in glyphs that resolved the prevailing dilemmas of the age. Like Rafinesque's *Walam Olum*, Smith's original tablets had vanished without a trace, leaving only his copies to posterity.

In an article entitled "The American Nations and Tribes Are Not Jews" (*Atlantic Journal* 1 [September 1832]: 98–99), Rafinesque railed against *The Book of Mormon* as a mistaken view of the peopling of America. He believed that Joseph Smith's tablets were a hoax based upon an incorrect theory, but they appear to have inspired him to create his own "long-lost tablets." In Rafinesque's version, the American Indians were not lost tribes of Israel but Central Asian tribes that had crossed into America over the Bering Strait land bridge. And the sophisticated Mound Builders of the Midwest had arrived years before from the east, after emigrating from their homeland—the legendary "lost continent" of Atlantis.

The Book of Mormon was surely fodder for the creation of Rafinesque's epic, but the immediate impetus for crafting it came later. During my research, I learned of a Rafinesque essay with an appended "supplement" that had been discovered in 1982 at the Royal Institute of France by historian Joan Leopold. Rafinesque had competed for the institute's prestigious Volney Prize (and its award of twelve hundred francs) for the best essay with new information about the languages of the "Lenni Lennape, Mohegan et Chippaway." Upon examining the essay and supplement, I found that they revealed both the timing and motivation for the hoax. Written in October of 1834, the essay makes no mention of ancient North American hieroglyphic records, noting only a traditional Shawnee Indian story about forebears who had crossed "over the ice." The essay betrays an abysmal lack of competence in the Delaware language and undermines Rafinesque's later claims that he had possessed the *Walam Olum* for years and mastered the Delaware language by 1833. In the supplement, written in December of 1834, he implores the French committee to accept "new philological and also graphic materials"—the *Walam Olum*. What could have prompted the abrupt appearance of this intriguing document in the two months after the essay was written?

David M. Oestreicher

Once again the answer lay in the archives of the American Philosophical Society. With the aid of the society's librarians, I found that Heckewelder's all-important list of Delaware names was first published in November of 1834. It is "Names which the Lenni Lenape . . . had given to Rivers, Streams, Places, &c. . . . and also Names of Chieftains . . . and Biographical Sketches," *Transactions of the American Philosophical Society* n.s. 4 (1834): 351–96. That Rafinesque was in possession of the list by December of 1834 is documented not only by internal linguistic evidence in the *Walam Olum* but also by one of his surviving letters to the Boston book dealer O. L. Perkins (see p. 267, n. 43). Later, in order to conceal any connection with Heckewelder's 1834 list, Rafinesque had to claim that he had already translated the Walam Olum by 1833.

Ironically, Rafinesque lost the French academy's contest to the philologist Peter Stephen Du Ponceau, president of the American Philosophical Society, who had edited most of the major linguistic materials Rafinesque plagiarized to construct his hoax. After losing the contest, Rafinesque tried to market the *Walam Olum* for a considerable sum, even petitioning the king of France for a pension in exchange for his handwritten copy of the chronicle, translated into English and French. Finally, when he published *The American Nations*, Rafinesque was able to inject his fraudulent epic into the growing body of literature on American Indians.

The *Walam Olum* consumed the energies of dozens of investigators who toiled for years over its mysteries. It promulgated a staggering amount of misinformation about American Indians and has confused and hurt younger Lenape seeking a clearer understanding of their heritage. Having suffered the loss of their language and earnestly seeking to reclaim their past, some Lenape were deeply disappointed when I presented my findings at a symposium in Oklahoma in May of 1995.

The history of science contains some remarkable hoaxes: Piltdown man, which gave British anthropologists comfort by providing them with "the first Englishman"; the gentle "stone age" Tasadays of the Philippines, who "proved" the inherent goodness of humanity in its natural state; and the stirring *Songs of Ossian*, which ascribed a glorious past to eighteenth-century Scottish nationalists. To this distinguished body we may add the remarkable *Walam Olum*, a work that sought to reconcile American Indian origins with what had been traditionally taught in the Bible. Despite its acclaim, the *Walam Olum* is not the "oldest native North American history," but one of the oldest hoaxes in North America.

Would that it were not so.

Indian Languages and
the Prix Volney

CHARLES BOEWE

A SUM OF MONEY left in the will of Constantin-François Chassebœuf, Comte de Volney (1757–1820), was intended to fund a series of prizes for essays on various aspects of the scientific study of languages. In its original form, the prize was a gold medal valued at twelve hundred francs, but by 1834 the medal awarded to the winner of the 1832 contest was worth only a thousand francs. By that time, however, the prestige of the prize was valued by scholars no less than its monetary worth. The selection of topics, as well as the judging of essays submitted in response to them, was conducted by a commission set up in the Institut de France. First conferred in 1822, the Prix Volney became the major award in the nineteenth century for research in comparative linguistics. The topic set for the 1835 *concours* was "detérminer le charactère grammatical des langues de l'Amérique du nord connues sous les noms de Leni-Lenape, Mohegan, et Chippaway." Only two authors responded: Peter Stephen Du Ponceau and C. S. Rafinesque, both living in Philadelphia. Du Ponceau and Rafinesque were acquainted with each other but, since submissions had to be anonymous, it is unlikely that either knew the other was a candidate. Du Ponceau won and was awarded a gold medal. After a delay of several years, his prize-winning essay was published as a 470-page book titled *Mémoire sur le système grammatical des langues de quelques nations Indiennes de l'Amérique du Nord* (Paris, 1838).

Charles Boewe

I

Rafinesque's earliest personal contact with indigenous Americans occurred in 1804, during his first sojourn in the United States, when he met a delegation of Osage Indians from the Arkansas Territory who were visiting the nation's capital. He reported (*A Life of Travels*, 21) that "Pauska or white head their chief, gave me a good vocabulary of their language by the help of the interpreter Mr. [Pierre] Chouteau."[1] This also is the earliest recorded instance of Rafinesque's practice of collecting vocabularies, which, like Jefferson and others of his time, he considered a proper part of a naturalist's endeavor. At this stage of his career, collecting words had about the same importance as collecting plants or other artifacts of the natural world. A few years later at Nicosia, in central Sicily, he found "a remnant of the Provençaux or Catalan conquerors of Sicily; I investigated and took a Vocabulary of their Language which indeed quite resembles old Provence or Catalan, being a Dialect of the Romance language."[2] And in the next breath he went on to mention finding huge fossil bones, possibly those of an extinct whale. Fossil language, fossil animals—they held much the same interest for Rafinesque.

It was after he had returned to the United States and was settled in Kentucky that Rafinesque had a need to put his knowledge of obscure languages to work. It was there, in collaboration with his friend John D. Clifford, that he set out to solve the mystery of the origin of the massive prehistoric earthworks that are found throughout the Ohio Valley; two prominent sites, which he mapped, were within a few miles of Lexington.[3] Now he began collecting "ancient monuments" with the same avidity he had been collecting plants, shells, fossils, and other natural history specimens. He preserved these records in the form of written descriptions and maps, which he exchanged with other investigators as freely as he did botanical specimens, only expecting a reasonable return for his trouble. By 1824 he was able to list 148 prehistoric sites in Kentucky (most of which he had visited) and 393 elsewhere (most of which he had not).

Through direct experience, he had been brought face to face with the historical problem posed by the American Indians and their handiwork. Unlike earlier ecclesiastical students of Indian languages—Cotton Mather, Jonathan Edwards, and John Heckewelder, for instance—he was untroubled by the theological riddle of how there could be living souls susceptible of salvation in an unknown part of the world if, in fact, the Flood had been universal.[4] Rather, Rafinesque, who had surveyed earthworks that appeared to be the product of a people in an advanced state of civilization, was preoccupied with the problem of reconciling these artifacts with the primitive and sometimes debauched Indians he had talked with himself. Maybe the puzzle could be solved by tracing the all-but-

Indian Languages and the Prix Volney

forgotten migrations of peoples by means of the "fossilized" record of their journeys left in the languages they had once spoken, a record which might still be recovered from living informants and documentary sources.

Rafinesque probably derived from Leibniz, whom he had read, the idea that languages can be seen as monuments of the past.[5] And it is clear that Rafinesque's concern with Indian languages was as a tool to accomplish other ends, essentially non-linguistic, never to study the nature of language itself.

Almost as though he were writing a grant proposal, Rafinesque next outlined a research program in a series of three letters, all addressed to Thomas Jefferson, but published in a Lexington newspaper.[6] He had just read Heckewelder[7] on the traditions of the Lenni-Lenape, called Delawares by the whites, and he took from this source a name for their predecessors, the "Alleghawees," the builders of the prehistoric mounds who had been defeated and driven southward by the Lenape and their allies. He concluded his third letter with the pious hope "that some able hand will soon rescue from oblivion the scattered materials relating to these ancient nations"—which is exactly what *he* was planning to do.

His first opportunity to do so came in the form of an invitation to write a brief introduction for the expanded second edition Humphrey Marshall was planning for his *History of Kentucky* (1824). Rafinesque was given fewer than forty pages in which to cover the history of Kentucky from the creation of the world to the coming of the whites, Marshall having reserved the remainder of the projected two volumes for himself, for he needed ample room to trounce his political foes. Copies of Rafinesque's contribution also were struck off in a separate pamphlet, and usually this is known under the pamphlet's second title, *Ancient Annals of Kentucky* (1824). Fatuous as the narrative was, Rafinesque did have the privilege here of inserting his lists of prehistoric sites,[8] an "Ethnological and Philological Table of the Primitive Nations and Languages" (an abridgment, he said, of "an elaborate survey of about 500 languages and dialects of both Continents"), and a bibliography listing more than 160 "Authors and Works Consulted." These last two sections prefigure major intellectual interests of the author during the remainder of his life.

The "Ethnological and Philological Table" took the concepts of *heaven* (God, sky, paradise), *land* (earth, world, ground, soil), *water* (sea, river, lake, rain), and *man* (male, nation, people) as root ideas and listed a word representing each of them in fifty languages, including fourteen languages of the New World. Rafinesque inserted a sign beside those Old World nouns that he believed to be cognate with any in a New World language. The New World languages are arranged into two groups: the "Eastern Branch," consisting of seven extending

Charles Boewe

from "Atalan" to "Peruvian"; and the "Asiatic or Western Branch" of seven more ranging from "Mexican" to "Caralit." The Old World languages are grouped as "European Languages, Asiatic Languages, Polynesian Languages, and African Languages." In addition, "dialects" are given as subentries to a number of the fifty languages, both in the New and Old World lists; for example, the language of the Cherokees being considered a dialect of Atalan in the Eastern Branch and the language of the Shawnees a dialect of "Lenapian" in the Western Branch. Thus, in 1824, Rafinesque began to develop the language typology that would guide his studies in comparative philology—and sometimes mislead him in his later historical research.

The bibliography of the pamphlet shows that Rafinesque had available a remarkable working library in the backwoods of Kentucky by virtue of his being at Transylvania University. Among authors and titles listed there in cryptic abbreviation are "Adelung, *Mithridates*" (which Du Ponceau, who had trouble getting it in Philadelphia, considered "the most astonishing philological collection that the world has ever seen"); "Barton, *Indian Languages*"; "Gebelin, *Monde Primitive*"; and "Reland, *American Languages*"[9]—although, understandably, most of the titles are those of histories and books of travel (some of which probably contained vocabularies). The bibliography also shows that he had available twelve volumes of *"Asiatick Researches,"* to which Sir William Jones contributed his research on Sanskrit; a *"Dictionnaire historique,"* otherwise unidentified; the *North American Review* and the British *Quarterly Review*, whose lengthy articles were almost the equivalent of reading the book under discussion; and long runs of such periodicals as the *Transactions of the American Philosophical Society* (where Du Ponceau's papers appeared), and the New-York Historical Society's *Collections* (where Edwin James later published on Algonquin mythology), as well as those of the Massachusetts Historical Society (which published the Indian-language studies of both John Eliot and Jonathan Edwards, Jr.). Since the bibliography cites "Pickering, *Indian Languages*," Rafinesque probably also had available the *Memoirs of the American Academy of Arts and Sciences*, where John Pickering's linguistic articles appeared, although that series is not cited by name in the list. However somnolent Transylvania became later in the century, in the 1820s it was about as good a place as any in the United States for an ambitious scholar to undertake linguistic research.

In addition to the authors listed in this printed bibliography, during the same year Rafinesque recorded in an unpublished manuscript[10] that among important "authors who have written on American languages" were Gottfried Wilhelm Leibniz, Pierre François Xavier de Charlevoix, Lorenzo Hervás y Panduro, Alexander and Wilhelm von Humboldt, Francisco Clavigero, Julius von Klaproth, and Juan

Indian Languages and the Prix Volney

Ignacio Molina—all of whom he must have read, or read about, in the Transylvania library.[11] Whatever he may have made of his reading, Rafinesque knew how to ferret out appropriate sources. After his return to Philadelphia in 1826, he also used the American Philosophical Society's library and others there, and he traveled to Harvard University especially to consult the Ebeling library of Americana shortly after its acquisition, when it was hardly known at all to his fellow Philadelphians.

Rafinesque also understood the advantage that could be gained by team research. On the trip to Philadelphia in 1825 preparatory to his final leave taking from Transylvania, he stopped in Washington to persuade the War Department's Office of Indian Affairs to dispatch a circular letter to all its Indian agents and to the superintendents of Indian schools requesting information about the languages spoken by those under their charge. In this letter he sought vocabularies of the same one hundred words in the languages of more than fifty Indian nations.

Albert Gallatin usually is credited for such pioneering linguistic research; his similar request was acted upon later the same year by the Office of Indian Affairs. Gallatin did seek larger vocabularies and, unlike Rafinesque, he also tried to ascertain something about grammatical forms. Both investigators struck a sympathetic chord with the superintendent of Indian Affairs, who had already sent out, three weeks before he dispatched Rafinesque's request to the field, an August 9 circular asking for an alphabet and grammar, plus a "chapter" on a subject of the respondent's choice in the language reported on.[12] Probably because so many requests flooded in to the Indian agents at once, the return was disappointing.

Perhaps the most noteworthy aspect of Rafinesque's request was the phonetic chart he devised to try to bring a degree of uniformity to transcriptions and thus "obviate the perplexing difficulties of the anomalies of modern written languages."[13] Because they were French and were having to use Indian vocabularies collected by speakers of English, French, German, Spanish, and even Swedish, both Rafinesque and Du Ponceau were among the early students of phonology. We do not know the contents of a Rafinesque manuscript titled "Elements of Universal Phonology," which has never been found, but it is of interest to note that for the Indian agents—all of whom it must be assumed were native speakers of English—he expressed the vowel sound in French *vivre* as "ih," leaving the unmodified "i" to represent the vowel sound in English *mill*, the "e" in French *parler* to be represented by "eh" and leaving the unmodified "e" as it is sounded in English *best*. What would have given his American informants more trouble was his discrimination of five "nasal vowels" as "an," "oen," "in," "on," and "un," to which he added (probably to make clear what he meant by nasal): "when not nasal, as *an* in *Fan,* write it *Fanh.*" He did not see a need to specify the sounds of consonants, despite his own tendency to omit the final one when writing such words as *savant* (a not uncommon practice of French writers at the time) and his long-standing problem of getting right the surname of the Saxon botanist F. T. Pursh, which sounded like "Bursh" to Rafinesque's ears.

His letter of August 1, 1825, specifying the information he needed, was printed and dispatched to fifty-six recipients under cover of another printed letter signed by Thomas L. McKenney, the superintendent of Indian Affairs. Rafinesque continued his contact with the Office of Indian Affairs during the following year without achieving a satisfactory return. The handwritten "Office of Indian Affairs, Letters Received Register (1 January 1824–31—December 1826)," at the National Archives, records five queries from him, the last being dated December 1, 1826. One of these, dated July 1, 1826, which does not appear to be extant, is summarized in the register as a report that Rafinesque "has visited the Seneca and Tuscarora nations and collected the traditions of the latter," showing that he continued field work among the Indians during his travels from Lexington to take up permanent residence in Philadelphia.

II

As we know, after his dispute with Transylvania University's president, Horace Holley, Rafinesque left Lexington in a huff in 1826 and returned to Philadelphia, where he remained the rest of his life except for short excursions up and down the East Coast. Aside from a brief teaching assignment at the Franklin Institute in Philadelphia, he was now obliged to make his living as best he could by capitalizing on his manifold intellectual interests. If most of the productions of his pen, usually printed at his own expense, failed to realize much of a profit, it must have occurred to him that an alternate opportunity lay in writing for prize contests.

Although these efforts were crowned with little more success, Rafinesque was no novice when he sat down to write his Prix Volney contribution. In 1825 he had contested for a thousand-dollar prize with an essay on the best way to rid the Ohio River of snags and for a hundred-dollar prize for "the best account of the materials existing on the history of the native tribes of America." The prize in the first competition was awarded, in his opinion, to another through political influence; and his second essay was "stolen or mislaid" by Abiel Holmes, the historian.[14] The substance of the second essay is summarized in twenty-one numbered paragraphs in *The American Nations* (2 vols.; Philadelphia, 1836), I: 42–45, where item seven states that "the philological solution of historical affinities, must be sought in the roots of the languages, their conformity or analogies, the number of similar sounds, roots and words; which are susceptible of a mathematical calculation, and referable to the theory of probabilities." He experienced no greater success in 1831, when he wrote for the five-hundred-dollar prize offered by the American Peace Society for a dissertation on "a congress of nations."[15] But returning to linguistic analysis, he did win a gold medal the next year, 1832, from the Société de Géographie at Paris for his "Mémoires sur l'Origine des Nations Nègres."[16]

Writing under the motto "Languages do not lie—Horne Tooke,"[17] Rafinesque drew extensively on lexical evidence to support his hypotheses, first, that the brown, white, and black nations had a common origin and, second, that the cradle of these peoples lay in the Himalaya mountains. Making tidy use of the same ideas elsewhere, he wrote in English that he had "demonstrated the affinities" of the Asian blacks' languages with those of "the African and Polynesian Negroes, as well as with the Hindus and Chinese," affinities which, in his view, "renders it probable that all the Negroes"—like all other human beings—"originated in the Southern Slopes of the Imalaya [sic] Mountains."[18] Although the Société de Géographie remained unconvinced by his argument, we may at least be impressed by his willingness to side-step the idea of a destructive universal flood sanctioned by the Bible, and by his willingness to embrace the concept of the unity of mankind regardless of skin color—neither of which was a popular view in the 1830s.

Perhaps misled by Klaproth, whom he had quoted with approval as having established that "languages are better guides than physical characters for researches on mankind, and roots more important than grammar,"[19] Rafinesque based this and his subsequent studies on the following axioms:

1. That words are the elements of languages.
2. That the names given to the most common and obvious objects are their first elements, and the least subject to variations.
3. That words resembling each other more or less are the links uniting the dialects and languages into groups or clusters.
4. That these words must be such as apply to the same objects, or are synonymous in many cases.
5. That syntax and grammar or the modes in which words are modified and combined are subservient to the radical or elementary words, and thus of much less relative importance.[20]

Rafinesque stated that the linguistic analysis in his prize essay was based upon his own rigorous and mathematical formula, which he discovered in 1822, announced in 1824, and published in 1828. Such a publication has not been found. The essence of it, as published elsewhere in 1832, the year of his "geographical" prize essay, has been discussed by Belyi (pp. 223–24). Rafinesque had a ready name for his "discovery"; "I call it the *Synoremic formula,* or the Numerical and Analogical Rule." His example of it compares the cardinal numerals in English and French (languages, he averred, which "are double in form, having each a written and spoken dialect"), and makes the sweeping conclusion that "the English and French languages compared merely by their 10 cardinal numbers, which are a very fair scale in many languages, evince a considerable analogy of 7 in 10 equal to 70 per cent."[21]

Jumping to vast conclusions on the basis of far too little evidence was the fatal flaw that marred much of Rafinesque's scholarship, including that in the natural sciences. It may have resulted from his irregular formal education; no skeptical teacher had ever hauled him up short and demanded rigorous evidence for his fanciful hypotheses. Rafinesque's reach invariably exceeded his grasp, and with a smattering of knowledge in nearly every field of learning, he launched glib analogies, many of which were intended to lead to the establishment of a new "science." If the paleontologist Cuvier could reconstruct an extinct fish from a few scraps of fossil bone, Rafinesque decided he could reconstruct lost languages from a few tattered cognate vocabularies; he called this science *palingenesy:* "Give me but a single *genuine* word of an ancient or extinct language, and I can find its analogies with all others. Give me 2 or 3 or a few, and I can trace its alliances. Give me several, and all its origins, parentage, filiation, claims, affinities, peculiarities &c can be traced."[22]

Most of the vocabularies used in Rafinesque's study of Asiatic Negroes had been obtained through his wide reading in travel books and histories, but references in the essay also show that he had come across Adriano Balbi's *Introduction à l'atlas ethnographique* (Paris, 1826) by this time. In the belief that he and Balbi might be helpful to each other, he opened "a philological correspondence" with him,[23] as he had done with many of the leading European natural scientists, some of whose replies are extant. During this period he continued to have personal contact with Philadelphians having kindred interests—among them the naturalists at the Academy of Natural Sciences—including Du Ponceau, who "lent me his manuscript collection of Vocabularies."[24]

Yet Rafinesque never quite fit in, anywhere. He began making donations to the American Philosophical Society (probably his own publications) as early as 1820. The last mention of him in its minutes, October 6, 1826, was "On the ancient languages of America; communicated by Mr. Rafinesque, and referred to the Historical Committee," of which Du Ponceau was chairman.[25] He was never

Indian Languages and the Prix Volney

elected to membership. With such treatment in mind, he wrote in the 1830s that "the scientific institutions and societies of Philadelphia are often disgraced by their tenacious errors, and by admitting unworthy members for the sake of mere fees!" He decided to "keep aloof from them," although he went on to say that he "cultivated chiefly the friendship of old friends or liberal savan[t]s," among whom he named Du Ponceau as one.[26] Their friendship must have been something of a trial for Du Ponceau, especially when Rafinesque, failing to get a hearing at the American Philosophical Society, began addressing Du Ponceau in print.[27] And in 1835, the year Du Ponceau won the Volney contest, Rafinesque struck out with some bitterness—albeit anonymously—at the errors of "this learned philologist" who "has become famous for his paradoxical opinions."[28]

Even though excluded from the select fellowship of the American Philosophical Society, by 1835 Rafinesque was in the thick of philological disputation in Philadelphia. In his *Knickerbocker* essay he defended the then prevailing opinion that Chinese characters could be understood by those unable to pronounce Chinese words, which Du Ponceau formally opposed three years later when he published his book-length *Dissertation on the Nature & Character of the Chinese System of Writing*.[29] Rafinesque went on to dispute Du Ponceau's assertion that there had been no monosyllabic languages indigenous to America, claiming Otomí as one. This Otomí dispute, which wracked the Philadelphia philologists for some time, resulted from the testimony of Don Manuel Nájera, who was personally known both to Du Ponceau and to Rafinesque. It resulted in the need Du Ponceau felt to clarify his position by including several pages (68–73) of otherwise extraneous material in the preface of the *Mémoire* (1838) that won for him the Prix Volney. And finally in his *Knickerbocker* essay Rafinesque threw in for good measure the accusation that Du Ponceau had overlooked two sounds in his "English Phonology or an Essay towards an Analysis and Description of the Component Sounds of the English Language."[30]

Yet for Rafinesque these forays into general linguistics, and even the essay on Asiatic Negroes, were but way stations on the great route he had chosen to follow in unraveling the origin of the indigenous peoples of the North America. For him, linguistic evidence was but one of several tools, and not a central object of study, as it was for Du Ponceau. Rafinesque drew with equal dexterity on history and mythology, plant and animal geography, geology and paleontology.

The earliest expression of his diffusionist thesis had been articulated just prior to the publication of *Ancient Annals of Kentucky* (1824). Modified somewhat in detail as his experience broadened and deepened through reading, this theme pervaded all of Rafinesque's subsequent writing on American prehistory. In summary, "the first and most ancient" wave of peoples "came to America from the East, and sprung from five ancient North-African nations. . . . Their [earthen] monuments are known by their great antiquity and circular shapes. One half of the American [Indian] population has sprung from this ancient stock, and has divided itself in[to] 1000 nations. . . ." The second wave—which he would later drop, perhaps because he learned that the distinction between circular and angular mounds really was not of diagnostic significance—"came from Asia, by the West" and "have furnished but few nations to America. . . . Their monuments are known by their angular shapes. . . ." "The third race sprung in Siberia . . . and came the last to America," where it divided into six "Lenapian" nations having four hundred tribes; "their monuments," if they can be said to have any at all, "are known by their recent state and rude structure." He went on to observe that "philological enquiries are of great use in ascertaining the affiliation and ancestry of those ancient nations" and, without giving a shred of evidence for it, he asserted that he had already reduced the two thousand known Indian dialects to twenty-five mother languages, which could be traced back to one or another of ten Old World grandmother languages, which in turn were reducible to a "single Primitive language divided into 3 branches, Iranic, Atlantic and Scythic."[31]

Although Rafinesque did not publish the names of the twenty-five mother languages, he must have mentioned them in correspondence no longer extant. For in his *Researches, Philosophical and Antiquarian, concerning the Aboriginal History of America* (Baltimore, 1829), 28, J. H. McCulloh, Jr., gave a "tabular view of the original languages of America, according to [Rafinesque's] nomenclature":

NORTH AMERICA	SOUTH AMERICA
1. *Uskih,* Esquimaux, &c.	15. *Aruac,* Arrowack, &c.
2. *Onguy,* Wyandot, &c.	16. *Calina,* Caraib, &c.
3. *Lenap,* Chippeway, &c.	17. *Puris,* Mayapuris, &c.
4. *Wacash,* North West Coast	18. *Yarua,* Betoy, Charua, &c.
5. *Skere,* Paunee, &c.	19. *Cuna,* Darien, Choco, &c
6. *Nachez,* Natchez, &c.	20. *Mayna,* Yameos, &c.
7. *Capaha,* Sioux, &c	21. *Maca,* Muhizca, &c.
8. *Chactah,* Chocktaw, &c.	22. *Guarani,* Guarani, &c.
9. *Otaly,* Cherokee, &c.	23. *Maran,* Peruvian, &c.
10. *Atalan,* Tarascan, &c.	24. *Lule,* Abipone, &c.
11. *Otomi,* Otomi, &c.	25. *Chili,* Araucanian, &c.
12. *Azteck,* Mexican, &c.	
13. *Maya,* Huasteca, &c.	
14. *Chontal,* Tzendal, &c.	

McCulloh further remarked that Rafinesque "thinks a more complete investigation, may possibly reduce them to but eighteen. Thus, for instance, 4 and 5 may be found to be the same, so also 6, 7, 8, and 9, 10, 11, as they have considerable analogies with each other, 15, 16, and 19, approximate also by gradual dialects with 17, 18, 20." (McCulloh's own contribution to linguistics receives further treatment in Stuart's essay, which follows.)

So much for Rafinesque's grandiose speculation; probably in the 1820s it was based on little more than a hunch that there had to be some kind of connection among ethnic types, the languages they spoke, and the physical artifacts of their making. Nevertheless, it was the theoretical scaffold he would spend the rest of his life seeking evidence to support—and, according to Oestreicher, pp. 239–40, inventing evidence when it could not be found. His great work on the subject, like so many Rafinesque projects, remained a truncated monument. This was his "Outlines of a General History of America," announced as a two-volume work in 1827. When the two volumes actually appeared in 1836 under the title *The American Nations,* they were described as part of what now had become a projected six-volume work. However, only the two small, privately published volumes of 1836 ever were issued.

Beginning with the publication of *Ancient Annals of Kentucky* (1824), Rafinesque's studies in the prehistory of the New World had attracted censure on both sides of the Atlantic. That pamphlet was lampooned in the *Cincinnati Literary Gazette* (2 [Dec. 25, 1824], 202–4) by an anonymous reviewer who made much fun of its absurd chronology and especially took umbrage at the pamphlet's pretentious motto, *numquam otiosus* (never idle).The reviewer also was cognizant of the bitter quarrel occasioned by Rafinesque's review of Caleb Atwater and enthusiastically sided with Atwater. But Rafinesque must have become inured to the many attacks on him that continued to issue from the "Queen City" on the Ohio. What stung more was a review in Paris by Georg Bernhard Depping, who hinted that the bibliography had been padded, for it contained a number of Greek and Latin authors who could not possibly know anything about America, much less Kentucky. The reviewer demanded to be told: What are "the unknown and authentic sources" Mr. Rafinesque says he drew on?[32]

Thus, after more than a decade, when Rafinesque finally could begin the publication of *The American Nations* his first task was to answer his French and American critics by printing some of the "sources inconnues et authentiques" that he believed supported the migration theories he only had space to outline in *Ancient Annals of Kentucky.* It was purely by accident that a few dozen pages of the illustrative material he chose to include in the first of these volumes brought him whatever lasting fame—or notoriety—he now enjoys in American prehistory and cultural anthropology.

This is the text of the *Walam Olum,* included to document the last wave of migration from Siberia of the Lenapian peoples. Also included in the same volume was the much longer "Haytian Annals," intended to document an earlier Atlantic migration from North Africa; but this has never received any attention at all. Having little interest in alleged prehistoric Atlantic migrations, twentieth-century scholars have ignored the "Haytian Annals"; but having arrived by other means at the strong conviction of a prehistoric migration via the Bering Strait, they have been much interested in the *Walam Olum.* Nobody, then or since, has ever seen fit to comment on the second volume of *The American Nations,* which dwelt on such topics as "Mosaic Nations," "Chinese Nations," "Early Japanese Annals," "Annals of the Atlantes," "Intercourse with America before Columbus," etc. As the notes to these volumes attest, all of this quaint lore had been gathered from the numerous libraries Rafinesque had frequented during the past decade and had recorded in about a hundred manuscript notebooks to which he gave the collective name *Tellus* (cf. Belyi, p. 222).

III

With such materials heaped around him and always in need of bringing in enough cash to live on, no wonder Rafinesque laid aside his historical studies to take up the challenge of the Institut Royal de France: "To determine the grammatical nature of the languages of North America known under the names of Leni-Lenape, Mohican, and Chippewa." One of two persons to respond to the challenge, Rafinesque submitted a 256-page essay followed by a 14-page supplement. His anonymous manuscript (as required by the rules of the contest) was filed with that of the winner, Peter Stephen Du Ponceau, and forgotten until discovered in Paris by Joan Leopold in 1982 and its author identified by Jean Rousseau in 1983.

Rafinesque's treatise, entitled "Examen Analytique des Langues Linniques de l'Amérique Septentrionale, et surtout des Langues Ninniwak, Linap, Mohigan &c avec leurs Dialectes," was a workmanlike effort that did address the subject of the competition but nevertheless managed also to include many of its author's related interests. There seems little question that Du Ponceau's contribution was superior. But if Du Ponceau's single most important observation about the Algonkian languages was their polysynthetic character,[33] Rafinesque's discussion of the same phenomenon in a chapter titled "Des Mots Amalgamés" (35–39) at least bears comparison, though he probably detracted from the singularity of the phenomenon by illustrating it with supposed parallels from Italian. He observed that what Sicilians accomplish in one "amalgamated" word *(damanuzza)* takes three words to say in Tuscany: *dami lua mania*. Nor was he astonished, as many observers were, by the length of some amalgamated expressions in Indian languages; he recalled that an Italian poet had once coined *precipitevolissimevolmente* to express "with the most rapid velocity" (36). At any rate, the anonymous report of the committee judging the two essays pounced on Rafinesque's amalgamation concept, declaring that his examples amounted to nothing more than contractions, and also—more justly—finding fault with his forced etymologies.[34]

Both authors needed a generic term for the family of languages under discussion. Du Ponceau accepted the name already in use by French missionaries, "Algonquine," while Rafinesque coined "Linnique," which, he said, had "the advantage of being derived from the language, and to be similar to Celtic, Arabic, Italic, &c."[35] After an introductory section of some eighty-five pages in which he discussed the nature of the Linnique languages, of which that of the Lenape people was said to be but one of twelve, Rafinesque dutifully devoted more than a third of his essay to the specific subject of grammar, writing five short chapters on substantives, six on verbs, and a chapter each on articles, adjectives, pronouns, adverbs, and particles. He appended fifteen pages of vocabularies as examples of the various dialects under consideration, and he included a whole section on syntax and idioms, widely construed to include such topics as style and eloquence, poetry, mythology, and graphic systems. Obviously, Rafinesque's interests in the Indians were much broader then Du Ponceau's and, verging on ethnology, probably gained him little credit in a competition focused on grammar. Since the languages and dialects under consideration were patently related to each other, the Institute was mercifully spared an explication of Rafinesque's "synoremic formula" for the numerical affinity among them.

And if Rafinesque cast his net more widely than Du Ponceau to include ethnological observations, he also cast it more widely linguistically—both in respect to the number of Indian languages treated and also in respect to his comparisons of these languages with those of Europe. In fact, he considered these European analogies one of his major innovations, stating with some pride that "the philologists have not yet perceived this fact, which I announce with pleasure."[36] Then he stated his conviction that Italian (whose forms are "highly variable and complicated") and English (whose forms are "very simple, since it is an analytical language and almost monosyllabic, like the other Teutonic languages") could provide valuable insight into the structure of the Linnique languages. As so often happened with Rafinesque's discoveries, concrete illustrations proved less engaging, however, than the avowed principle. The following, a fair example of his practice, recapitulated quite obvious characteristics of three European languages for the sake of reaching a far from certain conclusion about the relative youthfulness of Amerindian languages: "In Italian the adjectives may precede or follow the substantives, and the adverbs follow or precede the verbs. In French the position of the adjectives often changes the sense. In English and the Linnique languages all and sundry must almost always precede the terms which they modify, a form less natural than the other and consequently less ancient."[37]

While Du Ponceau concentrated his attention mostly on the grammatical structure of "Lénâpé" and "Chippéway," considering the latter "the true Algonquian," Rafinesque drew his examples from sixty "dialects"—as he denominated them—arranged in twelve family groups, each of which he called a language. For each of the sixty "dialects" he gave a word that he believed to be equivalent to "l'homme," "la terre," and "l'eau," and remarked that "one must recall that some synonyms mean male, the ground, and waters or the sea."[38] He acknowledged that his typology for these languages differed from that of other philologists, for earlier writers, such as Adelung and Balbi, had grouped them geographically. Thinking like a botanist, he arranged his samples phylogenetically according to affinities among the languages themselves. These affinities were demonstrated by twenty-one vocabularies given in an appendix (241–56). It was necessary to devote some attention to all of these languages, he thought, because the descent of the Linnique languages was obscure, unlike the pedigree of the better-known European tongues.

Nevertheless, his grammatical analysis was based largely on "Ninniwak" as Du Ponceau's was on "Chippéway"—in both cases different names for the same language, that of the Ojibwa people—and probably for the same reason: the availability of abundant lexical material for that language. The two candidates for the Prix Volney differed, however, in their sources of information about this language. Du Ponceau depended chiefly upon the works of Henry Schoolcraft, whose principal native informant was his wife, an Ojibwa woman. Rafinesque drew chiefly upon John Tanner's captivity narrative as reported by Edwin James. Marginal notes in his manuscript show, however, that Rafinesque was equally familiar with Schoolcraft's published works.

Moreover, Rafinesque further sought breadth of understanding by making extensive comparisons and contrasts with such geographically neighboring tongues as those of the Dakotas, Hurons, and Cherokees, and occasionally he reached out as far as Chile, Haiti, and Mexico for comparative materials—all of which he had, of course, in his own collection of vocabularies in the manuscript volumes called *Synglosson*.

Among the comparative word lists given in his appendix, the one of greatest historical interest is an analytical vocabulary for Lenape (254–56), not least because Rafinesque documented his sources for it as vocabularies given by Benjamin Smith Barton, David Zeisberger, George Henry Loskiel, William Penn, Tomas Campanius Holm, and Christian Frederick Dencke. All but the last of these were published sources (albeit Holm was available, when Rafinesque wrote, only in the Swedish original). Dencke's manuscript glossary of 3,700 words was unknown to other scholars until discovered in the Moravian Archives long after Rafinesque's death and used by Daniel G. Brinton as the basis for *A Lenape-English Dictionary*, published in 1889.[39] This section shows, too, the degree of sophistication Rafinesque possessed about the historical development of Lenape. For many of this vocabulary's eighty-one words (two of which were repeated with variant definitions), Rafinesque gave three Linnique forms: Unami, which he recognized as an older form of the language, as well as Lenape, and "Minsi or Munsey," two more recent forms which he considered to be dialects of Unami.

Believing that all spoken languages lacking the ballast of a written form are bound to change completely in about a thousand years (57), he considered it the task of philology, as the handmaiden of history, to chart the path of these changes. Although he believed change was less rapid among mountain people, such as the Basques, and among primitive people, his personal familiarity with Indian life provided a ready explanation for the cause of change in Indian languages. He noted the Indian practice of capturing women in wars, and the willingness of the captive women to become integrated into the tribe of their captors. Aware that in most tribes women spoke among themselves a variant of the tribal language, he believed such captives would in time enrich the vocabulary of their abductors (58).

The principal outcome of Rafinesque's analysis was a marked advance over his dogmatic certainty in 1824 that the myriad of Indian dialects could be reduced to twenty-five mother languages, and those recovered through palingenesy. Now, in 1834, he concluded that all the Linnique languages were sisters descended from an ancient language since lost; but, of them, Ninniwak appeared to be the eldest sister still extant. Like Du Ponceau, he also concluded that, far from being barbaric, these languages were highly expressive, even noble. In a three-page general conclusion, he summarized his observations on Linnique grammar, emphasizing its peculiarities. Among these, he noted as others had done the singular characteristic it had of dividing words between animate and inanimate modes. While he thought the bewildering conjugations of verbs a flaw, nevertheless he concluded that this characteristic made the languages "concise, spirited and copious," well adapted for the direct communication of ideas (221). He believed, too, that his analysis had corrected some of the accumulated errors of philologists concerning Indian languages, among them the opinion that these languages had never been monosyllabic. Since he had satisfied himself that all the composite substantives could be analyzed into constituent roots (the kind of exercise he later carried out with much misguided ingenuity on Hebrew), he declared it probable that the lost mother language of the Linnique tongues also had been monosyllabic.

Probably the greatest significance today of his Prix Volney essay lies in the contribution it makes to a better understanding of the method Rafinesque pursued in his historical studies. Taking his materials as usual from printed sources, manuscript vocabularies collected by himself and others, and personal communications with Indians and their interpreters, Rafinesque was much interested—as Du Ponceau was not—in how knowledge had been transmitted from generation to generation among ostensibly unlettered people. While others were studying Indian tongues as spoken languages recorded through the imprecise orthography of foreigners, Rafinesque was surely among the first to try to understand the meager records of the indigenous people themselves. Characteristically, he pushed his conclusions as far as—and perhaps a little farther than—the evidence would bear. In doing so he had something of historical importance to say about one of the major sources he chose to use; and, in view of the sources available to him at this time, what he chose not to say may be of equal importance.

Missing from the 256-page essay Rafinesque dispatched to France in October of 1834 was any reference to the *Walam Olum,* what was believed to be the great epic of the Lenape people whose substance he claimed had been in his possession since 1822. One would think that so extensive a document ought to yield capital evidence for the analysis of the language it was written in, especially since Rafinesque went out of his way to discuss Linnique poetry and mythology, and even treated graphic systems without mentioning any of the *Walam Olum*'s 184 pictographs recorded in his own manuscript copy dated 1833. The *Walam Olum* was discussed at some length, however, and several sample pictographs transcribed, in the fourteen-page supplement he fired off to Paris on December 24, 1834.

As we have seen, David Oestreicher understands this puzzle to mean that Rafinesque had not yet finished *inventing* the *Walam Olum* in 1834. Oestreicher has demonstrated[40] that the text of the *Walam Olum* is indeed a hodgepodge of half-understood Lenape roots and misused place names, a mishmash that would cause making sense of it all the more difficult if, as seems equally plausible, Rafinesque himself were once again the victim of someone else's hoax. But if we take him at his own word, Rafinesque long had trouble deciphering the *Walam Olum*, and he was not sufficiently satisfied with his work to include this material in his first Prix Volney submission.

Continuing his explication of the record-keeping systems used by Indians, Rafinesque alluded to wampum and petroglyphs and went on to explain that some tribes employed "notched sticks, and stakes with various grooves & paintings, to transmit and communicate ideas." He added that the "Lenape had similar sticks, which would give the events for the last 300 years."[41] This period of only three centuries suggests that he certainly had not finished his work on the *Walam Olum*—whether one understands his relationship as that of translator or fabricator—because, when he did publish it, he estimated that the ninety-six successive chiefs mentioned there implied a chronology of thirty-two centuries.[42] If his intention was to palm off a spurious Indian document of his own invention, it is hard to understand why he would confuse his French readers by telling them in the first instance that Lenape records stretched back only 300 years if he was planning to spring on them a chronology of 3,200 years later.

With the *Walam Olum* unavailable—for whatever reason—as he wrote his Prix Volney essay, Rafinesque based his language analysis on a document called the *Neobagun*, a text he had alluded to earlier in his published writing merely as one of several "painted tales and annals" of the Algonkian people. Now he discussed this text extensively—as well he might, because he owned the original, though his ownership was mentioned nowhere in his published writing.[43] As it is known today, the *Neobagun* appears as an appendix to the captivity narrative of John Tanner, edited for publication by Edwin James (1830).[44] James had served as botanist, geologist, and surgeon on the 1820 expedition to the Rocky Mountains led by Maj. Stephen H. Long, and he compiled its official report, *Account of an Expedition from Pittsburgh to the Rocky Mountains* (2 vols. and atlas; Philadelphia, 1822–23). Later, stationed in the Michigan Territory as an army surgeon, he became interested in Indian languages, befriended Tanner, and translated the New Testament into Chippewa (the language of the Ojibwas) in 1830. It was during James's short residence in Albany (1833–37) that Rafinesque met him and probably obtained there the *Neobagun*, which James no longer needed.[45]

It may be because the *Neobagun* was conveniently at hand that Rafinesque chose its vocabulary as central among the twelve Linnique languages and sixty dialects he identified. Calling this language neither Chippewa nor Ojibwa but Ninniwak, he said that it differed from the Lenape language as German differed from English (16). Elsewhere he made the somewhat puzzling observation that the word *Neobagun* means "male tool,"[46] a concept which is repeated in the Prix Volney manuscript, where the phallic allusion is clarified by this explicit description of how this kind of Indian writing, used both in the *Neobagun* and the *Walam Olum*, was done: "These Linnique characters are traced on narrow strips of wood, threaded at the end, and forming thus a written bundle. They are engraved with a stone or an awl, or sometimes painted with a stylus dipped in black or red paint. I have some before me, which show the two manners intermingled."[47] He coined the term synoglyphs for these pictographic characters "since they express ideas joined together." Synoglyphs, he said, are not employed "only for charms"—which is their chief function in the *Neobagun*—but also are used for "songs, fables, and to recall everything that appears very significant."[48] He explained that the synoglyphs could not be read without long instruction in the art.

Fig. 14.1. Top row: Neobagun *pictographs. That on extreme right was said to represent cannibalism. Bottom row: Samples of* Walam Olum *pictographs in the supplement.*

Since James had reproduced 104 pictographs, or synoglyphs, of the *Neobagun* with each followed by the appropriate Ninniwak words, and they in turn were translated into English, his book provided a ready source for an analysis of the language, as the *Walam Olum* did not as yet. Another reason for believing Rafinesque was not fully aware of the significance of the *Walam Olum* in 1834 is that all he had to say about Lenape traditions in his Volney Prize essay was that they implied that people's "origin or their immigration from the north west, and proximity to Asia," which is merely equivalent to Heckewelder's hazy report on the tradition of their origin. At this time, as far as Rafinesque was aware, only the Shawnees claimed "to have crossed the Bering Strait on the ice."[49] However, when he was satisfied with his translation of the *Walam Olum* and published it in 1836 in *The American Nations*, he was pleased to point out that it contained "a fine poem on the passage to America over the ice" (131). Again, why would anyone bent on creating an ethnogenic hoax set up in one document the crucial story about the Bering Strait passage only to have to contradict it in a later one? Nor does it appear that however Rafinesque may have been misled by unknown persons out to trick him, it was not in his own guileless nature to concoct poetry as a cruel joke on posterity.

James also had included comparative vocabularies in his book, a few of the examples being derived from Du Ponceau's published writing. While using James's lexical material, Rafinesque did not slavishly follow the interpretations of the Metai or medicine songs of the *Neobagun* as he found them in James. To James the Indians were degraded barbarians, and the character of the renegade cut-throat John Tanner, whose story he told, had been shaped by Tanner's forced association with them at an impressionable age. One piece of evidence revolting to James was a pictograph that he interpreted as revealing the practice of canni-balism. Rafinesque selected this pictograph from among 104 as one of five to reproduce in his Prix Volney essay, merely remarking with great sang-froid that it showed a "slave boy in the box that one might eat" (see fig. 14.1).[50] With a com-mendably objective attitude for his time, Rafinesque could conclude of all the Indian peoples that, despite a bit of cannibalism here and there, "this civilization thus is not as barbarian as that of many of the peoples of Africa & Australia; & thus we should not place the Linnique peoples among the absolutely barbarian nations; but they are like their languages, a mixture of the extremes in Philology & in Civilization."[51]

The lengthy supplement that Rafinesque dated December 24, 1834, and addressed to Silvestre de Sacy, Secrétaire de l'Académie des Inscriptions et Belles-Lettres, reached Paris in March of the following year.[52] His explanation for writing a supplement was that new evidence on a Linnique dialect, the "Rennapi" spoken by the Delawares in the vicinity of Philadelphia late in the seventeenth century, had just appeared in Philadelphia.[53] Titling this a "Premier Supplement," he also offered to forward six additional supplements—which were briefly described—if the works were valued and requested by the savants in Paris. It is a pity the request was not made, because the third supplement would have been an analytical translation of the *Walam Olum* manuscript in the Lenape language, with its "syno-glyphs" compared with those of the *Neobagun* in the Ninniwak language. To arouse interest, Rafinesque included drawings of seven *Walam Olum* pictographs (see fig. 14.1) and devoted a page to their analysis, to point out the philological, graphical, and historical importance of this document. But, alas, as far as we know the Paris savants never asked for more; they awarded the Volney prize to Du Ponceau at their annual meeting held on May 2, 1835.

Nevertheless, even in his abridged references to the *Walam Olum* in the supplement, Rafinesque clarified two issues about that document which have long been puzzling. One is the question of whether he possessed a physical artifact or merely a transcript of the pictographs. The other is whether the artifact consisted of a bundle of wooden sticks or a collection of birch bark strips (as had been confidently asserted by E. G. Squier). Here Rafinesque tells us that the figures were engraved on cedar wood tablets or sticks. Wherever they may have come from, and whoever may have made them, he says they were in his possession in 1834.

> There is in Philadelphia among several fragments of the *Neobagun* . . . a manuscript in sticks of cedar wood (the sacred tree of the Linnique people) which is covered with the figures, signs or hieroglyphs engraved or painted, similar with those of Tanner. . . . The name of the manuscript deserves attention, it is WALAM OLUM, two words meaning *painted sticks* in the French idiom but analytically WALAM indicates *picture* or *resemblance* employed here as an adjective, and derived from or analogous to Walam, the truth. OLUM indicates tablets, sticks or engraved stakes, and derives from OLO, one of the synonyms of all that is hollow or is engraved. So one could render Walam Olum analytically by "picture-engraving"![54]

Fig. 14.2. George Catlin's sketches of Indians holding the kind of sticks on which it is believed such documents as the Walam Olum *were recorded.*

Indian Languages and the Prix Volney

Not even hinted at is any explanation for why Rafinesque offered to forward an analysis of the *Walam Olum* only in his *third* supplement if, in fact, he already had finished "inventing" it. If he was itching to fool the Institut de France and, in turn, posterity, why wait?

IV

Rafinesque's failure to win the contest in no way dampened his interest in Indians and their languages. During the remaining six years of his life much of his published output was in the field of botany and, unable to publish the remaining projected volumes of *The American Nations,* he was left with vast amounts of raw material on hand. As usual, he was happy to share what he had with those who could use it. In 1838 he carried out an exchange of correspondence with the dramatist John Howard Payne—a writer remembered today for little more than his song "Home, Sweet Home"—and offered to turn over his unpublished "History of the Cherokees," if that would help Payne in his defense of Cherokees being expelled from their ancestral lands in Georgia. He wrote that this work consisted of three parts: "the ancient or original traditions of the *Tsalagi* [Cherokees] themselves"; "the Colonial & Modern history of the Nation"; and his "Remarks on their Origin, Language & Manners." In short, this was a study not much different in scope from what he had done earlier on Algonkian peoples.[55]

One source of information about the Cherokees was their newspaper, the *Cherokee Phoenix,* to which Rafinesque subscribed. With no hard feelings toward his French colleagues, he went to the trouble of sending his copies to France after he had read them. Assuming there would be no other examples in France of the eighty-five characters Sequoyah had invented to express in written form the language of the Cherokees, he suggested that if the Société de Géographie did not want the newspapers the copies should be handed over either to the Bibliothèque Nationale or to the Académie des Inscriptions et Belles-Lettres.[56]

The newspaper published his letter to its editor dated April 5, 1828, in which he framed eight detailed questions about Cherokee history and ten about the Tsalagi language. Those on language were answered very fully by the missionary Samuel Austin Worcester, assistant to the editor Elias Boudinot, although there was no printed response to those on history.[57] Somehow, perhaps through private correspondence no longer extant, Rafinesque obtained from the Cherokee Chief John Ross "the true names & meanings of their tribes, besides the Giants & Dwarfs of their mythology," he told Payne. Through his linguistic analysis of all this material, he had determined, he said, that the Tsalagi ("meaning Mountaineer and very akin to . . . the American Atlantes") were a remnant of the Talegawi nation, thus descendants of those people whom Heckewelder had called the Alleghawees, the "Mound Builders" driven south into Mexico by the ancestors of the Lenapes. The oldest known name of the Cherokees was Taliga, a form corrupted by the Lenapes; but properly it should be Tzuluki, a name related to that of the Tzuli bird of Mexican mythology. Rafinesque concluded that he had therefore connected the Cherokees with what was known of both the Lenni Lenape and the Aztecs.[58]

Thus Rafinesque's research had come full circle and accounted for the vastly different language groups represented by the Cherokees and the Lenapes. Just as he had hinted long ago in the *Ancient Annals of Kentucky*—when he "was not yet then, sufficiently advanced in this difficult study"—the ancestors of the first group had entered North America via the Atlantic, those of the second group via the Pacific. As far as he was concerned, their geographical propinquity in the United States had long masked the fact that these Indian nations had radically different origins, a difference borne out by the nature of their languages.

This keystone to the arch of Rafinesque's grand design in answer to his critics' demand for documentation for his theories can only be dropped into place because a few scattered documents in the hands of others at the time of his death have survived, while most of his personal papers either were pulped into newsprint or went to the Philadelphia city dump. A single letter exists as a file copy from Payne's side of the correspondence.[59] In it he agreed to publish Rafinesque's Cherokee material in a volume he was planning, and to summarize Rafinesque's theories about the peopling of the Americas. But Payne's book never appeared.

Notes

1. A copy of this vocabulary, consisting of seventy-three words and twelve numerals, was passed to Benjamin Smith Barton the same year. Another "Signs & Emblems used by the Osages" in the form of forty pictographs, with explanations, is in the Rafinesque MS titled "Graphic Systems of America" (1833). These pictographs are attributed to "Pahuska [*sic*] & Mr. Leblanc—1804 & 1818." Both manuscripts are at the American Philosophical Society.

2. Charles Boewe, Georges Reynaud, and Beverly Seaton, eds., *Précis ou Abrégé des Voyages, Travaux, et Recherches de C. S. Rafinesque* (Amsterdam, 1987), 47–48.

Claiming to be no more than a journeyman translator, I repeat here the French original: "un reste des Provençaux ou Catalans conquérans de la Sicile, j'allais les visitor et pris un Vocabulaire de leur Langue qui en effet ressemble assez à l'ancien Provençal ou

Indian Languages and the Prix Volney

Catalan, étant un Dialecte de la langue Romance." This observation is worth recovering as it occurred to Rafinesque originally, in French, because when he translated it for English-speaking readers he simplified it as follows: "*Nicosia* . . . has also a remnant of a Catalan or provencial colony, that has preserved its language of which I took a vocabulary" (*Life of Travels*, 35). With ancestral roots in Provence, he also had personal reasons to be interested in the history of its dialect.

3. Their collaboration is examined in Charles Boewe, ed., *John D. Clifford's Indian Antiquities: Related Material by C. S. Rafinesque* (Knoxville, 2000).

4. Nor did he have any patience with the Ten Lost Tribes of Israel theory, so favored at the time by ecclesiastics seeking to explain the peopling of the Western Hemisphere. See his "The [Native] Americans Are Not Jews," *Saturday Evening Post* 8 (Sept. 12, 1829): [1], and "Letter of Professor Rafinesque," *Troy Whig*, Jan. 20, 1835.

5. Giuliano Bonfante, "Ideas on Kinship of the European Languages from 1200 to 1800," *Cahiers d'Histoire Mondiale* 1 (1953–54): 692, attributes this idea to Leibniz.

6. These appeared in the *Kentucky Reporter*, Aug. 16, Aug. 23, Sept. 6, 1820, under the general title "Three Letters on American Antiquities, Directed to the Honorable Thomas Jefferson." Open letters on subjects of general interest, addressed to an appropriate patron and published in a newspaper, were a common device of the time, used, in part, to save the high cost of postage. Yet Rafinesque did expect the ex-president to read the published letters, and he inquired in a personal letter to Jefferson whether they had reached him. When Jefferson replied, on November 6, 1820, they had not. See Edwin M. Betts, "The Correspondence between Constantine Samuel Rafinesque and Thomas Jefferson," *Proceedings of the American Philosophical Society* 87 (1944): 372.

7. John Heckewelder, "An Account of the History, Manners, and Customs of the Indian Nations . . . ," *Transactions of the Historical & Literary Committee of the American Philosophical Society* 1 (1819): 1–348. Reprinted in 1876 as volume 12 of the *Memoirs of the Historical Society Pennsylvania* along with other Heckewelder contributions on the Lenape language, this reprint in turn was again made available by Arno Press in 1971.

8. An edited and annotated reprint of this now appears in *John D. Clifford's Indian Antiquities*, 117–28.

9. Johann Christoph Adelung, *Mithridates oder Allgemeine Sprachenkunde* (4 vols.; Berlin, 1806–17; vols. 2–4 edited from Adelung's MSS, with additions, by Johann Severin Vater); Benjamin Smith Barton, *New Views of the Origin of the Tribes and Nations of America* (Philadelphia, 1797); Antoine Court de Gébelin, *Monde primitif analysé et comparé avec le monde moderne* (9 vols.; Paris, 1773–82); *Hadriani Relandi Dissertationum Miscellanearum* (Utrecht, 1706). Reeland's Dissertatio XII, "De Linguis Americanis," was mostly on South American and Mexican languages, but it did have a five-page section on the "Lingua Algonkina" with a short vocabulary. His dissertations VI, VII, IX, XI, and XIII also treated languages, especially the Oriental ones.

Charles Boewe

A handwritten charge book still preserved in the Transylvania University Library shows that Rafinesque checked out Vols. 2, 3, 8, and 9 of Court de Gébelin's *Monde primitif* in 1820, several volumes of "Historical Collections," which cannot be further identified, in 1823–24, and "Barton on American Indians" four times in 1824, when he was writing *Ancient Annals of Kentucky*. When he himself became librarian, Rafinesque exercised the prerogatives of the office by charging no more books to his own name after May 3, 1824.

10. Rafinesque, "An Introduction to American Ethnology," a four-page essay sent to the American Antiquarian Society under cover of a letter, July 5, 1824, addressed to Isaiah Thomas. Neither the essay nor the letter has been published.

11. Something of the stature of the nineteenth-century Transylvania University Library is suggested by a description published in the *Transylvania College Bulletin* 11 (Nov. 1919): 1–51, where 238 books printed before 1700 were listed. One of the university's greatest book purchases occurred in 1821, when it sent a faculty member, Dr. Charles Caldwell, to Paris with $17,000 subscribed by the trustees and the citizens of Lexington, to buy books, which he acquired at bargain prices from the private holdings of aristocratic French families brought down by the Revolution. Unexpected donations continued to arrive also from European well-wishers. "His Britannic Majesty, William IV," sent in 1834 a gift of 81 folios, which included a facsimile of the Doomsday Book. During the year Rafinesque was writing, his fellow botanist François Andre Michaux gave the university a massive Chinese dictionary with definitions in French and Latin. Still treasured by the Transylvania library, it is one of two copies known to be in the United States. Consisting of 1,177 thick folio pages, it was compiled by Chrétien Louis Joseph de Guignes (1759–1845) when he was French Resident in China: *Dictionnaire Chinois, Français et Latin* (Paris, 1813).

12. McKenney's earlier effort appears in W. Edwin Hemphill, ed., *The Papers of John C. Calhoun* (Columbia, S.C., 1959–), vol. 9 (1976): 273–74.

13. C. S. Rafinesque to Col. Thomas L. McKenney, Washington City, Aug. 1, 1825, Historical Society of Pennsylvania. The Beinecke Library at Yale University has a copy of the printed version of this letter, as well as McKenney's covering letter dated August 22. At the National Archives in Washington, D.C., a register titled "Office of Indian Affairs, Letters Sent," II, 129, records the text of McKenney's letter as well as the names of the persons to whom it was sent. McKenney's role in these efforts is well covered in Herman J. Viola, *Thomas L. McKenney, Architect of America's Early Indian Policy* (Chicago, 1974).

14. *A Life of Travels*, 92. Rafinesque believed that his 240-page essay, although the best submitted, was too long for the American Academy of Arts and Sciences, sponsor of the contest. Asking that the manuscript be deposited at the American Antiquarian Society, of which he was a member, he jumped to the conclusion that the manuscript had been stolen when he went to the society in Worcester in 1827 and failed to find it there. Nor has it been found since.

15. This contest is not mentioned in *A Life of Travels,* because the award was still pending when that book was published in 1836. It had been alluded to in the French original of that book (85), and it is discussed in several unpublished letters. The manuscript has not been found, but probably the substance of his views were repeated poetically in lines 3366–82 of Rafinesque's *The World, or Instability* (Philadelphia, 1836).

16. See Rafinesque, "The Primitive Black Nations of America," *Atlantic Journal* (Sept. 1832): 85–86; "Reward of Merit," *Atlantic Journal* 1 (Spring 1833): 157. Even this honor was tarnished. Having received but a single entry in its contest for a prize of fr.1,000, the society awarded Rafinesque "une médaille d'encouragement du prix de cent francs"—a medal of encouragement of the value of a hundred francs (*Bulletin de la Société de Géographie* 17 [1832], 185)—although one would never guess from his account of it that Rafinesque's was a consolation prize. A draft of the essay under a slightly different title is at the American Philosophical Society; with a supplement it consists of a total of 105 closely written pages.

17. A search of John Horne Tooke's *Epea pteroenta, or Diversions of Purley* (2nd ed., 2 vols.; London, 1798), has failed to identify these words as a quotation. Probably what Rafinesque had in mind was the total thrust of this study undertaken because the author "very early found it . . . impossible to make many steps in the search after . . . the nature of *human understanding* . . . without well considering the nature of language" (I, 12). Rafinesque's familiarity with this book may have inspired his own fanciful speculations on etymology.

18. *Atlantic Journal* 1 (Sept. 1832): 85. The Himalayas as the cradle of humankind was not an original idea. It had been advanced by Sir William Jones, whose contributions to *Asiatick Researches* were familiar to Rafinesque.

19. *The American Nations,* I: 8–9.

20. Rafinesque, "The Fundamental Base of the Philosophy of Human Speech, or Philology and Ethnology," *Atlantic Journal* 1 (Summer 1832): 48–51.

21. Ibid., 50.

22. Rafinesque, "Historical and Ethnographical Palingenesy," *The Good Book* 1 (Jan. 1840): 68–70. From a store of Latin and Greek roots, Rafinesque denominated new sciences with abandon. Ironically, though, the only one attributed to his invention, *malacology* (the study of invertebrates that inhabit a shell, as distinguished from *conchology,* of the shell itself) was picked up by later naturalists from his *Principes fondamentaux de somiologie* (Palermo, 1814) after Rafinesque himself had come to favor a different coinage, "apalogy."

23. "J'avais déjà ouvert une Corresp. philologique avec Balbi," *Précis ou Abrégé,* 84. *A Life of Travels* names Balbi as a correspondent without mentioning the nature of the correspondence. Their letters have not been found.

24. *A Life of Travels,* 84.

25. *Proceedings of the American Philosophical Society* 22, pt. 3 (1885): 558. No notice of the committee's action appeared, nor was Rafinesque mentioned anywhere in the three volumes

of the MS. Letterbooks of the committee. Exercises in "palingenesy," the two essays of the contribution are preserved at the American Philosophical Society and are titled "Vocabulary of the Extinct Haytian or Taino Language," 3 pp., and "Vocabulary of the Chontal Language and Its Dialects," 7 pp. When the society took no notice of his contributions, Rafinesque expanded the Haytian paper to occupy a full chapter in *The American Nations,* I: 215–59.

26. *A Life of Travels,* 95–96.

27. The earliest example is his "Important Historical and Philological Discovery, to Peter Duponceau, Esq.," *Saturday Evening Post* 6 (Jan. 13, 1827), [2]. The significance of this in deciphering the Mayan glyphs is discussed in George Stuart's article, p. 274. In his Chontal language essay the previous year, Rafinesque pointed out that the Société de Géographie had offered a prize for vocabularies of the language of the inscriptions at Palenque. Du Ponceau must have read these essays and discussed their substance with Rafinesque, for in his "Second Letter to Mr. Champollion on the Graphic Systems of America" (*Atlantic Journal* 1 [Summer 1832]: 40–44) Rafinesque remarked that Du Ponceau "was struck by the analogy" he had detected.

28. In an essay signed only "R": "The Chinese Nations and Languages," *The Knickerbocker* 5 (May 1835): 365–76.

29. *Transactions of the Historical & Literary Committee of the American Philosophical Society* 2 (1838): ix–xxxii, 1–376.

30. Du Ponceau, *Transactions of the American Philosophical Society* n.s. 1 (1818): 228–64.

31. Rafinesque, "Clio No. 1: Ancient History of North America," *Cincinnati Literary Gazette* 1 (Feb. 21, 1824), 59–60. In Paris, the *Bulletin Universal* was willing to print, under the title "Ethnologie Américaine," Rafinesque's names of the twenty-five Indian nations that flourished "between 1400 and 1492" and the number of "hordes" or branches comprising each (*Septième Section* 6 [1826]: 168–69). But in an editorial note, probably written by Champollion, who had deciphered the Rosetta Stone, the magazine refused to publish his lists of indigenous peoples prior to Columbus until "M. Rafinesque nous fasse connaître les sources où il a puisé ces renseignemens nouveaux"—Mr. Rafinesque makes known to us the sources where he obtained this new information. He did not do so.

32. Depping wrote: "dans ce nombre les uns sont des auteurs grecs et latins qui n'ont pu avoir connaissance de l'existence de l'Amérique," and asked, "Quelles sont donc les sources inconnues et authentiques où M. Rafinesque a puisé?" *Bulletin Universel (Septième Section)* 1 (1824): 280–82. Rafinesque's lame explanation was the admission that the pamphlet was "far from being perfect, as I was not yet then, sufficiently advanced in this difficult study" (*A Life of Travels,* 73–74).

33. This is the view of Murphy D. Smith, "Peter Stephen Du Ponceau and His Study of Languages," *Proceedings of the American Philosophical Society* 127 (1983): 143–79, where the

significance of the concept of polysynthesis is based on the conclusions of linguistics scholars Daniel G. Brinton, Peter P. Pratt, and Mary R. Haas. The phenomenon of polysynthesis in Indian languages was called "agglutination" by Wilhelm von Humboldt, "coalescence" by Lewis Cass, and both "encapsulated" and "holophrastic" by Francis Lieber. Henry Schoolcraft described it as "clustered or botryoidal, thought exfoliating thought, as capsule within capsule, or box within box."

34. To take a flagrant example, when Rafinesque explained that *kich* or *gich* means vast, and illustrated it with the one Algonkian expression known to all readers of Longfellow, *gichi gumi* (vast water, i.e., lake), he was not content to stop there but felt compelled to add that *gumi* stems from the same source as our (French) word humid—"dérive de la même source que notre mot humide" (112).

35. He wrote: "l'avantage d'être dérivé du Langage, et d'être analogue à Celtique, Arabique, Italique, &c." (96).

36. He wrote: "les philologues n'ont pas encore aperçu ce fait, que j'annonce avec plaisir" (24).

37. He wrote: "En italien les adjectifs peuvent précéder ou suivre les substantifs, et les adverbes suivre ou précéder les verbes. En français la position des adjectifs change souvent le sens. En Anglais et les Langues Linniques les uns et les autres doivent presque toujours précéder les termes qu'ils modifient, forme moins naturelle que l'autre et par conséquent moins antique" (30).

38. He wrote: "Il faut se rapeller que quelques synonymes signifient mâle, le Sol, et les Eaux ou la mer" (81).

39. C. A. Weslager, "A New Look at Brinton's Lenape-English Dictionary," *Pennsylvania Archaeologist* 42 (Dec. 1972): 23–25. Dencke, who was a Moravian missionary, shared other interests with Rafinesque. The American Philosophical Society has a Dencke notebook containing a "Catalogue of Plants Found in Burlington and Gloucester Counties, N[ew] J[ersey], by C. S. Rafinesque."

40. David M. Oestreicher, "Unmasking the *Walam Olum:* A 19th-Century Hoax," *Bulletin of the Archaeological Society of New Jersey* No. 49 (1994): 1–44, and "Text Out of Context: The Arguments that Created and Sustained the *Walam Olum,*" *Bulletin of the Archaeological Society of New Jersey* No. 50 (1995): 31–52.

41. He wrote: "des baguettes déchiquetées, et des Pieux avec des entailles diverses & peintures, pour transmettre et communiquer des Idées," and added: "Linapis avaient de pareilles baguettes, qui apportaient des Evénemen[t]s depuis 300 ans" (216).

42. Rafinesque, *The American Nations,* I: 124.

43. Rafinesque to O. L. Perkins, Philadelphia, Dec. 5, 1834, Gray Herbarium Library. In this leter he offered for sale to the Boston bookseller the "M[anuscri]pt & original *Neobagun* or Written books on Wood in scratched signs or glyphs—by the Ninniwak, Ottawas &

Charles Boewe

Miamis Indians. 5 M[anuscri]pt[s]. Very curious—some are in color—of some I have the explanation or transl[ation.]" In the same letter he offered for sale "250 Vocabularies of Languages of North & South America" as well as a manuscript "on the structure & analogies of all the Languages akin to the Mohigan, Linapi & Ninniwak or Chipeway," probably the file copy of his recently completed Prix Volney essay. He did not offer to sell Perkins the *Walam Olum* because he was still using it to prepare his Prix Volney supplement (Dec. 24, 1834) and to write *The American Nations* (1836). However, it, too, was offered for sale—at $100 for the "painted copy"—in the Spring 1837 issue of his *Bulletin of the Historical and Natural Sciences* No. 4. It is not known whether he succeeded in selling any of these items.

44. Edwin James, *A Narrative of the Captivity and Adventures of John Tanner . . . during Thirty Years Residence among the Indians in the Interior of North America* (New York, 1830).

45. Now at the University of Kansas, an unpublished Rafinesque travel journal transcribed by T. J. Fitzpatrick recorded their meeting on July 31, 1833, when Rafinesque wrote: "Dr Edwin James—Bible in Ojibway! lang[uage.]" Four years later James went West again and spent the rest of his life in Nebraska and Iowa. Du Ponceau was little acquainted with James's work, mentioning only a Chippewa spelling book by James in the body of his *Mémoire* and drawing almost entirely on Henry R. Schoolcraft's *Narrative of an Expedition thro' the Upper Mississippi* (New York, 1834) for that language. He had learned of the Tanner captivity narrative, however, before his *Mémoire* was published in 1838, and cited it in an appendix, where he also made use of vocabulary material given there by James.

46. *The American Nations*, I: 123.

47. He wrote: "Ces Caractères Linniques sont tracés sur des Planchettes étroites de Bois, enfilées par un bout, et formant ainsi un faisceau Ecrit. Ils sont gravés avec un Caillou ou une Alêne, ou quelquefois peints avec un style trempé dans de la peinture noire ou rouge. J'en ai sous les yeux, qui offrent les deux manières entremêlées" (217). Cf. James (*Narrative*, p. 341), who wrote that "these rude picture are carved on a flat piece of wood, and serve to suggest to the minds of those who have learned the songs, the ideas, and their order of succession; the words are not variable, but a man must be taught them, otherwise, though from an inspection of the figure he might comprehend the idea, he would not know what to sing."

48. He wrote: "on pourrait les nommer Caractères Synoglyphes puisqu'ils expriment des idées réunies" and that "on ne les employe pas seulement pour les charmes, mais aussi pour les chansons, fables, et pour se rapeller de tout ce qui parait très important" (217).

49. He wrote: "origine ou leur Immigration du Nord Ouest, et voisinage de l'Asie" and "avoir passe le détroit de Behring sur la glace" (229).

Whatever traditions of the Shawnees may have been in Rafinesque's possession at this time have since been lost. In Lexington he had lectured "On the History and Traditions of the Shawanoe Nation" (*Cincinnati Literary Gazette* 1 [Mar. 13,1824], 87), and he later

Indian Languages and the Prix Volney

offered for sale his "M[anuscri]pt history of the Shawanis, with the Laws of Lolloway the prophet" (C. S. Rafinesque to O. L. Perkins, Dec. 5, 1834). At one time he must have had similar Shawnee pictographs, for he said in the Prix Volney MS that "the great sacred song of Shawanees for becoming a good hunter contains only 27 signs & verses or couplets" ("La Grande Chanson Sacrée des Shawanis pour devenir Bon Chasseur contient seulement 27 Signes & Vers ou Couplets"), yet an Indian who knew it well is able "to charm the wild beasts like an Orpheus" ("charmer les Bêtes fauves comme un Orphée") (217). He said also that the Shawnees once had "an historical society to preserve the memory of their emigrations and exploits" ("une Société historique pour conserve la Mémoire de leurs Emigrations & Exploits") (228).

50. He wrote: "garçon esclave dans la Case que l'on peut manger" (218).

51. He wrote: "cette civilization n'est donc pas si barbare que celle de beaucoup de Peuples de l'Afrique & de l'Australie; & Il ne faut donc pas placer les peuples Linniques parmi les Nations absolument Barbares; mais ils sont comme leurs Langues, un Mélange des Extrêmes en Philologie & en Civilisation" (228–29).

52. At the American Philosophical Society is a February 24, 1835, letter by Alexandre Eyriès (a merchant at Le Havre and brother of Jean-Baptiste-Benoît Eyriès, the French geographer and founder of the Société de Géographie), advising Rafinesque that the packet sent to him under cover of a January 5 letter had been transmitted, anonymously, to Silvestre de Sacy. This was the supplement. Receipt of the longer manuscript and its mode of transmission also can be documented, for, in a letterbook now at the Maryland Historical Society, David Bailie Warden, an American living in Paris, copied his December 29, 1834, letter to Rafinesque where he reported handing over the manuscript without a name on it to the secretariat of the academy on December 26, which is in fact the date docketed on the 256-page manuscript, also marked No. 1 as the first submission received. Both of Rafinesque's contributions were identified by the motto: "L'analyse & la comparison sont les bases de la philosophie des Langues," and his name did not appear on them.

53. This was Tomas Campanius Holm, "A Short Description of the Province of New Sweden . . . Translated from the Swedish . . . By Peter S. Du Ponceau," *Memoirs of the Historical Society of Pennsylvania* 3 (1834): 1–166. The name "Rennapi" equals Lenape; Holm reported that the language of this clan did not have the sound of the letter L.

54. He wrote: "Il y a à Philadelphie parmi plusieurs fragmens des *Neobagun* . . . un Manuscript en tablettes de Bois de Cèdre (l'arbre sacré des peuples Linniques) qui est couvert des figures, signes ou Hiéroglyphes gravés ou peints, pareils à ceux de Tanner. . . . Le nom du M[anuscri]pt mérite attention, c'est WALLAM OLUM, deux mots signifiant *Peintes tablettes* en Idiome français mais Analytiquement WALLAM désigne *Peinture* ou *Res[s]emblance* employé ici comme un Adjectif, et dérivé ou analogue à Wallam, la vérité. OLUM désigne tablet[t]es, bâtons ou pieux gravés, et dérive de OLO un des synonymes de tout ce qui

est creux ou gravé. En sorte que l'on pourrait rendre analytiquement Wallam Olum par Peinture gravure!" (266).

55.   Rafinesque to John Howard Payne, Philadelphia, Aug. 28, 1838, Columbia Univ. Library.

56.   Rafinesque to Edme François Jomard, Philadelphia, Oct. 22, 1834, Archives of the Société de Géographie, Paris.

57.   This exchange may be read in *New Echota Letters*, ed. by Jack Frederick Kilpatrick and Anna Gritts Kilpatrick (Dallas, 1968), 16–31.

58.   Rafinesque to John Howard Payne, Philadelphia, Oct. 31, 1838, Columbia Univ. Library.

59.   Payne to C. S. Rafinesque, New York, Oct. 23, 1838; Harvard Univ. (Theater Collection). Payne also expressed only guarded interest in Rafinesque's proposal that the two of them collaborate in writing an "American Iliad" and an "American Odessey" based upon Rafinesque's collections.

The Beginning of Maya Hieroglyphic Study:
Contributions of Constantine S. Rafinesque
and James H. McCulloh, Jr.

GEORGE E. STUART

NO ERA OF MAYA RESEARCH is more fascinating than the period between 1820 and 1840. Yet those two decades, which immediately preceded the epochal explorations of John Lloyd Stephens and Frederick Catherwood, remain the least known in detail of any period in the history of Maya investigation.

At the outset, it should be noted that the time span in question was one of tremendous geopolitical change in both Europe and the Americas: Napoleon's invasion of Spain and the resulting abdication of Charles IV in 1808 set the stage for the patriotic movements that (despite the restoration of the Spanish Bourbons in 1814) culminated in the political independence of most of Spanish America by 1825. This chain of events opened up a huge and relatively unknown area for travelers, merchants, and scientists, who began early on in the process to take advantage of the opportunity. The German naturalist Alexander von Humboldt had helped initiate the trend with a prodigious output of scientific publications on America that began in 1805. And those who followed his example found they had a ready market on both sides of the Atlantic for any publication dealing with the new American nations. As noted by McNeil and Deas (1980: 23), the number of travel books on various parts of Spanish and Portuguese America reached a volume between 1815 and 1830 that would not be equaled again until the end of the century.

George E. Stuart

Given the time and the circumstances, it is hardly surprising that the articles and books published before 1840 on the subject of the Maya and their remains were overwhelmingly descriptive and often laced with pseudoscientific speculation. Those works, however, constitute the first "modern" publications on the subject and helped to shape the subsequent century and a half of investigation.

In this brief essay, I will focus on the work of Constantine Samuel Rafinesque (1783–1840), one of the most intriguing characters in the annals of early American science, whose publications between 1827 and 1833 represent the first attempts to analyze Maya hieroglyphic writing. To my knowledge, Günter Zimmermann (1964) was the first modern scholar of Mesoamerica to formally recognize Rafinesque's pioneering role in Maya epigraphic studies. The present study relates the work of Rafinesque to that of his contemporary, the remarkable Dr. James H. McCulloh, Jr., and to the periods that both preceded and followed their brief but productive collaboration on the investigation of the Maya (see fig. 15.6).

Published Sources Available in the Mid–1820s

Given the present abundance of published works on American archaeology, it is all but impossible to fully appreciate the relative paucity of published literature available to an antiquarian of Rafinesque's era.[1] Nor was there a conceptual framework to aid scholars in the appraisal of the few data that *were* available.

Among the major published sources on the Maya available to Rafinesque and his contemporaries we may count the works of Bernardo de Lizana (1633), Pedro Sánchez de Aguilar (1639), and Diego López Cogolludo (1688) on Yucatán; and that of Juan de Villagutierre Soto-Mayor (1701) on the conquest of Tayasal. Much valuable material had also appeared in wider ranging general histories of the conquest and colonization of New Spain. Notable among these were the various editions of *De Orbe Novo* by Peter Martyr (1530; etc.); the story of Cortés and the conquest of Mexico by Francisco López de Gómara (1552; etc.); and Antonio de Herrera y Tordesillas's monumental *Historia General* (1601–15; etc.).

In 1739, Johann Christian Götze, of the Royal Saxon Library of Dresden, purchased in Vienna a screenfold manuscript "book" full of figures and hieroglyphic texts for his institution. Some twenty years afterward, and an ocean away, the first hints appeared of a large ruined city in the forest near the town of Palenque in northern Chiapas, then part of the Kingdom of Guatemala.[2]

In 1787, Antonio del Río, accompanied by the artist Ricardo Armendáriz, conducted the first intensive investigation of the ruins.[3] Del Río's account did not appear in print, however, until 1822, when Henry Berthoud of London published it in English, along with a lengthy essay by Dr. Paul Felix Cabrera on the connections between Palenque and the Old World.[4] The 1822 *Description of the Ruins of an Ancient City*, the first illustrated account of a Maya ruin to be published anywhere, featured seventeen plates engraved from copies of the Armendáriz drawings by "J. F. W."—none other than Jean Frédéric Waldeck, who thus made his own debut in the history of Maya studies (fig. 15.2).[5]

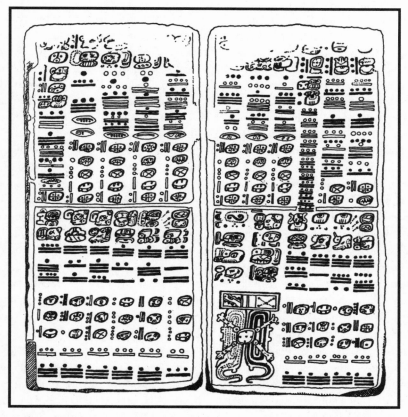

Fig. 15.1 Two pages of the Dresden Codex *from among the five illustrated in Plate 45 of Humboldt 1810 as a "Fragment d'un Manuscrit hiéroglyphique conservé à la Bibliothéque Royale de Dresde." These pages, from a tracing of the manuscript, show the numbers—combinations of bars (fives), dots (ones), and ovate shells (zeros) deciphered by Rafinesque in his pioneering study of the ancient Maya writing system. Courtesy of the Center for Maya Research.*

George E. Stuart

Meanwhile, in 1810, five of the seventy-nine pages of the Dresden Codex (fig. 15.1), along with a drawing of a Palenque bas-relief, had been reproduced by Humboldt in his monumental folio edition of *Vues des Cordillères* (fig. 15.7).[6]

Thus, by the end of 1822, in the midst of Rafinesque's first period of intensive interest in American antiquities, both the Dresden Codex and the ruins of Palenque had appeared, at least in part, in illustrated reports. And while the cultural relationship between these two seems patently obvious to us today, the connection had not been made in the early 1820s. The stage was set, however, for the studies that would bring the unprovenienced and enigmatic manuscript and the remarkable architecture and sculpture of mysterious Palenque together.

On New Year's Day 1827, Constantine Rafinesque wrote an open letter to the prominent philologist Peter Stephen Du Ponceau, which he submitted to the *Saturday Evening Post*.[7] It appeared on January 13 under the headline "Important Historical and Philological Discovery," and it stands as the first known interpretative work in print dealing with ancient Maya hieroglyphic writing (fig. 15.3).

The letter deals with the hieroglyphs of Otolum (Palenque), which Rafinesque had seen in the plates of the 1822 publication of Del Río and Cabrera (see glyphs to the right of the male image, fig. 15.2). Although the letter to Du Ponceau is laden with labored arguments connecting the writing with "Old Lybian" and suggesting the outside origins of civilization in the New World, parts are of more than routine interest: "The characters of OTOLUM are totally different from any other we are acquainted with, since they are formed by many curvilinear figures, compactly connected or blended together, and forming square groups in vertical series" (Rafinesque 1827). A few paragraphs later, he wrote: "But the letters instead of being rows, form compact groups, each group being a word, or short sentence," and "the main letters are generally larger, and succeed each other from right to left. Appearances of syllabic combinations are often evident, and numbers are perspicuously delineated by long ellipsoids marking 10 with little balls for unities, standing apart." The letter continues: "These OTOLUM characters, are totally different from the Azteca or Mexican paintings, which are true symbols, and also from every other American mode of expressing ideas by carving, painting, or quipos. They appear besides to belong to a peculiar language, distinct from the Azteca, probably the TZENDAL, (called also Chontal, Celtal, &c.) yet spoken from Chiapa to Panama, and connected with the Maya of Yucatan."

Rafinesque had made three important points in this landmark essay of January 1827: First, that Maya hieroglyphic writing was distinct from the Central Mexican scripts; second, that the language of the Maya script was related to modern Mayan languages; and third, that the bars and dots represented numbers. He also noted in passing the apparent presence of syllables.

The Beginning of Maya Hieroglyphic Study

Fig. 15.2. The Tablet of the Cross, Waldeck's engraving after Armendáriz, in Del Río and Cabrera 1822. Courtesy of the Center for Maya Research.

James Hugh McCulloh and the 1829 *Researches*

James Hugh McCulloh, Jr., (1791–1869) was one of Maryland's most distinguished citizens. A graduate of the University of Pennsylvania's School of Medicine, he served as a military surgeon in the War of 1812. From 1822 until 1850 he served as deputy collector of customs at the Port of Baltimore, and afterward served for twelve years as president of the Bank of Baltimore.

Sometime before his first war service, McCulloh began a lengthy essay on the aborigines of America "under the disadvantages of youth, occupation, and a limited library" (McCulloh 1817: 2). The 1817 edition of the work (which replaced a badly flawed version issued earlier) is entitled *Researches on America*. There is little of note about the work except perhaps for its astonishing scope. In just over two hundred pages, McCulloh ranges the literature of geography, history, zoology, and scripture for data on human customs, on distributions of mammals, birds, and insects, and on the narratives of real and mythical history from the Pacific Islands to India—all in an effort to explain the origins and history of the ancient Americans. As if overcome by the sheer weight and diversity of the data he sampled, the young

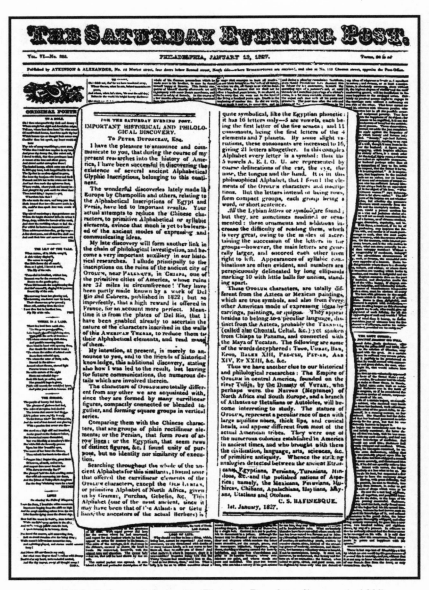

Fig. 15.3. Rafinesque's Saturday Evening Post *letter, January 1, 1827. Courtesy of the Center for Maya Research.*

The Beginning of Maya Hieroglyphic Study

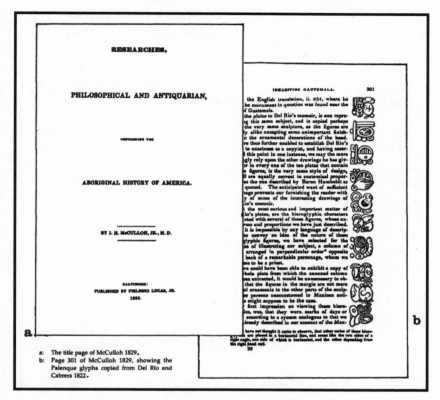

a: The title page of McCulloh 1829.
b: Page 301 of McCulloh 1829, showing the Palenque glyphs copied from Del Río and Cabrera 1822.

Fig. 15.4. McCulloh's 1829 Researches, *with its page containing the Palenque text; (a) title page, (b) page 301 showing Palenque glyphs copied from Del Río and Cabrera 1822 (cf. fig. 15.2). Courtesy of the Center for Maya Research.*

author concluded "that all the light which will ever be thrown upon the subject will be through the uncertain medium of conjecture" (McCulloh 1817: 220).

At some point, perhaps after seeing Rafinesque's letter to Du Ponceau, McCulloh solicited more details on the researches into the history of Otolum. Four letters from Rafinesque to McCulloh appeared in the *Saturday Evening Post* over the summer of 1828 (Rafinesque 1828a–d).

The *First Letter* is a superficial narrative dealing mainly with the contrasting racial types in the Americas. The *Second Letter* is a minor masterpiece of the historiography of Otolum, citing the varied works that Rafinesque had consulted up to that time, including the all-important accounts of López Cogolludo [cited as "Ayeta," who was actually the "editor" of that 1688 work]; Villagutierre de Soto-Mayor

George E. Stuart

(1701); and Del Río and Cabrera (1822). These sources led Rafinesque to speculate on the existence of ancient American "alphabetical writing" and the general belief in a triad of gods. The *Third Letter* "proves" that the ultimate origins of the American "nations" lie in North Africa and the Atlantic islands. The *Fourth Letter* briefly surveys the polities of ancient America, laments the lack of data, and ends on a note of optimism regarding Rafinesque's planned opus, *Outlines of a General History of America*—"I find new material every day" (Rafinesque 1828d).

The Rafinesque–McCulloh correspondence almost certainly continued outside the pages of the *Saturday Evening Post* during the remainder of 1828 and well into 1829, for at the time McCulloh was engaged in writing a volume to replace his 1817 *Researches on America*. The new work, submitted for copyright on October 31, 1829, bore the title *Researches, Philosophical and Antiquarian, concerning the Aboriginal History of America*. In its preface, McCulloh noted that his earlier effort "has been almost entirely forgotten" (McCulloh 1829: v), and that the new work owes much to Professor Rafinesque of Philadelphia "for an acquaintance with some valuable books and communications of great interest" (McCulloh 1829: viii).

The 1829 *Researches* (fig. 15.4) dwarfs its earlier version in all respects. It is larger in format and more than twice as long—535 pages vs. 220—and, more importantly, it is pervaded by a cautious rigor in its survey of virtually every available bit of data available on ancient America. On the matter of the hieroglyphic writing at Palenque, McCulloh lays the hand of reason on Rafinesque's speculations and makes many new points, creating what stands as the earliest rigorous discussion of the subject to reach print. His woodcut of ten glyphs from the Tablet of the Temple of the Cross, copied from one of the plates in the 1822 London edition of the Del Río narrative (McCulloh 1829: 301) is to my knowledge the first illustration of Maya glyphs to be published in the Americas (fig. 15.4).

In the course of ten pages, McCulloh drew upon Juarros (1808–18), Humboldt (1810; 1814), Del Río and Cabrera (1822), and others to demonstrate one crucial link that Maya epigraphists now take for granted—that the hieroglyphic inscriptions of Palenque and the text of the Dresden Codex were *both* products of the ancient Maya. He also tied Peter Martyr's famed description of the books of Yucatan to the Dresden Codex, lamenting that Humboldt "had not sufficient time to study this singular manuscript at full leisure" (McCulloh 1829: 305).

Even though some of McCulloh's conclusions—that the Maya writing system was ideographic and did not represent sounds or words, or that books were in use among the ancient Peruvians—were later rejected, many of the flaws in his work can be ascribed to the inadequacy of the data available to him rather than to his methods of inquiry. All in all, the 1829 *Researches* retains a remarkable integrity as a rigorous study of the cultures of ancient America.[8]

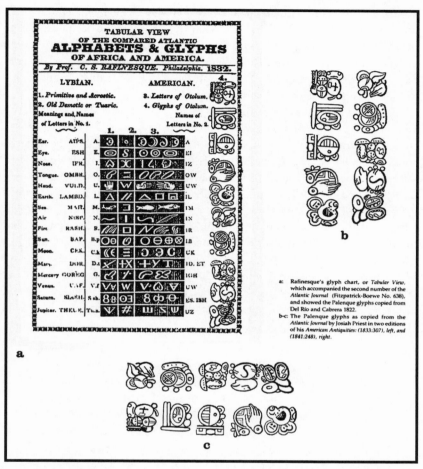

Fig. 15.5. Rafinesque's Atlantic Journal *chart, and its derivatives. (a) Rafinesque's "Tabular View," which accompanied the second number of his* Atlantic Journal *(Fitzpatrick-Boewe No. 638) and contains the glyphs copied from Del Río and Cabrera 1822. Palenque glyphs (b & c) as copied from the* Atlantic Journal *by Josiah Priest in two editions of his* American Antiquities: *b, 1833: 307; c, 1841: 248. Courtesy of the Center for Maya Research.*

Rafinesque's *Atlantic Journal*

Constantine Rafinesque was far from idle in the three years or so that followed the appearance of his letters to McCulloh. During that time he produced some sixty publications, mostly short excerpts from his *Medical Flora* of 1828–30, along with a work on a cure for consumption, a treatise on wine making, and a monograph on bivalves (Boewe 1982: 162–78). Early in 1832, Rafinesque began issuing the quarterly *Atlantic Journal, and Friend of Knowledge*, a small magazine to which he was virtually the sole contributor. The *Atlantic Journal* lasted for eight issues, and contained a grand total of nearly 160 brief articles of truly "rafinesquian" scope, including some dealing with Maya writing.

In the very first number appeared his "First Letter to Mr. CHAMPOLION, [*sic*] on the Graphic systems of America, and the Glyphs of OTOLUM or PALENQUE, in Central America." In it, Rafinesque recapitulates much of the material contained in his *Saturday Evening Post* letters to Du Ponceau and McCulloh, and divides ancient American writing into twelve types. According to those categories, the texts of Palenque fell into the Seventh Series—"alphabetical symbols, expressing syllables, or sounds, not words," and the Maya manuscripts fell into the Eighth Series "cursive symbols in groups, and the groups in parallel rows, derived from the last" (Rafinesque 1832a). In a second letter to the French Egyptologist, Rafinesque (1832b) reiterated his work with "Demotic Libyan" and other Old World scripts, and included with it a chart comparing Libyan and American glyphs (fig. 15.5a), using the ten-glyph Palenque text derived from Del Río and Cabrera (1822). He ended his discussion with a summary statement:

> Besides this monumental alphabet, the same nation that built Otolum, had a Demotic alphabet belonging to my 8th series; which was found in Guatimala and Yucatan at the Spanish conquest. A specimen of it has been given by Humboldt in his American Researches, plate 45, from the Dresden Library, and has been ascertained to be Guatimalan instead of Mexican, being totally unlike the Mexican pictorial manuscripts. This page of Demotic has letters and numbers, these represented by strokes meaning 5 and dots meaning unities, as the dots never exceed 4. This is nearly similar to the monumental numbers.
>
> The words are much less handsome than the monumental glyphs; for they are also uncouth glyphs in rows formed by irregular or flexuous heavy strokes, inclosing within in small strokes, nearly the same letters as in the monuments. It might not be impossible to decypher some of these manuscripts written on metl paper: since they are written in languages yet spoken, and the writing was understood in Central America, as late as 200 years ago. If this is done, it will be the best clue to the monumental inscriptions. (Rafinesque 1832b: 43–44)

At the end of this letter, Rafinesque appends the notice of the death of the learned Champollion, received as the text went to press, and the news of the publication of Dupaix's *Antiquités Mexicaines* in Paris.[9]

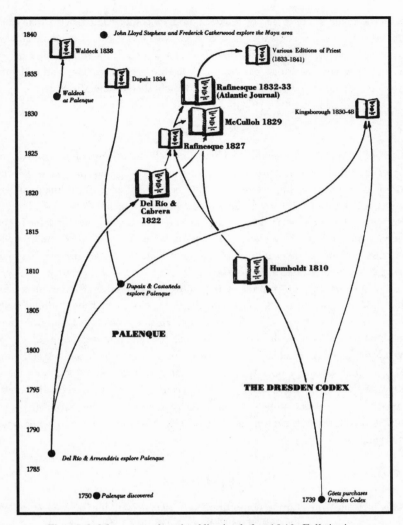

1840

John Lloyd Stephens and Frederick Catherwood explore the Maya area

Waldeck 1838

Various Editions of Priest (1833-1841)

1835

Dupaix 1834

Rafinesque 1832-33 (Atlantic Journal)

Waldeck at Palenque

1830

Kingsborough 1830-48

McCulloh 1829

Rafinesque 1827

1825

1820

Del Río & Cabrera 1822

1815

1810

Humboldt 1810

Dupaix & Castañeda explore Palenque

1805

PALENQUE

1800

1795

THE DRESDEN CODEX

1790

Del Río & Armendáriz explore Palenque

1785

1750 Palenque discovered

Göetz purchases
1739 Dresden Codex

Fig. 15.6. Maya research and publication before 1840. Full citations appear in References. Courtesy of the Center for Maya Research.

For all practical purposes, the letters to Champollion represent the last of Rafinesque's pioneering contributions to Maya epigraphy, just as McCulloh's *magnum opus* of 1829 marked his final published statement on the subject. There is no doubt, however, that Rafinesque continued to keep up with happenings in the Maya area through his correspondence. Early in 1835, the enthusiastic eccentric published

George E. Stuart

extracts of a letter sent to him by Waldeck, from Palenque. "The reliefs and inscriptions are very complicated and difficult to copy," wrote Waldeck, "it took me 20 days to copy 114 glyphs" (Rafinesque 1835: 6). And although Rafinesque's major survey of ancient American history, *The American Nations,* appeared in 1836, it added nothing of substance to that which he had already published on the Maya glyphs.

Summary

Between them, Rafinesque and McCulloh took the scanty and often inconsistent evidence of their time and produced several major conclusions that have stood the test of time. The most important appears to be McCulloh's explicit linking of the Dresden Codex and the archaeological site of Palenque as Maya. Rafinesque's contention that the best path to the decipherment of the ancient script lay in the study of the modern Mayan languages was far, far ahead of its time, and his decipherment of the bar-and-dot numbers as combinations of ones and fives anticipated the work of Brasseur de Bourbourg by more than half a century.

Unfortunately, this promising beginning was diluted by Rafinesque's penchant for the "shotgun" approach and the distinction, which he seldom made, between careful analysis and sheer speculation. However that may be, the very fact that both he and McCulloh were right some of the time is nothing short of amazing, given the almost total lack of any sort of intellectual framework for the results of their analyses. We can only speculate that, had Rafinesque possessed more glyphic texts with which to work (and ones accurately drawn), and had he maintained the interest that impelled him to the productivity of the period between 1827 and 1832, both the nature and the pace of subsequent research would have been quite different.

Epilogue

Many of Rafinesque's articles from his *Atlantic Journal,* including the important letter to Champollion (illustrated with the now-familiar Palenque glyphs [see fig. 15.5b & fig. 15.5c]), enjoyed a decade or so of reprinting in a perennial anthology of natural and historical curiosities entitled *American Antiquities, and Discoveries in the West,* by Josiah Priest, an Albany harness maker and carriage fitter turned "author."[10] The final appearance of Rafinesque's work on Palenque, again derived from the old *Atlantic Journal,* appeared in the appendix of Benjamin F. Norman's *Rambles in Yucatan,* 1843. By then, any accomplishments of Rafinesque and McCulloh (not to mention those of Norman himself) in the realm of Maya studies had

The Beginning of Maya Hieroglyphic Study

been eclipsed by the work of Stephens and Catherwood, whose spectacularly successful published travels (Stephens 1841; 1843) mark the beginning of a whole new era of Maya investigation. Ironically, just as Stephens and Catherwood, fresh from their initial trip into the Maya area, basked in triumph in New York in the late summer of 1840, the end came for Constantine Rafinesque. On September 18, 1840, he died in relative obscurity in Philadelphia, still seeking recognition for his contributions to Maya hieroglyphic study.[11] According to Von Hagen (1947: 187–88), John Lloyd Stephens received a letter from Rafinesque a month after Stephens and Catherwood returned from Central America. In it, the dying scientist reminded Stephens of his priority in discovering the nature of Maya hieroglyphs. In a letter mentioned but unverified by Von Hagen, Stephens apparently acknowledged the priority. Under the terms of Rafinesque's will, James Hugh McCulloh, Jr., was named an executor.[12] McCulloh himself died in Baltimore on December 21, 1869.[13]

Notes

1. Many of the key descriptions of the eyewitnesses of the Spanish Conquest and of the Maya and their antiquities remained in manuscript for centuries. Indeed, some remain unpublished to the present day. Among the important sources of data that were unavailable to Rafinesque, McCulloh, and their contemporaries we can list Diego de Landa's 1566 *Relación de las cosas de Yucatán*, published in 1864 (see G. Stuart 1989); Diego García de Palacio's now-famous 1576 letter to Philip II describing Copán (Squier 1858); and the reports on Palenque by Calderón and Bernasconi (Angulo Iñiguez 1934; and see n. 3, below).

2. The circumstances surrounding the modern discovery of the ruins of Palenque are unknown in precise date or detail. The earliest source on the episode is the Guatemalan Historian Domingo Juarros, who attributes the discovery, around the middle of the eighteenth century, to "some Spaniards having penetrated the dreary solitude" (Juarros [1808–18] 1823: 18). According to Brasseur de Bourbourg (1866: 3–4), some relatives of Antonio de Solís, licentiate of Tumbalá (and no relation to the earlier historian of the same name), moved to the town of Santo Domingo de Palenque (founded 1567) and came upon the ruined city around 1746. In 1773, after word of the "stone houses" had spread, a small group organized by Ramon Ordoñez y Aguiar of Ciudad Real (now San Cristóbal de las Casas) visited the ruin and reported it to José Estacheria, Governor General of Guatemala (Bernal 1980: 87). The manuscript account by Ordoñez y Aguiar is in the library of the National Museum of Anthropology, Mexico City (Castañeda Paganini 1946: 17–20).

3. Antonio del Río's 1787 expedition to Palenque was the fourth of record. In 1884, long after the visit by Ordoñez y Aguiar (see n. 2 above), a second exploring party was sent

by Estacheria. It was led by José Antonio Calderón, who spent three days at the site. The resulting report was accompanied by four drawings—crude images of part of the Tablet of the Temple of the Sun, two standing figures from the Palace piers, and the tower. The originals of these are in the Archives of the Indies, Seville, and were published by Angulo Iñiguez (1934: pls. 137 & 138) and Castañeda Paganini (1946: 22–29).

A third expedition, led by the architect Antonio Bernasconi in 1785, produced his renderings of a site map, building plans and elevations, reliefs, and a throne. Copies of these are in the Archives of the Indies (Angulo Iñiguez 1934: pls. 133–38), the British Museum (Graham 1971: 50), and the library of the National Museum of Anthropology, Mexico City (Maricruz Pailles, personal communication, 1985).

4. Biographical data on Dr. Paul Felix Cabrera may be found in the brief sketch by Heinrich Berlin-Neubart (1970: 108–11).

5. The artist Jean Frédéric Waldeck (ca. 1768–1875) himself entered the Maya field about a decade after he engraved the plates for the Del Río and Cabrera publication of 1822. Waldeck lived at Palenque for about a year in 1832–33. His contributions to Maya research lie mainly in the realm of illustration. Aside from the 1872 engravings, he made numerous drawings and paintings of the architecture and sculpture of both Palenque and Uxmal. Waldeck's published drawings have been rightly criticized for their over-romanticized European style, and for the occasional inclusion of elephant heads among the glyphs and sculptures he copied. However, his original field drawings of Palenque—now part of the Ayer Collection in the Newberry Library, Chicago—are relatively accurate and quite useful for research on details of the stucco carving at that site.

The story of the 1787 Armendáriz drawings of Palenque and copies derived from them is, in itself, worthy of separate treatment. The earliest known extant set appears to be that made in 1789 for Charles III, now in the library of the Casa Real, Madrid. That set consists of thirty figures drawn on twenty-six separate sheets, thus matching perfectly the illustration references in Antonio del Río's account.

At some point between 1807 and about 1820, copies of the Armendáriz drawings began to be distributed: A single image made its way to Alexander von Humboldt, who incorporated it into his "atlas pittoresque," the 1810 folio edition of the *Vues des Cordillères* (see n. 6 below). A partial set of the drawings apparently reached Luciano Castañeda, the artist on Guillermo Dupaix's expedition to Palenque in 1807–8, for as many as fifteen of Castañeda's renderings are clearly direct copies of them (see Berlin-Neubart 1970: 111–18). Another partial set—and the perfect match-up suggests that it was the very same one that Castañeda copied—found its way to London, where it served as the basis for Waldeck's engravings for the 1822 Del Río and Cabrera book on Palenque. This is why the figure references in the Del Río text published in that work, which match the *original* Armendariz set, bear absolutely no relation to the engravings that appear with it in the 1822 imprint.

Fig. 15.7. From Plate 11 of Humboldt 1810. Courtesy of the Center for Maya Research.

6. The drawing in question (Humboldt 1810: pl. 11) was copied from Armendáriz's rendering of Pier E of House A of the Palace at Palenque.

Humboldt's caption reads *Relief Mexicaine trouvé à Oaxaca*. The description of the "Oaxaca" bas-relief appears on pages 47–51 (in signatures 12 and 13) of the same work. Humboldt corrected this mis-identification in a short comment buried among the notes in the final section of the book (p. 320, signature 80).

7. Peter Stephen Du Ponceau is perhaps best known for his 1838 *Mémoire sur le système grammatical des langues de quelques nations indiennes de l'Amèrique du Nord*. The noted philologist was

George E. Stuart

president of the American Philosophical Society, and a member of the Pennsylvania Historical Society and the Philadelphia Athenaeum. He also served as correspondent of the Institute of France and the Paris Society of Geography. In 1835 the famed French resident of Philadelphia received the linguistics prize of the French Royal Academy, for which Rafinesque was the only other candidate.

[*The Saturday Evening Post* issued in 1827 from the press of Samuel C. Atkinson and Charles Alexander, publishers of the first volume of Rafinesque's *Medical Flora*, would not be recognized by later readers of the magazine of that name which was edited by George Lorimer for the Curtis Publishing Company and often was embellished by Norman Rockwell cover paintings. Atkinson and Alexander's *Post* resembled a newspaper in size, format, and makeup. It took its name from the fact that in was published weekly to provide recreational reading on Sunday for Philadelphians still opposed to any kind commercial activity on the Sabbath. **Ed.**]

8. For a detailed critique of early American archaeological research, including comments on Rafinesque and McCulloh, see Haven (1856).

9. Although the three explorations of Dupaix took place from 1805 to 1809, the publication of the results, including the illustrations of Castañeda (see n. 5 above) did not begin until 1830, when Kingsborough included them in his massive, multivolume *Antiquities of Mexico*. The first "official" edition of Dupaix's work appeared under the editorship of Baradere, and its title page bears the date 1834. Rafinesque's reference to that specific publication in his February 1832 article in the *Atlantic Journal* is therefore somewhat mystifying. Perhaps the work, like many others of its size during that period (i.e., Humboldt "1810"), was issued signature by signature over a period of months, even years, with the title date marking, in this case, the end of the publication span. Alcina Franch (1949) evidently has grounds for believing the final work did not appear until a decade *after* 1834, for he consistently cites the work as "Dupaix 1844."

10. The work appeared in five editions and at least ten states between 1833 and 1841. Boewe (1982: 15–17) provides a cogent discussion of the bibliographical confusion created by Priest's publishing zeal. He also details the turbulent relationship between Rafinesque and Priest—a relationship that created "the only known instance where Rafinesque actively resisted publication" (Boewe 1982: 15).

11. Boewe (1985: n. 24) cites a letter written by Rafinesque to the dramatist John Howard Payne, asking Payne to ascertain whether Stephens had received his pamphlet, *The Ancient Monuments of North and South America* (Philadelphia 1838). On the day the letter was written—October 10, 1839—Stephens and Catherwood were already a week into their first journey to Central America.

12. Rafinesque's will is published by Call (1895: appendix).

13. There is inconsistency in the rendering of the name McCulloh. Sources in the archives of the Maryland Historical Society sometimes use the spelling "McCulloch." I have retained "McCulloh" throughout this essay, in keeping with the name as it appears in his publications of 1817 and 1829.

References

Alcina Franch, José. 1949. *Guillermo Dupaix: Expediciones acerca de los monumentos de la Nueva España, 1805–1808.* 2 vols. [text and pates], comprising Nos. 27 and 28 of the *Colección Chimalistac de libros y documentos acerca de la Nueua España.* Madrid: Ediciones José Porrúa Turanzas. [Perhaps the most exhaustive treatment of the complicated history of Luciano Castañeda's drawings, including those of Palenque, and the fate of the various manuscript copies derived from the original set.]

Angulo Iñiguez, Diego. 1934. *Planos de monumentos arquitectónicos de América y Filipinas existentes en el Archivo de Indias. Catálogo, II.* Seville: Universidad de Sevilla, Laboratorio de Arte. [Plates 133–136 contain the map, drawings, and plans of Palenque drawn by Bernasconi; plates 137 and 138, the Calderón drawings.]

Baily, J., trans. 1823. *A Statistical and Commercial History of the Kingdom of Guatemala, in Spanish America . . .by Don Domingo Juarros, a Native of New Guatemala.* London: J. F. Dove. [An English translation of Juarros 1808–18.]

Berlin-Neubart, Heinrich. 1970. Miscelánea Palencana. *Journal de la Société des Américanistes* 59: 107–28. Paris.

Bernal, Ignacio. 1980. *A History of Mexican Archaeology: The Vanished Civilizations of Middle America.* London and New York: Thames and Hudson.

Boewe, Charles. 1982. *Fitzpatrick's Rafinesque: A Sketch of His Life with Bibliography.* Weston, Mass.: M&S Press.

———. 1985. A Note on Rafinesque, the Walam Olum, the Book of Mormon, and the Mayan Glyphs. *Numen* 32, fasc. 1 [July]: 101–13. Amsterdam.

Brasseur de Bourbourg, E. Charles. 1866. *Recherches sur las ruines de Palenqué et sur las origines de la civilisation du Mexique.* Paris: Arthus Bertrand. [This 84-page folio accompanied the *Monumens anciens du Mexique,* the 56 lithographic plates—40 of Palenque subjects —by Jean Frédéric Waldeck. The whole was issued under the latter title.]

Call, Richard Ellsworth. 1895. *The Life and Writings of Rafinesque.* Filson Club Publication 10. Louisville, Ky.: John P. Morton & Co.

Castañeda Paganini, Ricardo. 1946. *Las Ruinas de Palenque: su descubrimiento y primeras exploraciones en el siglo xviii.* Guatemala, C. A. [A very useful compendium of the accounts of the expeditions to Palenque by Calderdón, Bernasconi, and Del Río, along with

George E. Stuart

poorly reproduced but adequate illustrations of the manuscript drawings (or, in the case of the Armendáriz set, a contemporary copy) produced on each expedition. For some reason, the rendering of the Tablet of the Temple of the Sun (64) is that of Catherwood as published by Stephens (1841, 2: frontispiece); not the Armendáriz version. This short but important work is dedicated to the Carnegie Institution of Washington.]

y

Del Río, Antonio, and Dr. Paul Felix Cabrera. 1822. *Description of the Ruins of an Ancient City, discovered near Palenque, in the Kingdom of Guatemala, in Spanish America: translated from the original manuscript report of Captain Antonio del Río: followed by Teatro Critico Americano; or, a Critical Investigation and Research into the History of the Americans, by Dr. Paul Felix Cabrera, of the City of New Guatemala.* London: Henry Berthoud. [German translations of this work appeared in Meiningen (1823) and Berlin (1834). A French version by David Bailie Warden was published in Paris in 1835. In its first American edition, the Del Río report, with some of the illustrations, appeared in parts in the *Family Magazine, or Weekly Abstract of General Knowledge* (New York, 1833), 1: 266–67; 275–76; 283; 290–91.]

Dupaix, Guillermo. 1834. *Antiquités Mexicaines. Relation des trois expéditions du Capitaine Dupaix, ordonées en 1805, 1806, et 1807, pour la recherche des antiquités du pays, notamment celles de Mitla et de Palenque* . . . Paris: Jules Didot l'Aîné.

Graham, Ian. 1971. *The Art of Maya Hieroglyphic Writing. January 28–March 28, 1971: An Exhibition in the Art Gallery, Center for Inter-American Relations, Sponsored Jointly by the Peabody Museum of Archaeology and Ethnology, Harvard University, Cambridge, Mass., and Center for Inter-American Relations, Inc., New York, N.Y.* Cambridge: Harvard Univ. Printing Office.

Haven, Samuel F. 1856. *Archaeology of the United Sates, or Sketches, Historical and Bibliographical, of the Progress of Information and Opinion Respecting Vestiges of Antiquity in the United States.* Smithsonian Contributions to Knowledge, vol. 8. Washington, D.C.: Smithsonian Institution.

Herrera y Tordesillas, Antonio. 1601–15. *Historia general de los hechos de los Castellanos en las islas y tierra-firme del mar océano.* 8 vols. in 4. Madrid.

Humboldt, Alexander von. 1810. *Vues des cordillères, et monumens des peuples de l'Amérique.* Paris: F. Schoell. [Octavo editions of this work were published in English (London 1814) and French (Paris 1816). Each contained a single plate (No. 16) showing portions of pages 47 and 51 of the Dresden Codex, labeled as Aztec hieroglyphic paintings.]

Juarros, Domingo. 1808–18. *Compendio de la historia de la Ciudad de Guatemala.* 2 vols. Guatemala, C. A.: Ignacio Beteta. [See Baily 1823.]

Kingsborough, Edward King, Viscount. 1830–48. *Antiquities of Mexico: Comprising Facsimiles, of Ancient Mexican Paintings and Hieroglyphics, Preserved in the Royal Libraries of Paris, Berlin, and Dresden; in the Imperial Library of Vienna; in the Vatican Library; in the Borgian Museum at Rome; in the Library of the Institute at Bologna; and in the Bodleian Library at Oxford. Together with the Monuments of New Spain, by M. Dupaix; with their Respective Scales of Measurement and Accompanying Descriptions. The Whole Illustrated by Many Valuable Inedited Manuscripts, by Augustine Aglio.* London: James Moynes (vols. 1–7) and Colnagi, Son, & Co. (vols. 8 and 9). [This important work bears various dates, as follows: vols. 1 and 2: 1829, 1830, or 1831; vols. 3–7, 1830 or 1831; and vols. 8 and 9, 1848. The Dresden Codex appears in vol. 3. The original Aglio tracing of the manuscript is in the Museum of Mankind, London. Kingsborough's basic thesis—and the *raison d'etre* of the work—was that the Americas had originally been peopled by the "Lost Tribes" of Israel. The publication of vols. 1–7 drew a vehemently critical review by Rafinesque (1832c)!]

Lizana, Bernardo de. 1633. *Historia de Yucatan, devocionario, de nuestra Señora de Iz[a]mal, y Conquista Espiritual.* Valladolid, Spain: Gerónimo Morillo.

López Cogolludo, Diego. 1688. *Historia de Yucathan.* Madrid: Juan Garcia Infanzón.

López de Gómara, Francisco. 1552. *Historia de las Indias y conquista de Mexico.* Zaragoza.

Martyr, Peter. 1530. *De Orbe Novo Petri Martyris ab Anglaia Mediolanensis protonotarij.* Alcalá de Henares.

McCulloh, James H., Jr., M.D. 1817. *Researches on America; being an Attempt to Settle Some Points Relative to the Aborigines of America, &c.* Baltimore: Joseph Robinson. [McCulloh's preface notes an earlier edition of this work (1816, Baltimore: Coale and Maxwell), which I have not seen.]

———. 1829. *Researches, Philosophical and Antiquarian, concerning the Aboriginal History of America.* Baltimore: Fielding Lucas Jr.

McNeil, R. A. and M. D. Deas. 1980. *Europeans in Latin America: Humboldt to Hudson.* Catalogue of an exhibition held in the Bodleian Library, Dec. 1980–Apr. 1981. Oxford: Bodleian Library.

Norman, B. M. 1843. *Rambles in Yucatan, including a Visit to the Remarkable Ruins of Chi-Chen, Kabah, Zayi, Uxmal, &c.* New York: J. & H. G. Langley. [The appendix contains two articles by Rafinesque, one on "Otolum"; the other, the first letter to Champollion. Both were apparently excerpted from Josiah Priest (1841?), who, in turn, had lifted the articles from the *Atlantic Journal* (see Rafinesque 1832).

Priest, Josiah. 1833. *American Antiquities, and Discoveries in the West: Being an Exhibition of the Evidence that an Ancient Population of Partially Civilized Nations, Differing Entirely from Those of the Present Indians, Peopled America, Many Centuries before Its Discovery by Columbus. And Inquiries into Their Origin, with a Copious Description of many of their Stupendous Works, now in Ruins. With Conjectures Concerning What May Have Become of Them. Compiled from Travels, Authentic Sources, and the Researches of Antiquarian Societies.* Albany, New York: Hoffman and White. [This book is labeled "Second Edition, revised." Other editions or issues are dated at various times between 1833 and 1841. These differ mainly in the

George E. Stuart

typography of the title pages and, for the 1841 edition, a change of printer. The later editions of this work carry the additional note that "22,000 volumes of this work have been published for subscribers only."]

Rafinesque, Constantine Samuel. 1827. Important Historical and Philological Discovery. To Peter Duponceau, Esq. *The Saturday Evening Post,* vol. 6, no. 285, p. [2], cols. [2–3], Jan. 13, 1827. Philadelphia. [Fitzpatrick-Boewe No. 492.]

———. 1828a. Four Letters on American History by Prof. Rafinesque, to Dr. J. H. M'Culloh, of Baltimore: First Letter. *The Saturday Evening Post,* vol. 7, no. 358, p. [1], cols. [3–4], June 7, 1828. Philadelphia. [Fitzpatrick-Boewe No. 979.]

———. 1828b. Four Letters on American History by Prof. Rafinesque, to Dr. J. H. M'Culloh, of Baltimore: Second Letter. *The Saturday Evening Post,* vol. 7, no. 360, p. [1], cols. 1[–5], June 21, 1828. Philadelphia. [Fitzpatrick-Boewe No. 979.]

———. 1828c. Four Letters of Professor Rafinesque, to Dr. J. M'Culloh, of Baltimore, on American History: Third Letter. *The Saturday Evening Post,* vol. 7, no. 364, p. [2], cols. [1–2], July 19, 1828. Philadelphia. [Fitzpatrick-Boewe No. 979.]

———. 1828d. Four Letters of Prof. Rafinesque, to Dr. J. H. M'Culloh, of Baltimore: Fourth Letter. *The Saturday Evening Post,* vol. 7, no. 371, p. [1], cols. [4–5], Sept. 6, 1828. Philadelphia. [Fitzpatrick-Boewe No. 979.]

———. 1832a. Philology. First Letter to Mr. Champol[l]ion, on the Graphic systems of America, and the Glyphs of Otolum, or Palenque, in Central America. *Atlantic Journal, and Friend of Knowledge,* vol. 1, no. 1 [spring], pp. 4–6. Philadelphia. [Fitzpatrick-Boewe No. 614.]

———. 1832b. Philology. Second Letter to Mr. Champollion on the Graphic Systems of America, and the Glyphs of Otolum, or Palenque, in Central America—Elements of the Glyphs. *Atlantic Journal and Friend of Knowledge,* vol. 1, no. 2 [summer], pp. 40–44. Philadelphia. [Fitzpatrick-Boewe No. 640.]

———. 1832c. The American Nations and Tribes Are Not Jews. *Atlantic Journal and Friend of Knowledge,* vol. 1, no. 3 [autumn], pp. 98–99. Philadelphia. [Fitzpatrick-Boewe No. 683.]

———. 1835. Otolum, near Palenque. In *Bulletin No. 1 of the Historical and Natural Sciences, by Prof. C. S. Rafinesque,* pp. 5–6. Philadelphia. [Fitzpatrick-Boewe No. 851c.]

———. 1836. *The American Nations; or Outlines of their General History, Ancient and Modern.* Two volumes. Philadelphia. [Fitzpatrick-Boewe Nos. 858–59.]

———. 1838. *The Ancient Monuments of North and South America.* Philadelphia. [Fitzpatrick-Boewe No. 887.]

Sánchez de Aguilar, Pedro. 1639. *Informe contra idolorum cultores del obispado de Yucatan.* Madrid: Viuda de Juan González.

Squier, Ephraim George. 1858. *The States of Central America; Their Geography, Topography, Climate, Population, Resources, Productions, Commerce, Political Organization, Aborigines, etc., etc.,*

The Beginning of Maya Hieroglyphic Study

Comprising Chapters on Honduras, San Salvador, Nicaragua, Costa Rica, Guatemala, Belize, the Bay Islands, the Mosquito Shore, and the Honduras Inter-Oceanic Railway. New York: Harper & Brothers. [The famous 1576 letter of García de Palacio describing the ruins of Copán appears on pp. 242–44.]

Stephens, John Lloyd. 1841. *Incidents of Travel in Central America, Chiapas, and Yucatan.* New York: Harper & Brothers.

———. 1843. *Incidents of Travel in Yucatan.* New York: Harper Brothers.

Stuart, George. 1989. *Glyph Drawings from Landa's* Relación: *A Caveat to the Investigator.* Research Reports on Ancient Maya Writing, No. 19. Washington, D.C.: Center for Maya Research.

Villagutierre Soto-Mayor, Juan de. 1701. *Historia de la conquista de la provincia de el Itza, reducción, y progresos de la de el Lacandon, y otras naciones de Indios bárbaros, de la mediación de el reyno de Guatimala, a la provincias de Yucatán, en la America Septentrional. Primera parte.* Madrid: Lucas Antonio de Bedmar y Narvaez. [The second part was never issued.]

von Hagen, Victor Wolfgang. 1947. *Maya Explorer: John Lloyd Stephens and the Lost Cities of Central America and Yucatan.* Norman: Univ. of Oklahoma Press.

Waldeck, Jean Frédéric. 1838. *Voyage pittoresque et archeologique dans la Province d'Yucatan, pendant les années 1834 et 1336.* Paris: F. Didot-frere.

Zimmerman, Günter. 1964. La escritura jeroglifica y el calendario como indicadores de tendencias de la historia cultural de los Mayas. In *Desarrollo Cultural de los Mayas,* ed. Evon Z. Vogt and Alberto Ruz L., pp. 243–56. México: Universidad Nacional Autónoma de México. [Zimmermann, having only the 1832 "second letter" to Champollion from the *Atlantic Journal,* erroneously ascribes the date 1828 to Rafinesque's letter to Du Ponceau, and also assumes that the letter in question connects the Dresden Codex and the Palenque texts. As noted in the essay above, this crucial connection was first stated explicitly by McCulloh (1829).]

Part IV

The Writer

The Medicine and Medicinal Plants
of C. S. Rafinesque

MICHAEL A. FLANNERY

IN RECENT YEARS there has been a concerted effort to examine historical texts as potential sources for pharmaceutical prospecting. A noteworthy example is that of Bart K. Holland, of the New Jersey Medical School, who assembled a team of scholars to pursue this provocative theme (Holland 1996). While such an integrated approach is bound to be fruitful, the utilization of classic and traditional literature can by its very nature only address the examination of European medicinal plants. Patient and persistent inquiry has uncovered indigenous species with therapeutic promise from every part of the globe, but it has been suggested that the plant drugs historically used in the United States may hold greater phytomedicinal promise than rare species from the rain forest and other distant climes (Tyler 1996). While Holland and his colleagues study the texts of Theophrastus, Dioscorides, Galen, Paracelsus, and others, surviving texts in New World medicinal plant investigation are suggested in Johann David Schoepf's *Materia Medica Americana* (1787), William P. C. Barton's *Vegetable Materia Medica of the United States* (1817), C. S. Rafinesque's *Medical Flora* (1828–30), and Wooster Beach's *American Practice of Medicine* (1833).

By far one of the most colorful representatives of this group is Constantine Rafinesque (1783–1840). This "erratic genius" presents himself as an especially interesting and valuable study in the history of medical botany for three basic reasons: (1) because of a well-defined (albeit controversial) medical career that comes together in one major work, Rafinesque is relatively easy to bring into sharp historical focus; (2) because he was considered a maverick by the medical establishment, he became ipso facto a seminal figure within the American botanico-medical movement; and (3) because of his importance to these sectarians, inquiry into the validity and endurance of his work will also shed equal light on the materia medica of nineteenth-century American medicine and perhaps the future of some North American plant species in pharmacognostic inquiry.

Rafinesque and Medicine

The importance of Rafinesque as a medical botanist has long been recognized (Berman 1952; Weakes 1945; Wilder 1904), a point made self-evident by his work in the healing arts and his close relationship with some of America's leading medical men of the age. In Sicily as early as 1808 he engaged profitably in the drug trade, having "established a manufacture of prepared dry Squills [i.e., the bulb of a sea onion, *Urginea maritima* (L.) Baker] on a large scale" (Rafinesque [1836] 1944: 307). Upon his arrival in America he met the Congregational minister-turned-botanist/physician Manasseh Cutler (1742–1823) and was promoted and supported by Dr. Samuel Latham Mitchill (1764–1831), co-founder of the *United States Pharmacopoeia* in 1820 (hereafter referred to as *USP*). In 1819 he began teaching at the progressive Transylvania University in Lexington, Kentucky, until strained relations forced his departure early in 1826. Despite detractors, like the botanist Lewis David von Schweinitz (1780–1834), who called Rafinesque "crack-brained" (Schweinitz [1832] 1921) and others who characterized him as a ne'er-do-well eccentric quack (Drake [1830] 1970: 198; Hanley 1977: 140–41), he did have a strong background in botany and medicine.

It is largely on three fronts that Rafinesque's work in medicine must be examined: (1) his relationship to the early botanico-medical movement, specifically through Wooster Beach (1794–1868) and Alexander Wilder (1823–1908), who attempted to link the naturalist with the eclectic medical movement; (2) his invention of Pulmel, an alleged cure for tuberculosis; and (3) his publication of the two-volume *Medical Flora*.

The first of these issues has been previously examined, and despite the strained efforts of eclectics (the best organized and most enduring of the nineteenth-century botanical cults) to tie Rafinesque to the early movement, little direct relationship can be demonstrated (Berman 1956). While Rafinesque may have generally supported Wooster Beach's ideas and his new medical sect, he was never actively affiliated with it, despite claims of Alexander Wilder that this medical botanist "gave his adhesion" to Beach's medical practices (Wilder 1904: 438).

The Medicine and Medicinal Plants of Rafinesque

The second important component of Rafinesque's work in medicine relates to his theories on the diagnosis and treatment of consumption (tuberculosis) and the development, promotion, and sale of his product called Pulmel. Believing that he had contracted the disease, Rafinesque claimed in his tract titled *The Pulmist* that his "previous skill in practical and medical botany" enabled him to compound a mixture of "some active medical plants" that restored him "to perfect health and sound constitution" (1829: 7). This botanical preparation, which he called Pulmel, was made into several different dosage forms: most commonly as a syrup for internal use, but also as a lotion for external use and as a balm for inhalation. Each dosage form had a prescribed use under certain conditions.

Rafinesque's treatment included not only Pulmel but a restorative therapy that included lifestyle change and careful attention to diet (1829: 48–59). Compared with the harsh heroic measures of the regular physicians and their preference for blistering, bleeding, and dosing with calomel (mercurous chloride), it is easy to concur with one twentieth-century analyst that the "treatment recommended by Rafinesque was in advance of his period" (Weaks 1945: 48). Unfortunately, the ingredients of his Pulmel are not known and so it is impossible to assay it or to conduct the clinical trials necessary to determine its efficacy. Although Rafinesque had no idea he was battling *Mycobacterium tuberculosis*, it is conceivable that one or more of his botanicals contained antibacterial or immune-stimulant properties that acted against a variety of respiratory ailments. A potentially valuable phytopharmaceutical may lurk within the long-lost Pulmel preparation; if so, it is likely to be found in his major work in medicine—the primary focus of this chapter—the *Medical Flora, or Manual of the Medical Botany of the United States of North America*, a two-volume work, the first volume of which was published in 1828.

The *Medical Flora*

Reviews were mixed. While modern analysts readily admit the importance of the *Medical Flora* (Berman 1952; Merrill 1949: 38–39; Packard 1931, 2: 1229), Rafinesque's contemporaries, especially physicians, were less appreciative. Dr. Asahel Clapp insisted that Rafinesque "was not a physician, and is not entitled to much confidence in regard to the properties of plants" (Berman 1952). Likewise, Dr. R. Eglesfeld Griffith (1798–1850) called the Medical Flora "so mingled with wild hypothesis and unsubstantiated assertions, as to render it an unsafe guide . . ." (1832). Another view was held by D. F. L. von Schlechtendal (1794–1866), the famous German botanist at the University of Halle. Schlechtendal devoted five pages to the *Medical Flora* in his *Litteratur-Bericht zur Linnaea für das Jahr 1834*, extracting a large portion of the monograph on *Aristolochia serpentaria* L. for his readers. He found adequate the woodcut illustrations printed with green ink and called the work "a very convenient handbook" (1835).

Whatever the opinions of Rafinesque's contemporaries, larger historical developments in medicine destined his book for temporary obscurity. The steady decline of the botanico-medical movement following a peak of activity among the eclectics in the late 1860s and 1870s saw a general waning of interest in medicinal plants among the medical community. Concomitant with this trend was the emergence of more sophisticated laboratory techniques and the rise of synthetic drugs, causing a dramatic and progressive shrinkage of vegetable products with official *USP* status (Boyle 1991: 57). When medical botany was discussed at all in the opening decades of the twentieth century, the unorthodox Rafinesque was often forgotten.

Nevertheless, the twentieth century has seen a gradual reawakening of interest in Rafinesque's *Medical Flora*. The centennial symposium that convened at Transylvania College on October 30, 1940, to honor its former faculty member did much to revive this American medical classic. It was also an opportunity to call for a broader approach to medicine, one that went beyond the professional fascination with the magic bullet synthetics dominating the day's research projects (Haag 1941).

Rafinesque's own view of medical progress was multidisciplinary, resting upon rather sophisticated ideas about the interplay of taxonomic botany, medical botany, ethnobotany, phytochemistry, and pharmacy. He acknowledged the previous work of his contemporaries Jacob Bigelow (1787–1879) and William P. C. Barton (1786–1856) and called their books on medical botany "imperfect" but "useful assistants to those who can afford them." Rafinesque's qualifications for writing his *Medical Flora* included "fifteen years of botanical and medical observations and researches, and 8000 miles of botanical travels, wherein he diligently enquired and elicited from the learned and the illiterate, the result of their practical experience" (Rafinesque 1828: viii). Taking his cues from any and all sources showing merit, he was unmistakably attuned to the ethnobotanical uses of New World plants by Native Americans. His appreciation of Native American medicinal-plant use in particular has been noted by modern analysts (Mignone 1975). Daniel Drake (1785–1852), who ridiculed Rafinesque for "inquiring into the arts and sciences of these savage hordes" (Horine 1961: 234), strikes today's reader as narrow, Eurocentric, and anachronistic.

Interest in the *Medical Flora* today has been invigorated since the Rafinesque centennial symposium. Recently the book has been deemed of enough significance to be microfilmed as part of the American Institute of the History of

The Medicine and Medicinal Plants of Rafinesque

Pharmacy's ambitious facsimile editions project (King 1987: 42). The reasons for this change in attitude reflect a broader shift in contemporary medicine itself. A recent study revealed that in a national survey over one-third of those sampled used herbal medicines or other alternative therapy (Eisenberg 1993). The popularity of herbal products prompted the passage of the Dietary Supplement Health and Education Act (DSHEA) in 1994, which permits some limited structure and function claims for herbals and requires that the Food and Drug Administration submit proof that the product is unsafe before removing it from over-the-counter sale. Thus the renewed interest in medicinal plants and their growing prominence in the American health care scene suggest this present appraisal.

Methodology

Two questions underpin this examination of Rafinesque's *Medical Flora:* (1) What was the character of the *Medical Flora* within its own historical context? And (2) How many of the plants selected therein are used therapeutically today? I have used a summary analysis to address these issues. Each plant entry is first listed under the original monograph number with its family and binomial. (I have reduced Rafinesque's one hundred monographs to ninety-nine because his split of the Vitis spp. into two separate numbers is inexplicable and inconsistent with the rest of the work.) If there is no variance between Rafinesque's name and the current name a distinction is not necessary and the binomial is simply given with the common name. If there is a difference between the Rafinesque name and the current name, both are given. In either case, a common name is provided followed by its inclusion in the *United States Pharmacopoeia* (defined in the summary as "*USP* status"), and finally whether or not the plant has had any past or present medicinal use.

Rafinesque's notorious penchant for altering the names of plants and the sheer passage of time have caused many taxonomic changes since his writing. While many of the binomial names need updating, most of the species in the *Medical Flora* are readily identifiable through the accompanying illustrations in each monograph. I sought the assistance of various taxonomists in the few cases of those species whose identification was uncertain. In some cases the name used by Rafinesque is synonymous with the current name; in others they are Rafinesque's inventions; in still others a current name has been provided that is more in keeping with Rafinesque's taxonomic description and illustration. In addition, numerous authoritative sources have been used as guides to the nomenclature (Fernald 1950; Kartesz 1994; Merrill 1949; Torkelson 1996). The common names for the plants in Rafinesque's *Medical Flora* are taken from the *USP* or from *Gray's Manual of Botany* (Fernald).

The *USP* status contains inclusive dates that appeared in either the primary or secondary lists of the *USP*'s various editions (exclusive of the *National Formulary*) for each plant described in Rafinesque's *Medical Flora*. These dates were largely taken from earlier studies (Boyle 1991; Hershenson 1964).

The last entry in the summary shows the medicinal use of each plant in the *Medical Flora*. It is very important to point out that medicinal use should *not* be considered synonymous with efficacy. This study is a synthesis of secondary literature and seeks only to point the way toward some of the more promising candidates for phytomedicinal inquiry; it in no way presumes to be more than a summary account of each plant listed. Certain authoritative benchmarks were used as a guide in reporting references to the current medicinal use of each plant. These include the following: British Herbal Medicine Association 1992; British Pharmacopoeia Commission 1990, 1993; Bruneton 1995; Duke 1985; Foster and Duke 1990; Lawrence review 1989; Martindale 1996; Merck 1996; Tyler 1993, 1994; United States Pharmacopoeial Convention 1995; Werbach and Murray 1994; Wichtl 1994. Multiple references to the medicinal use or uses of a plant may indeed be relying upon a single primary source, but the fact that these several citations exist serves to underscore the designation of a plant as in "current use." A code has been devised to delineate the medicinal use of each plant: CU = current medicinal use; HU = historical medicinal use; and NUR = no medicinal use reported. For a plant to be coded CU, it had to appear in *at least one* of the benchmark authorities. Furthermore, that use has to be in the context of current scientific research and/or known pharmacological activity; folk usage alone was insufficient for a plant to earn a CU designation. A plant coded HU means that it appears to have been used in the past by virtue of its inclusion in former editions of the *USP* or other compendia but is no longer to be found in any of the benchmark literature used in this study. In cases of HU plants without *USP* status, a brief description of the use and the authority for it is given. The "no use reported" designation goes to those plants showing no indications of past or present medicinal use. Although some of these NUR plants may have been accorded some medicinal use ethnobotanically, there is no indication that any of these particular species have been given serious consideration by the medical community past or present. Since the emphasis of this summary is to serve as a guide to those plants showing the most therapeutic promise, extended discussion is given only to those plants designated CU.

Finally, as will be apparent to anyone familiar with the *Medical Flora*, the present study only examines Rafinesque's full monographs in volumes I and II, not his fairly sizable compilation of "medical equivalents" found at the end of

The Medicine and Medicinal Plants of Rafinesque

volume II (1830: 181–276). Plants in this category represent those "omitted" from the main list and are given much briefer analyses; nor are botanical descriptions given for them, and they are not illustrated by woodcuts. It is hard to determine Rafinesque's criteria for inclusion in this section because some (e.g., Salix spp., the bark of which he notes is used as a febrifuge and "6 grains of Salicine [the active constituent] have cured agues") seem well founded, while others (e.g., "wild chervil" he simply lists as "Roots eaten like Chervil [an herbaceous plant of the parsley family] in Canada") seem like mere filler (1830: 260; 211). Overall, the impression is that Rafinesque added this "equivalents" section as an afterthought and that he spent little time on these entries.

Summary Analysis

1. Acoraceae. Rafinesque's name: *Acorus calamus* L. Current name: *Acorus americanus* (Raf.) Raf. Common name: calamus/sweet-flag. *USP* status: 1820–1900. CU. *A. calamus* and *A. americanus* are synonymous. Rafinesque lists calamus as "stomachic, tonic, corroborant [i.e., invigorating] and carminative [i.e., antiflatulent]." Modern uses also consider it as stomachic and carminative (Duke 1985: 14–15). There is some evidence that calamus lowers serum cholesterol in rabbits (Foster and Duke 1990: 86); and "The oil has a strong sedative and antispasmotic action . . ." (Lawrence review, s.v. "Calamus," Mar. 1996). A "bitter and carminative" widely used in Europe (Martindale 1996: 1682).

2. Adiantaceae. *Adiantum pedatum* L. Common name: maidenhair-fern. *USP* status: none. HU. Used in "chronic catarrhs and other pectoral affections" (Wood and Bache 1880: 1565).

3. Rosaceae. *Agrimonia eupatoria* L. Common name: agrimony. *USP* status: none. CU. Rafinesque lists agrimony as "A mild astringent, tonic and corroborant. Useful in coughs and bowel complaints." Modern usage accords with that of Rafinesque (Foster and Duke, 1990: 108; Lawrence review, s.v. "Agrimony," Aug. 1995; Wicht 1994: 49). "Plant is gargled for inflammation of the throat and mouth. Mixed with fat, it is used as a poultice to draw out indolent ulcers" (Duke 1985: 23).

4. Liliaceae. *Aletris farinosa* L. Common name: aletris/unicorn-root. *USP* status: 1820–80. CU. Rafinesque lists aletris as "tonic, stomachic, narcotic and repercussive [i.e., a reduction of swelling or tumor]." Steroidal compounds may underlie the estrogenic activity of aletris (Lawrence review, s.v. "Aletris," October 1993). It is also considered an antiflatulent (Merck 1996: 234).

5. Ericaceae. Rafinesque's name: *Andromeda arborea*. Current name: *Oxydendrum arboreum* (L.) DC. Common name: sorrel-tree. *USP* status: none. HU. Used topically in foot ulcerations and internally as a febrifuge (Wood and Bache 1880: 1577).

Michael A. Flannery

6. Asteraceae. *Anthemis cotula* L. Common name: mayweed. *USP* status: 1820–70. HU.

7. Apocynaceae. *Apocynum andros[a]emifolium* L. Common name: dogbane. *USP* status: 1820–70. HU.

8. Araliaceae. *Aralia nudicaulis* L. Common name: false sarsaparilla. *USP* status: 1820–70. HU.

9. Ericaceae. Rafinesque's name: *Arbutus uva-ursi.* Current name: *Arctostaphylos uva-ursi* (L.) Spreng. Common name: bearberry. *USP* status: 1820–1920. CU. Rafinesque lists bearberry as "Astringent, tonic and diuretic." Bearberry is used in urinary tract infections (Bruneton 1995: 217; Merck 1996: 1688; Wichtl 1994: 51) and is an ingredient in over eighty preparations (Martindale 1996: 1678). Diuretic effect has been demonstrated in rats (Bruneton 1995: 217).

10. Aristolochiaceae. *Aristolochia serpentaria* L. Common name: Virginia snakeroot. *USP* status: 1820–1930. CU. Rafinesque describes the active properties of Virginia snakeroot as "diaphoretic, tonic, anodyne [i.e., analgesic], antispasmodic, cordial, antiseptic, vermifuge, exanthematic [i.e., ameliorates skin eruptions], alexitere [i.e., antidotal (esp. snakebites)]." He also writes that it is an "excellent auxiliary to Peruvian Bark." Today this plant is used for fevers, indigestion, suppressed menses, and snakebite (Foster and Duke 1990: 224). *Aristolochia* spp. appear in the Chinese pharmacopoeia and are commonly employed as bitters (Martindale 1996: 1750). [Rafinesque's woodcut can be seen in fig. 17.3.]

11. Araceae. Rafinesque's name: *Arum triphyllum.* Current name: *Arisaema triphyllum* (L.) Schott. Common name: arum root. *USP* status: 1820–60. HU.

12. Aristolochiaceae. *Asarum canadense* L. Common name: wild ginger. *USP* status: 1820–70). HU.

13. Asclepiadaceae. *Asclepias tuberosa* L. Common name: butterfly-weed/ pleurisy root. *USP* status: 1820–90. CU. Rafinesque listed pleurisy root as a "diaphoretic, expectorant, diuretic, laxative, esc[h]arotic [i.e., caustic], carminative, antispasmodic, &c. It is a valuable popular remedy, and a mild sudorific [diaphoretic causing perspiration], acting safely without stimulating the body." Currently used as a poultice for bruises, swellings, and rheumatism (Foster and Duke 1990: 136). *A. tuberosa* contains glycosides used in treating cardiac insufficiency (Bruneton 1995: 573–74).

14. Fabaceae. *Baptisia tinctoria* (L.) R. Br. Common name: wild indigo. *USP* status: 1830. CU. Rafinesque lists wild indigo as an "astringent, antiseptic, febrifuge, diaphoretic, purgative, emetic and stimulant." This plant was extremely popular among the eclectic practitioners. The historical uses for this plant largely

The Medicine and Medicinal Plants of Rafinesque

accord with Rafinesque's, and modern studies suggest that wild indigo stimulates the immune system, thus explaining its "antiseptic" reputation (Foster and Duke 1990: 116). Wild indigo is an ingredient of Esberitox N. Ojad, and Toxi-loges C (tonics made in Germany); and Galium Complex (a tonic made in Australia) (Martindale 1996: 1970, 2171, 2339, 2002).

15. Berberidaceae. *Berberis canadensis* P. Mill. Common name: American barberry. *USP* status: none. HU. Listed as similar to *B. vulgaris* and used as a tonic or cathartic depending on dosage and "formerly given in jaundice" (Wood and Bache 1880: 177–79; Millspaugh 1887: 15:2).

16. Ranunculaceae. Rafinesque's name: *Botrophus serpentaria*. Common name: black cohosh Current name: *Cimicifuga racemosa* (L.) Nutt. *USP* status: 1820–1920. CU. Rafinesque describes the properties of black cohosh as "astringent, diuretic, sudorific, anodyne, repellent, emenagogue, subtonic [i.e., mild tonic?], &c." He also recommends the plant in the treatment of "acute and chronic rheumatism." Current European uses of black cohosh are found in premenstrual syndrome, dysmenorrhea, and menopause (Tyler 1993: 46; Tyler 1994: 136–37). Studies show estrogenic, hypoglycemic, sedative, and anti-inflammatory activity. The root extract has a strengthening effect in the female organs of rats (Bruneton 1995: 607; Foster and Duke 1990: 56; Werbach and Murray 1994: 236).

17. Cabombaceae. Rafinesque's name: *Brasenia hydropeltis*. Current name: *Brasenia schreberi* J. F. Gmel. Common name: water shield/purple wen-dock. *USP* status: none. NUR.

18. Fabaceae. *Cassia marilandica* L. Common name: American senna. *USP* status: 1820–70. CU. Rafinesque writes, "All the Sennas are simple cathartics. . . . They may enter into compound laxatives and cathartics, &c." Their use as a cathartic is due to their dianthrone glycosides (1.5 to 3.0 percent); it is an anthraquinone-containing laxative (Tyler 1994: 298). Senna is a common part of both African and Indian materia medicas (Foster and Duke 1990: 118). *C. acutifolia* and *C. angustifolia* are official in the *USP* (1995: 1407).

19. Berberidaceae. *Caulophyllum thalictroides* (L.) Michx. Common name: blue cohosh. *USP* status: 1880–90. CU. Rafinesque describes the therapeutic properties of blue cohosh as "demulcent, antispasmodic, emenagogue, sudorific, &c. . . . It appears to be particularly suitable for female diseases. . . ." Blue cohosh is a uterine stimulant and emenagogue, but the glycoside caulosaponin has a toxic effect on the cardiac muscle, making it a plant of questionable safety (Tyler 1993: 47–48). "Powdered rhizomes sold in limited quantities in U.S. herb stores, for use in uterine disorders" (Duke 1985: 108).

20. Rubiaceae. *Cephalanthus occidentalis* L. Common name: button bush. *USP* status: none. HU. Used in decoction or infusion as a laxative, tonic, and in periodic fevers (Wood and Bache 1880: 1611)

21. Chenopodiaceae. Rafinesque's name: *Chenopodium anthelminticum.* Current name: *Chenopodium ambrosioides* L. Common name: wormseed. *USP* status: 1820–1940. CU. Rafinesque calls wormseed "a powerful vermifuge." The essential oil of wormseed is highly toxic and can cause dermatitis or allergic reactions (Foster and Duke 1990: 216). Despite the dangers associated with its use, chenopodium oil is still listed in the pharmacopoeia of Italy.

22. Apiaceae. *Cicuta maculata* L. Common name: spotted cowbane. *USP* status: none. CU. Rafinesque describes spotted cowbane as "a strong narcotic." He also notes that "[a] few grains of the dried leaves or extract have been given in schirrose [i.e., scirrhous or hard tumors] and scrofulous [i.e., glandular] tumors and ulcers, with equal advantage." Interestingly, Takao Konoshima and Kuo-Hsiung Lee conclude that "*C. maculata* was found to show significant in vitro cytotoxicity in the 9 KB (human nasopharyngeal carcinoma) cell structure assay" (1986).

23. Lamiaceae. *Collinsonia canadensis* L. Common name: richweed/horse balm. *USP* status: none. HU. Described as useful in "catarrh of the bladder, leucorrheoea [vaginal discharge], gravel [kidney stones], dropsy, and other complaints" (Wood and Bache 1880: 1626).

24. Myricaceae. Rafinesque's name: *Comptonia asplenifolia.* Current name: *Comptonia peregrina* (L.) Coult. Common name: sweetfern. *USP* status: none. HU. Used as a tonic and astringent (Wood and Bache 1880: 1626).

25. Apiaceae. *Conium maculatum* L. Common name: hemlock. *USP* status: 1820–1900. CU. "A powerful acrid narcotic and resolvent," writes Rafinesque, "but the uncertainty of its action lessens its value." He further observes that it primarily "acts only as a palliation to pain, like opium, to which it is often preferable, as less constipating." Folk uses for poison-hemlock largely concur with Rafinesque's (Foster and Duke 1990: 58). Conium fruit is described as having "antispasmodic activity" (Merck 1996: 424).

26. Convolvulaceae. Rafinesque's name: *Convolvulus panduratus.* Current name: *Ipomoea pandurata* (L.) G. F. W. Mey. Common name: wild potato-vine. *USP* status: 1820–50. CU. Rafinesque admits, "The genera of this family had not been well fixed, and *Ipomea* particularly was so little distinguished from *Convolvulus* that many species were considered as belonging to both!" He refers to the medicinal properties of the plant as "cathartic, diuretic and pectoral." The root of this plant has been used in teas as a diuretic, laxative, and expectorant (Foster and Duke:

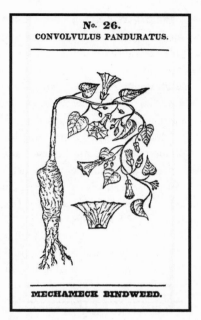

Fig. 16.1. "It is asserted that the Indians can handle Rattle-snakes with impunity, after wetting their hands with the milky juice of the root of this plant" (Medical Flora: *I, 126*).

20). Ipomea is used today as a cathartic, especially *I. orizabensis* that yields 15 percent ipomea resin (Merck 1996: 5088–89).

27. Ranunculaceae. Rafinesque's name: *Coptis trifolia*. Current name: *Coptis groenlandica* (Oeder) Fern. Common name: goldthread. *USP* status: 1820–70. CU. Rafinesque lists the properties of goldthread as "tonic and stomachic, promoting digestion, strengthening the viscera, useful in dyspepsia, debility, convalescence from fevers, and whenever a pure bitter is required. . . ." A very astringent member of the buttercup family. The plant contains berberine, thus giving it "anti-inflammatory and antibacterial effects" (Foster and Duke 1990: 38). It is currently listed therapeutically as a "bitter tonic" (Merck 1996: 427).

28. Cornaceae. *Cornus florida* L. Common name: flowering dogwood. *USP* status: 1820–80. CU. The flowering dogwood, according to Rafinesque, is useful as a "tonic, astringent, antiseptic, corroborant and stimulant." In addition, he describes it as useful "in intermittent and remittent fevers also, typhus and all febrile disorders." Dogwood was a common substitute for quinine during the Civil

War (Foster and Duke 1990: 270). At present it has found use in homeopathic tinctures for dyspepsia, intermittent fevers, and pneumonia (Duke 1985: 145).

29. Lamiaceae. *Cunila mariana* L. Common name: dittany. *USP* status: none. HU. Used in "flatulent colic" and in colds and fevers (Wood and Bache 1880: 1636).

30. Orchidaceae. Rafinesque's name: *Cypripedium luteum*. Current name: *Cypripedium pubescens* (Willd.) Correll. Common name: yellow lady's-slipper. *USP* status: 1860–1900. CU. Rafinesque lists this member of the orchid family as "sedative, nervine, antispasmodic, &c. and the best American substitute for Valerian in almost all cases." Rafinesque's claim for this plant as a sedative accords with its common usage throughout nineteenth-century American medical practice (Foster and Duke 1990: 94). *Cypripedium* is currently an ingredient in a homeopathic preparation made by Pfuger of Germany. *C. pubescens* is listed under the therapeutic category of sedatives (Merck 1996: 468).

31. Solanaceae. *Datura stramonium* L. Common name: jimson weed/thorn apple. *USP* status: 1820–1945. CU. Rafinesque treats "this loathsome weed" with therapeutic caution. He suggests that jimson weed is psychotropic and can be used

Fig. 16.2. "It is with some satisfaction that I am enabled to introduce, for the first time, this beautiful genus into our Materia Medica" (Medical Flora: *I, 143).*

The Medicine and Medicinal Plants of Rafinesque

"to lessen sensibility and pain," but when "taken internally in too great quantity" it can be lethal. Modern analysis accords with Rafinesque's findings. One of the nightshades, this plant contains atropine and other alkaloids (Foster and Duke 1990: 20). It is listed among the tropane alkaloids of belladonna *(Atropa belladonna)* and henbane *(Hyoscyamus niger)* (Bruneton 1995: 653–65). It is official in the Australian, Belgium, British, European, French, German, Hungarian, Italian, Dutch, Portuguese, and Swiss pharmacopoeias (Martindale 1996: 507).

32. Ebenaceae. *Diospyros virginiana* L. Common name: persimmon. *USP* status: 1820–70. HU.

33. Thymelaeaceae. *Dirca palustris* L. Common name: leatherwood/moosewood. *USP* status: none. HU. A strong emetic (Wood and Bache 1880: 1640).

34. Asteraceae. Rafinesque's name: *Erigeron philadelphicum*. Current name: *Erigeron philadelphicus* L. Common name: Philadelphia flea-bane. *USP* status: 1840–70. HU.

35. Liliaceae. Rafinesque's name: *Erythronium flavam*. Current name: *Erythronium americanum* Ker-Gawl. Common name: adder's tongue/trout-lily. *USP* status: 1820–50. HU.

36. Asteraceae. *Eupatorium perfoliatum* L. Common name: boneset. *USP* status: 1820–1900. CU. Rafinesque calls boneset "a valuable sudorific, tonic, alterative ["producing a change in the whole system"], antiseptic, cathartic, emetic, febrifuge, corroborant, diuretic, astringent, deobstruent and stimulant." It is still used by herbalists as a cold remedy and febrifuge and was carried in the U.S. *National Formulary* until 1950 (Tyler 1993: 49–50). "German research suggests nonspecific immune system-stimulating properties, perhaps vindicating historical use in flu epidemics" (Foster and Duke 1990: 78). Boneset is an ingredient in several different preparations marketed in Britain and Australia (Martindale 1996: 1866, 1989, 2339). [Rafinesque's woodcut can be seen in fig. 17.2.]

37. Euphorbiaceae. *Euphorbia corollata* L. Common name: flowering spurge. *USP* status: 1820–1900. HU.

38. Rosaceae. *Fragaria vesca* L. Common name: woodland strawberry. *USP* status: none. CU. Besides noting their food value, Rafinesque writes that wild strawberries "are useful in fevers" and "give relief in diseases of the bladder and kidneys." Related species, such as *F. vesca* and *F. virginiana,* have been used to make a tea as a diuretic and for kidney stones (Foster and Duke 1990: 38). *F. vesca* is an ingredient in several British, German, and French preparations (Martindale 1996: 1998, 2027, 2107).

39. Gentianaceae. Rafinesque's name: *Frasera verticillata*. Current name: *Swertia carolinensis* (Walt.) Ktze. Common name: American columbo. *USP* status: 1820–70. HU.

40. Ericaceae. Rafinesque's name: *Gaultheria repens.* Current name: *Gaultheria procumbens* L. Common name: teaberry/wintergreen. *USP* status: 1820–1900. CU. Rafinesque lists the medicinal properties of wintergreen or teaberry as "stimulant, anodyne, astringent, menagogue [i.e., promotes menstruation], antispasmodic, diaphoretic, lacteal [i.e., enhances lactation], cordial, &c." It has been shown to be "experimentally analgesic, carminative, anti-inflammatory, antiseptic," and "small amounts have delayed the onset of tumors" (Foster and Duke 1990: 28). Methyl salicylate, the essential oil found in the leaves of wintergreen, "is applied topically as a counterirritant in the form of liniments, gels, lotions, or ointments containing concentrations of 10 to 60 percent" (Tyler 1994: 147). Oil of wintergreen is currently listed in the *National Formulary* and numerous European pharmacopoeias (USNF XVIII 1995: 2266; Martindale 1996: 62).

41. Gentianaceae. *Gentiana catesb[a]ei.* Common name: blue gentian. *USP* status: 1840–70. CU. Rafinesque states that blue gentian is "very little inferior to the officinal Gentian [*Gentiana lutea*] in strength and efficacy, it invigorates the stomach. . . ." *Gentiana* spp. have found wide use as "bitters" for the stomach and were a common ingredient in nineteenth-century proprietary medicines (Tyler 1994: 145–46; Merck 1996: 745). Gentian has pharmacopoeial status in thirteen countries (Martindale 1996: 1709).

42. Geraniaceae. *Geranium maculatum* L. Common name: wild geranium/cranesbill. *USP* status: 1820–1900. CU. Rafinesque lists the wild or spotted geranium as a "powerful astringent, vulnerary [i.e., a healing agent for wounds], subtonic and antiseptic." The tannin-rich root is very astringent (Foster and Duke 1990: 146). *G. maculatum* is an ingredient in Acacia Complex, a preparation for diarrhea made by Blackmores of Australia (Martindale 1996: 1770).

43. Rosaceae. *Geum virginianum* L. Common name: Virginian avens. *USP* status: none. HU. Listed as "tonic and astringent" (Felter and Lloyd 1909, II: 931).

44. Rosaceae. Rafinesque's name: *Gillenia stipulacea.* Current name: *Porteranthus stipulatus* (Muhl. *ex* Willd.) Britt. Common name: Indian physic/western dropwort. *USP* status: 1820–70. HU.

45. Hamamelidaceae. Rafinesque's name: *Hamamelis virginica.* Current name: *Hamamelis virginiana* L. Common name: witch hazel. *USP* status: 1880–1910, 1995. CU. The medicinal uses of witch hazel are well known. Rafinesque describes its properties as "sedative, astringent, tonic, discutient [i.e., an agent which causes dispersal of a tumor or other lesion]. . . . The bark affords an excellent topical application for painful tumors and piles, external inflammations, and inflamed eyes, &c. in cataplasm or poultice or wash." This common OTC product is used in "hemorrhoids, irritations, minor pain and itching" (Foster and Duke 1990: 256). After

an eighty-five-year absence, witch hazel has found its way back into the U.S. pharmacopoeia (*USP* XXIII 1995: 1637).

46. Lamiaceae. *Hedeoma pulegoides* (L.) Pers. Common name: American pennyroyal. *USP* status: 1820–1900. CU. Rafinesque lists American pennyroyal as "carminative, resolvent, pectoral, diaphoretic, antispasmodic, menagogue, pellent [i.e., promoting discharge], stimulant, &c." He further notes that it is "a popular remedy for female complaints." Traditionally this member of the mint family has found use as an abortifacient, but its dangers are evident in its volatile oils, which contain 85–92 percent of the toxin pulegone (Tyler 1993: 243). It is also "a popular insect repellent" (Foster and Duke 1990: 190). Despite the dangers associated with ingesting pennyroyal, its continued "use in a variety of herbal self-treatment practices" make it therapeutically active (Lawrence review, s.v. "Pennyroyal," Jan. 1992).

47. Asteraceae. *Helenium autumnale* L. Common name: sneezeweed. *USP* status: none. HU. Listed a "tonic, diaphoretic, and errhine [causes sneezing]" (Felter and Lloyd 1909, 2: 979).

48. Ranunculaceae. Rafinesque's name: *Hepatica triloba*. Current name: *Hepatica nobilis* var. *obtusa* (Pursh) Steyermark. Common name: round-lobed liverwort. *USP* status: 1840–70. HU.

49. Saxifragaceae. Rafinesque's name: *Heuchera acerifolia*. Current name: *Heuchera villosa* Michx. Common name: alumroot. *USP* status: none. NUR.

50. Cannabaceae. *Humulus lupulus* L. Common name: common hops. *USP* status: 1820–1910. CU. "The whole plant," writes Rafinesque, "but chiefly the Strobiles and the Lupulin are tonic, narcotic, phantastic, anodyne, sedative, alterative, astringent, antilithic [i.e., an agent preventing the formation of calculi, especially uric-acid calculi], diuretic, corroborant, &c." The sedative properties of hops have been verified experimentally in rats (Foster and Duke 1990: 204; Tyler 1993: 176). It is currently used in conditions of "excitability, restlessness and disorders of sleep; lack of appetite" (*British Herbal Compendium* 1992: 128–30; *British Herbal Pharmacopoeia* 1990: 59). It is official in the German pharmacopoeia (Martindale 1996: 1722).

51. Ranunculaceae. *Hydrastis canadensis* L. Common name: goldenseal. *USP* status: 1830, 1860–1920. CU. Rafinesque describes the properties of goldenseal as tonic and ophthalmic and recommends it "as a bitter tonic, in infusion or tincture in disorders of the stomach, the liver, &c." The berberine content is antibacterial (Foster and Duke 1990: 50). Experimental study shows goldenseal may be effective in the treatment of alcoholic liver disease, cancer, and infection, and in the prevention of ventricular arrhythmias caused by ischemia (Werbach and Murray 1994: 23–24, 114–15). Goldenseal is official in the French pharmacopoeia (Martindale 1996: 1714).

52. Solanaceae. Rafinesque's name: *Hyosciamus niger.* Current name: *Hyoscyamus niger* L. Common name: black henbane/henbane. *USP* status: 1820–1945. CU. Rafinesque describes henbane as "narcotic, phantastic, phrenitic [i.e., causing frenzied, maniacal behavior], anodyne, antispasmodic, repellent, discutient, &c." Henbane is official in the Austrian, Belgian, British, European, French, German, Greek, Italian, Dutch, and Portuguese pharmacopoeias. It has found use in tachycardiac arrhythmia and glaucoma (*British Herbal Compendium* 1992: 131–32; *British Herbal Pharmacopoeia* 1990: 60).

53. Aquifoliaceae. *Ilex opaca* Ait. Common name: American holly. *USP* status: none. HU. Said to possess the same qualities as *I. Aquifolium* L. Used in "catarrh, pleurisy, smallpox, and gout" (Wood and Bache 1880: 1670).

54. Illiciaceae. *Illicium floridanum* Ellis. Common name: purple anise/tree anise. *USP* status: none. HU. A possible substitute for anise (*Pimpinella anisum* L.), an aromatic carminative (Wood and Bache 1880: 1671).

55. Berberidaceae. Rafinesque's name: *Jeffersonia bartoni.* Current name: *Jeffersonia dyphylla* (L.) Pers. Common name: twinleaf. *USP* status: none. HU. Reported to be emetic in large doses and tonic or expectorant in small doses (Wood and Bache 1880: 1681).

Fig. 16.3. "Samuel Thompson claims . . . to have discovered the properties of this plant towards 1790; but the Indians knew some of them; it was their puke weed, used by them to clear the stomach and head in their great councils" (Medical Flora: *II, 23).*

The Medicine and Medicinal Plants of Rafinesque

56. Cupressaceae. *Juniperus communis* L. Common name: common juniper. *USP* status: 1820–1940. CU. Rafinesque lists the properties of the common juniper as "stimulant, diaphoretic, diuretic, carminative [a mild laxative], anthelmintic, emmenagogue [*sic*], &c." He also warns that "pregnant women ought never to use them [junipers]; but they are very useful in dropsical complaints, menstrual suppressions, also in rheumatism, gout, worms, &c. in powder, conserve, or tincture." These uses (excepting "dropsy") accord with the current literature (Foster and Duke 1990: 226). Although juniper can have a harmful effect on the kidneys, it was carried in the *National Formulary* until 1960 and is still recommended by German Commission E for indigestion (Tyler 1994: 77). Juniper is official in the Austrian, French, Hungarian, and Swiss pharmacopoeias (Martindale 1996: 1717).

57. Ericaceae. *Kalmia latifolia* L. Common name: mountain laurel. *USP* status: none. HU. Reportedly "antisyphilitic, sedative to the heart, and somewhat astringent" (Felter and Lloyd 1909, II: 1094).

58. Asteraceae. Rafinesque's name: *Leontodon taraxacum*. Current name: *Taraxacum officinale* G. H. Weber *ex* Wiggers. Common name: dandelion. *USP* status: none. CU. Rafinesque lists the properties of dandelion as "deobstruent, diuretic, hepatic, subtonic, corroborant, aperient [i.e., a laxative or mild cathartic], &c." The leaves are rich in vitamins A and C. "Experimentally, [the] root is hypoglycemic, weak antibiotic against yeast infections *(Candida albicans)*, stimulates flow of bile and weight loss" (Foster and Duke 1990: 130). *T. officinale* is available in four proprietary and numerous multi-ingredient preparations (Martindale 1996: 1757). [Rafinesque's woodcut can be seen in fig. 17.5.]

59. Scrophulariaceae. Rafinesque's name: *Leptandra purpurea*. Current name: *Veronicastrum virginicum* (L.) Farw. Common name: Culver's-root. *USP* status: 1820, 1860–80. CU. Rafinesque writes of Culver's root, "The root alone is medical; it is bitter and nauseous, has never been analyzed, and is commonly used in warm decoction as purgative and emetic. . . . The safest way is to use it in weak and cold infusion. Employed also for rheumatism, spasms, and bilious complaints." Listed currently as a cathartic (Merck 1996: 928).

60. Campanulaceae. *Lobelia inflata* L. Common name: Indian tobacco. *USP* status: 1820–30. CU. Also known as puke weed, the "virtues" of this emetic plant were touted by Samuel Thomson and many of the nineteenth-century botanico-medical movement. Rafinesque referred to lobelia as a "powerful and efficient emetic, narcotic, expectorant, antispasmodic, suvorific [i.e., sudorific] diuretic, anti-asthma[t]ic, and sialagogue [i.e., causing profuse salivation]." Experimental study suggests a bronchodilation effect in guinea pigs that might be reproducible in humans (Werbach and Murray 1994: 84–85). Official in the Austrian pharmacopoeia and available in numerous multi-ingredient preparations (Martindale 1996: 1552).

311

61. Lamiaceae. *Lycopus virginicus* L. Common name: bugleweed. *USP* status: 1830–70. CU. Rafinesque describes bugleweed as "an excellent sedative, subtonic, subnarcotic, and subastringent." "Science has confirmed the potential value of this plant in treating hyperthyroidism" (Foster and Duke 1990: 70), although "the principles responsible for bugle weed's antithyrotropic function have not been identified" (Tyler 1994: 140–41).

62. Magnoliaceae. *Magnolia macrophylla* Michx. Common name: great-leafed magnolia. *USP* status: none. HU. Described as equivalent to the medicinal properties of other *Magnolia* spp.; used as a tonic for dyspepsia and typhoid fever (Felter and Lloyd 1909, II: 1226–28).

63. Menyanthaceae. Rafinesque's name: *Menyanthes verna*. Current name: *Menyanthes trifoliata* L. Common name: buckbean. *USP* status: 1820. CU. Rafinesque describes the properties of the buckbean or bogbean as "tonic, stomachic, febrifuge, purgative, asthritic [arthritic], diaphoretic, anthelmintic, &c." "Science confirms phenolic acids may be responsible for bile-secreting, digestive tonic, and bitter qualities" (Foster and Duke 1990: 14). It is one of the commonly encountered "minor bitter herbs" (British herbal compendium 1992: 41–42; British herbal pharmacopoeia 1990: 24–25; Tyler 1994: 45). Buckbean is official in several European pharmacopoeias (Martindale 1996: 1725).

64. Lamiaceae. Rafinesque's name: *Monarda coccinea*. Current name: *Monarda didyma* L. Common name: Oswego tea/bee-balm. *USP* status: none. HU. Described as related to *Monarda punctata* L. used as a "stimulant, carminative, sudorific, diuretic, and anti-emetic" (Felter and Lloyd 1909, II: 1275).

65. Brassicaceae. Rafinesque's name: *Nasturtium plaustra*. Current name: *Rorippa palustris* (L.) Bess. Common name: yellow cress. *USP* status: none. HU. Used in visceral obstructions (Wood and Bache 1880: 1707).

66. Nelumbonaceae. Rafinesque's name: *Nelumbium luteum*. Current name: *Nelumbo lutea* Willd. Common name: yellow nelumbo/water-chinquapin. *USP* status: none. NUR.

67. Nymphaeceae. *Nymph[a]ea odorata*. Common name: fragrant water-lily. *USP* status: none. HU. Listed as "astringent, demulcent, anodyne, and antiscrofulous" (Felter and Lloyd 1909, II: 1318).

68. Oxalidaceae. *Oxalis acetosella* L. Common name: wood-sorrel. *USP* status: none. HU. Described as "cooling and diuretic. Useful in febrile diseases, hemorrhages, gonorrhoea, chronic catarrh, urinary affections, and in scurvy" (Felter and Lloyd 1909, II: 1424).

69. Ericaceae. Rafinesque's name: *Oxycoca macrocarpa*. Current name: *Vaccinium macrocarpon* Ait. Common name: cranberry. *USP* status: none. CU. Among the medicinal properties of the cranberry Rafinesque lists its "anti-putrid [i.e., preservative]"

qualities and its usefulness in fevers. This is undoubtedly due to its ability in urinary tract infections "to prevent the microorganisms from adhering to the epithelial cells that line the urinary tract" (Tyler 1994: 80). Clinical studies suggest that cranberry extract may be of value in treating not only urinary tract infections but also in reducing ionized calcium in the urine (Werbach and Murray 1994: 206–7, 232).

70. Polygonaceae. Rafinesque's name: *Oxyria reniformis*. Current name: *Oxyria digyna* (L.) Hill. Common name: mountain sorrel. *USP* status: none. NUR.

71. Araliaceae. *Panax quinquefolius* L. Common name: American ginseng. *USP* status: 1840–70. CU. Rafinesque considered American ginseng to possess "nearly the same properties" as the Chinese and other varieties. He ascribed a wide range of curative and tonic properties to ginseng by indicating that "it renovates the vital spirits" and "invigorates old people" by removing "all the disorders of weakness and debility." "Research suggests it may increase mental efficiency and physical performance, aid in adapting to high or low temperatures and stress (when taken over an extended period)" (Foster and Duke 1990: 50). There are differences between various species "but in general, their effects are similar" (Tyler 1994: 172). It is considered a general tonic and is available in fifteen proprietary products and in numerous multi-ingredient preparations (Merck 1996: 751–52; Martindale 1996: 1710–11)

72. Rubiaceae. Rafinesque's name: *Pinckneya pubens*. Current name: *Pinckneya bracteata* (Bart.) Raf. Common name: fever tree. *USP* status: none. HU. Listed among the "important medicinal plants of this family." Suggested as a possible equivalent to the cinchonas (Millspaugh 1887, 1: 76–72).

73. Berberidaceae. Rafinesque's name: *Podophyllum montanum*. Current name: *Podophyllum peltatum* L. Common name: mayapple. *USP* status: 1820–1940, 1960– . CU. *Podophyllum montanum* is a Rafinesque invention. The mayapple was a popular remedy with eclectics; so frequently substituted where allopathic practice indicated calomel (mercurous chloride) that it received the common name "vegetable calomel." Rafinesque considered it "one of the best native cathartics," but warned that "two ounces of the leaves in decoction killed a dog." *P. peltatum* has current official status in the U.S. (*USP* XXIII 1995: 1238–39). "Etoposide, a semisynthetic derivative of this plant, is FDA-approved for testicular and small-cell lung cancers" (Foster and Duke 1990: 46).

74. Capparidaceae. Rafinesque's name: *Polanisia graveolens*. Current name: *Polanisia dodecandra* (L.) DC. Common name: stinkweed/false mustard. *USP* status: none. NUR.

75. Polygalaceae. *Polygala paucifolia* Willd. Common name: milkwort/flowering wintergreen. *USP* status: 1820–1920. CU. Rafinesque lists the properties of flowering wintergreen as "similar" to seneca snakeroot *(P. senega)*, describing it as

"stimulant, sudorific, restorative, &c." The therapeutic equivalence of these two species has been long established (Felter and Lloyd 1909, II: 1746). "It may be used in tea or decoction: being milder than either [wintergreen or seneca snake-root]; it may be very useful when the Senega would be too stimulant, and it may perhaps answer all its effects in asthma, rheumatism, dropsy, &c." The methyl salicylate found in the root of *P. senega* (also a constituent of wintergreen) "suggests a rationale behind use of this plant's root to relieve pain, rheumatism, etc." (Foster and Duke 1990: 72). Official in Austrian, Belgian, British, European, French, German, Italian, Japanese, Dutch, Portuguese, and Swiss pharmacopoeias (Martindale 1996: 1074).

76. Polygonaceae. *Polygonum aviculare* L. Common name: knotweed. *USP* status: none. CU. Rafinesque refers to knotweed or knotgrass as "astringent, vulnerary, diuretic, subtonic, &c." *Polygonum* spp. are still put to these therapeutic uses today, but it is used especially as a remedy in upper respiratory complaints. Knotweed has active ACE (angiotensin converting enzyme) inhibitors (Wichtl 1994: 386). It is an active ingredient in Antussan-Kombi, Dr. Boether Bronchitten S., Elisir Depuativo Ambrosiano, Silphoscalin, and Tussiflorin (Martindale 1996: 1806, 1945, 1958, 2282, 2353).

77. Polypodiaceae. Rafinesque's name: *Polypodium vulgare*. Current name: *Polypodium virginianum* L. Common name: common polypody. *USP* status: none. HU. Used as a purgative and in upper respiratory complaints (Wood and Bache 1880: 1736).

78. Monotropaceae. *Pterospora andromedea* Nutt. Common name: giant bird's nest. *USP* status: none. NUR.

79. Pyrolaceae. Rafinesque's name: *Pyrola maculata*. Current name: *Chimaphila maculata* (L.) Pursh. Common name spotted pyrola/ground holly. *USP* status: none. HU. Described as similar to *C. umbellata* (L.) W. Bart.—"diuretic, tonic, and astringent" (Wood and Bache 1880: 265).

80. Ranunculaceae. *Ranunculus acris* L. Common name: common buttercup/crowfoot. *USP* status: 1820, 1840–70. CU. Rafinesque treats *R. acris* and *R. bulbosus* (and several other species) as equivalent in therapeutic action: "They act on the skin as rubefacient [i.e., a mild counterirritant] and escharotics [i.e., caustic]." These two species have been described as "possessing similar properties" (Felter and Lloyd 1909, II: 1638). "The plant, especially the roots, has been used for cancers, especially of the breast, cervix, etc., corns, warts, and wens, from Chile to California" (Duke 1985: 400). The plant contains protoanemonin, a toxin which produces contact dermatitis. When the plant is dried, however, it yields anemonin, a Gram-positive/Gram-negative antibacterial (Duke 1985: 400; Bruneton 1995: 597).

81. Rutaceae. *Ruta graveolens* L. Common name: common rue/garden rue. *USP* status: 1830–80. CU. Rafinesque describes the properties of the common rue as "antispasmodic, deobstruent, stimulant, heating, rubefacient, and blistering, useful in spasmodic affections, hysteria, hypoc[h]ondria, obstructions, obstructed secretions; also in rheumatism of the joints, feet, and loins, applied externally." Rue oil is used in homeopathic preparations (Martindale 1996: 1749).

82. Gentianaceae. Rafinesque's name: *Sabbatia angularis.* Current name: *Sabatia angularis* (L.) Pursh. Common name: American centaury. *USP* status: 1820–70. HU.

83. Papaveraceae. *Sanguinaria canadensis* L. Common name: bloodroot. *USP* status: 1820–1920. CU. Rafinesque lists the medicinal properties of bloodroot as "an acrid narcotic, emetic, deobstruent, diaphoretic, expectorant, vermifuge, escharotic, and at the same time stimulant, tonic." He considers sanguinaria "one of the most valuable medical articles of our country." This member of the poppy family has had a long history of medicinal use. Currently it is used in plaque-inhibitor toothpastes and mouthwashes such as Viadent (Bruneton 1995: 745; Foster and Duke 1990: 48; Martindale 1996: 2374). It is also an ingredient in Lexat, an Australian product for digestive disorders (Martindale 1996: 2082).

84. Lamiaceae. *Scutellaria lateriflora* L. Common name: skullcap. *USP* status: 1860–1910. CU. Rafinesque describes the medicinal properties of skullcap as "used chiefly of late, in all nervous diseases, convulsions, tetanus, St. Vitus' dance, tremors, &c." He refers to the plant as a proven "tonic, astringent, antispasmodic, and anti-hydrophobic [rabies]." "Recent studies indicate that it may possess anti-inflammatory activity related to its ability to inhibit the enzyme sialidase" (Lawrence review, s.v. "Scullcap," Jan. 1993). Scutellarin, a constituent of scull-cap, "has confirmed sedative and antispasmodic qualities" (Foster and Duke 1990: 186). Scutellarin is an ingredient in several herbal preparations to relieve tension and stress (Martindale 1996: 841, 2028, 2153, 2193, 2240).

85. Liliaceae. Rafinesque's name: *Sigillaria multiflora.* Current name: *Polygonatum biflorum* var. *commutatum* (J. A. & J. H. Schultes) Morong. Common name: Solomon's seal. *USP* status: none. HU. Used internally for piles and externally for certain skin irritations (Wood and Bache: 1627).

86. Solanaceae. *Solanum dulcamara* L. Common name: nightshade/bitter-sweet. *USP* status: 1820–90. CU. Rafinesque lists bittersweet as "depurative [purifier], deobstruent, antiherpetic [i.e., remedy for ringworm], narcotic, diuretic, anodyne, repellent, &c." Today bittersweet has been shown to possess "significant anti-cancer activity" and is used as a base material for steroids (Foster and Duke 1990: 182).

Michael A. Flannery

Fig. 16.4. "[F]amous as a cure and prophylactic against hydrophobia.
This property was discovered by Dr. Vandesveer, towards 1772 . . .
and is said to have . . . prevented 400 persons and 1000 cattle
from becoming hydrophobous, after being bitten by mad dogs"
(Medical Flora: *II, 82*).

87. Loganiaceae. *Spigelia marilandica* L. Common name: Carolina pink/
pinkroot. *USP* status: 1820–1910. CU. Rafinesque describes pinkroot as "narcotic,
vermifuge, sedative, cathartic, and febrifuge." *S. marilandica* is still considered
anthelmintic (Merck 1996: 1496). Although it has toxic effects similar to strych-
nine, it continues to be used homeopathically (Duke 1985: 456).

88. Rosaceae. *Spir[a]ea tomentosa*. Common name: hardtack/steeplebush.
USP status: none. HU. "Spirea is tonic and astringent, and may be used in
diarrhoea, cholera infantum, and other complaints . . ." (Wood and Bache
1880: 869).

89. Plumbaginaceae. Rafinesque's name: *Statice tomentosa*. Current name:
Limonium nashii Small. Common name: marshroot/marsh rosemary. *USP* status:
1820–70. HU.

90. Boraginaceae. *Symphytum officinale* L. Common name: comfrey. *USP*
status: none. CU. Rafinesque lists comfrey as "beneficial in dysentery, nephritis,

316

haematuna [hematuria?], hemoptysis [i.e., bleeding from the lungs], strangury [i.e., difficulty and pain in urination], and many other diseases internally, while externally they are useful bruised and applied to ruptures and sprains." Rafinesque's claims for the topical application of comfrey are undoubtedly due to its allantoin content, which promotes healing (Foster and Duke 1990: 180). Despite the presence of toxic pyrrolizidine alkaloids (Tyler 1994: 158–59), comfrey is still found in several proprietary and multi-ingredient preparations (Martindale 1996: 1694).

91. Liliaceae. Rafinesque's name: *Trillium latifolium*. Current name: *Trillium flexipes* Raf. Common name: birthroot. *USP* status: none. HU. Virtually all the Trilliums were considered of medicinal value. Primarily used as an astringent, tonic, and antiseptic (Felter and Lloyd 1909, II: 1997).

92. Asteraceae. Rafinesque's name: *Tussilago frigida*. Current name: *Petasites sagittatus* (Banks *ex* Pursh) Gray. Common name: coltsfoot. *USP* status: none. CU. Rafinesque fails to make clear taxonomic differentiations here. The plant he describes in this section along with the accompanying plate is clearly *Petasites sagittatus*. There is, in fact, *Tussilago petasites* or butterbur (Torkelson 1996) used as a "spasmolytic with analgesic effects" (Wichtl 1994: 367). Yet in Rafinesque's discussion of the medicinal properties of "*T. frigida*" [*sic*] he cites Samuel Henry's *American Family Herbal* (1814), which refers to *T. farfara*. Since Rafinesque does not make a therapeutic distinction between any of these plants, *T. farfara* (the most commonly utilized) is retained for purposes of discussion. Rafinesque states that coltsfoot is commonly used in "coughs, complaints of the breast and lungs, asthmatic affections, [w]hooping cough, and also scrofula: either in tea or decoction, conserve or powder." It is found in several European cough remedies such as Antibron and Médiflor Tisane Pectorale d'Alsace (Foster and Duke 1990: 130; Martindale 1996: 1803, 2107). Like comfrey, however, coltsfoot contains toxic pyrrolizidine alkaloids (Tyler 1993: 95–96).

93. Pontederiaceae. Rafinesque's name: *Unisema deltifolia*. Current name: *Pontederia cordata* L. Common name: pickerelweed. *USP* status: none. NUR.

94. Scrophulariaceae. *Veronica beccabunga* L. Common name: speedwell/brooklime. *USP* status: none. CU. Rafinesque writes of *Veronica* spp., "They all purify the blood and humors, act as mild stimulants, strengthen the stomach, promote diuresis, and are said to correct the secretions of the liver, so as to remove melancholy or hypochondrical affections." *V. beccabunga* and *V. officinalis* "possess somewhat similar properties" (Felter and Lloyd 1909, II: 2058). This member of the figwort family finds numerous uses in Europe that include its application in gout, rheumatism, jaundice, coughs, asthma, other lung ailments, and as a "blood purifier" (Foster and Duke 1990: 174).

95. Fabaceae. *Vicia faba* L. Common name: broad bean. *USP* status: none. NUR.

96. Rutaceae. Rafinesque's name: *Xanthoxylon fraxineum.* Current name: *Zanthoxylum americanum* P. Mill. Common name: prickley-ash. *USP* status: 1820–1930. CU. Rafinesque writes, "[I]t is sialagogue, stimulant, pellant ["repelling the morbid fluids"], astringent, sudorific, antisyphylitic, odontalgic [i.e., a toothache remedy], &c." *Zanthoxylum* spp. have been used for toothache, rheumatic disorders, and circulatory insufficiency and as a diaphoretic (*British Herbal Compendium* 1992: 177; *British Herbal Pharmacopoeia* 1990: 72–73; Martindale 1996: 1768).

97. Scrophulariaceae. *Chelone glabra* L. Common name: balmony/turtlehead. *USP* status: none. HU. Described as "tonic and aperient" (Wood and Bache 1880: 1612).

98. Rubiaceae. *Galium verum* L. Common name: true bedstraw. *USP* status: none. CU. Rafinesque considers *G. verum* and *G. aparine* to be therapeutically equivalent. All the Galiums possess "similar medicinal virtues," although *G. verum* contains more galitannic acid than does *G. aparine* (Felter and Lloyd 1909, II: 909).

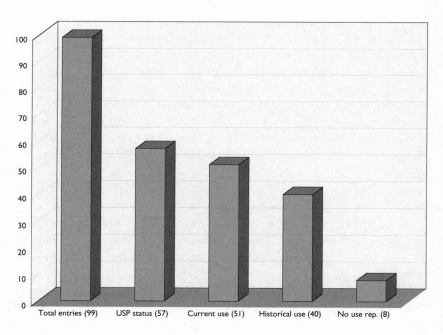

Fig. 16.5. USP *status and historical use of Rafinesque's* Medical Flora. *Courtesy of the New York Botanical Garden Press.*

"Externally applied in poultice," he writes, "it is a good discutient [i.e., heals skin sores] for indolent tumors, strumous [i.e., goitrous or scrofulous] swellings and tumors of the breast. Internally it is used in decoction sweetened with honey, for suppression of urine and gravely complaints [i.e., kidney stones], in scurvy, dropsy, hysterics, epilepsy, gout, &c." It has been shown experimentally to possess hypotensive and anti-inflammatory properties (Foster and Duke 1990: 36). It is also considered a mild diuretic (*British Herbal Compendium* 1992: 61–62).

99. Vitaceae. *Vitis* spp. L. CU. Rafinesque devotes nearly sixty pages to grapes and wine. He suggests that wine be given to "temperate persons" for debility, scrofula, scurvy, rickets, anemia, vaginal discharge, digestion, and as a heart stimulant. Rafinesque's praise of wine as a medicinal is in keeping with centuries-old practice. In Biblical times Paul told Timothy, "Drink no longer water, but use a little wine for thy stomach's sake." There is some evidence that moderate wine intake boosts high-density lipoproteins, thus improving serum lipid levels. "A large body of evidence has accumulated regarding the benefits of moderate wine intake in the management of a variety of diseases" (Lawrence review, s.v. "Wine," July 1993).

Conclusions

The findings of the preceding list are illustrated in fig. 16.5 and suggest several conclusions. First, if Rafinesque's work was as unreliable as some of his contemporaries insisted, one would expect to find a much higher number of plants with no use reported and few with pharmacopoeial status. With the vast majority of the *Medical Flora* plants (91) showing some medicinal use at some time, this is clearly not the case. Rafinesque's list includes 48 plants that had some pharmacopoeial status in the 1820 and 1830 *USP* editions before or during his writing. On this basis the regular physicians could hardly claim that the author was engaging in wild speculation on untried species. It is true that Rafinesque could be rather expansive in his various therapeutic claims (one of the most remarkable being that "good wines" were of such medicinal value that "they may restore to life new born and very weak children, likely to die, by merely rubbing it on their stomach"!); but it is also true that many of the medicinal properties that he described were in keeping with the practices of his times. A second conclusion borne out by this study is that over half of the medicinal plants (51) described by Rafinesque are still used today, a surprisingly high number that casts a more favorable light on the vegetable materia medica of Rafinesque. Finally, there clearly is no correlation between official *USP* status (past or present) and current medicinal use. Indeed, many of the species monographed in the *Medical Flora* show no *USP* status but current usage. Those who seek to prospect for phytomedicines in this old and venerable compendium are restricting themselves to a very small portion of the historic vegetable materia medica.

This does not suggest that Rafinesque's *Medical Flora* can or should be used as a therapeutic guide today. Yet, as Bart K. Holland and others have recently indicated of certain European texts, it can perhaps be used for "prospecting"— systematic phytopharmaceutical research of selected species. Rafinesque's *Medical Flora* is especially valuable for medicinal plant investigators in the United States because he primarily focused upon indigenous American species (over 80 percent of his total list) and he understood the importance of ethnobotanical considerations in delineating plants with therapeutic promise. The implication, of course, is that considerably more work needs to be done in the area of medicinal plant research. The requests of numerous scholars in the field for the establishment of a comprehensive plan for this to occur in the United States, for the most part, remains to be filled (Foster 1993; DerMarderosian, Tyler, and Blumthenthal 1996; Tyler 1994: 17–31). The findings of this study suggest that the *Medical Flora* still may be useful for phytomedicinal inquiry today. Indeed, based upon their past and present medicinal use, most of the plants listed in Rafinesque's compendium would make viable candidates for funded in vitro and clinical studies.

References

Berman, Alex. 1952. C. S. Rafinesque (1783–1840): A Challenge to the Historian of Pharmacy. *American Journal of Pharmaceutical Education* 16: 409–18.

———. 1956. A Striving for Scientific Respectability: Some American Botanics and the Nineteenth-Century Plant Materia Medica. *Bulletin of the History of Medicine* 30: 7–31.

Boyle, Wade. 1991. *Official Herbs: Botanical Substances in the United States Pharmacopoeias, 1820–1990.* East Palestine, Ohio: Naturopathic Press.

British Herbal Medicine Association. 1992. *British Herbal Compendium: A Handbook of Scientific Information on Widely Used Plant Drugs.* Dorset: BHMA.

British Pharmacopoeia Commission. 1990. *British Herbal Pharmacopoeia, 1990.* Dorset: BHMA.

———. 1993. *The British Pharmacopoeia.* 2 vols. London: HMSO.

Bruneton, Jean. 1995. *Pharmacognosy. Phytochemistry, Medicinal Plants.* Trans. Caroline K. Hatton. Paris: Lavoisier.

Bynum, W. F. 1994. *Science and the Practice of Medicine in the Nineteenth Century.* New York: Cambridge Univ. Press.

Drake, Daniel. [1830] 1970. The People's Doctors: A Review by "The People's Friend." Pp. 195–202 in Henry D. Shapiro and Zane L. Miller, eds., *Physician to the West: Selected Writings of Daniel Drake on Science and Society.* Lexington: Univ. Press of Kentucky.

The Medicine and Medicinal Plants of Rafinesque

Duke, James A. 1985. *CRC Handbook of Medicinal Herbs*. Boca Raton, Fla.: CRC Press.

Eisenberg, David, et al. 1993. Unconventional Medicine in the United States, *New England Journal of Medicine* 328: 246–52.

Felter, Harvey Wickes, and John Uri Lloyd. 1909. *King's American Dispensatory*. 19th ed., 4th rev., 2 vols. Cincinnati: Ohio Valley Co.

Fernald, Merritt Lyndon. 1950. *Gray's Manual of Botany*. 8th ed. New York: American Book Co.

Foster, Steven. 1993. *Herbal Renaissance: Growing, Using and Understanding Herbs in the Modern World*. Salt Lake City: Gibbs-Smith.

Foster, Steven, and James A. Duke. 1990. *A Field Guide to Medicinal Plants: Eastern and Central North America*. Boston: Houghton Mifflin.

Griffith, R. Eglesfeld. 1832. On *Frasera walteri*. *Journal of the Philadelphia College of Pharmacy* 3: 272–73.

Haag, H. B. 1941. Rafinesque's Interests—A Century Later: Medicinal Plants. *Science* 94: 403–6.

Hanley, Wayne. 1977. *Natural History in America: From Mark Catesby to Rachel Carson*. New York: New York Times Book Co.

Hershenson, Benjamin R. 1964. A Botanical Comparison of the United States Pharma-copoeias of 1820 and 1960. *Economic Botany* 18: 342–56.

Holland, Bart K. 1996. *Prospecting for Drugs in Ancient and Medieval Texts: A Scientific Approach*. Amsterdam: Harwood Academic Publishers.

Horine, Emmet Field. 1961. *Daniel Drake (1785–1852): Pioneer Physician of the Midwest*. Philadelphia: Univ. of Pennsylvania Press.

Kartesz, John T. 1994. *A Synonymized Checklist of the Vascular Flora of the United States, Canada, and Greenland*. 2nd ed., 2 vols. Portland, Ore.: Timber Press.

King, Nydia. 1987. *A Selection of Primary Sources for the History of Pharmacy in the United States*. Madison, Wisc.: AIHP.

Konoshima, Takao, and Kuo-Hsiung Lee. 1986. Cicutoxin, an antileukemic principle from *Cicuta maculata*, and the cytotoxicity of the related derivatives. *Journal of Natural Products* 49: 1117–21.

The Lawrence Review of Natural Products. Facts and Comparisons. 1989– . St. Louis, Mo.: Lippincott.

DerMarderosian, Ara, Varro E. Tyler, and Mark Blumenthal. 1996. Milestones of Phar-maceutical Botany. *Pharmacy in History* 38: 15–28.

Martindale: The Extra Pharmacopoeia. 31st ed. 1996. Ed. James E. F. Reynolds. London: The Pharmaceutical Press.

Merck. 1996. *The Merck Index: An Encyclopedia of Chemicals, Drugs, and Biologicals*. 12th ed. Rah-way, N.J.: Merck.

Merrill, Elmer D. 1949. *Index Rafinesquianus: The Plant Names of C. S. Rafinesque with Reductions, and a Consideration of His Methods, Objectives, and Attainments*. Jamaica Plain, Mass.: The Arnold Arboretum of Harvard Univ.

Mignone, Cherie. 1975. Materia Medica in Early American Literature. *The Cornell Plantations* 31: 3–7.

Millspaugh, Charles F. 1887. *American Medicinal Plants: An Illustrated and Descriptive Guide to the American Plants Used as Homeopathic Remedies.* 2 vols. New York: Boericke & Tafel.

Packard, Francis R. 1931. *History of Medicine in the United States.* 2 vols. New York: Paul B. Hoeber.

Rafinesque, C. S. 1828–30. *Medical Flora, or Manual of Medical Botany of the United States of North America.* 2 vols. Philadelphia: Atkinson & Alexander (1828); Samuel C. Atkinson (1830).

———. 1829. *The Pulmist, or Introduction to the Art of Curing and Preventing the Consumption or Chronic Phthisis.* Philadelphia: The Author.

———. [1836] 1944. A Life of Travels. *Chronica Botanica* 8: 294–360. Reprint of a work first published as *A Life of Travels and Researches in North America and South Europe.* Philadelphia: F. Turner.

Schlectendal, D. F. L. von. 1835. Review of Medical Flora by C. S. Rafinesque. *Litteratur-Bericht zur Linnaea für das Jahr 1834:* 95–100.

Schweinitz, Lewis David von. [1832] 1921. The Correspondence of Schweinitz and Torrey. Ed. C. L. Shear and Neil E. Stevens. *Memoirs of the Torrey Botanical Club* 16: 269–71.

Torkelson, Anthony R. 1996. *The Cross Name Index to Medicinal Plants.* 3 vols. Boca Raton, Fla.: CRC Press.

Tyler, Varro E. 1993. *The Honest Herbal: A Sensible Guide to the Use of Herbs and Related Remedies.* 3rd ed. New York: Pharmaceutical Products Press.

———. 1994. *Herbs of Choice: The Therapeutic Use of Phytomedicinals.* New York: Pharmaceutical Products Press.

———. 1996. Pharmacognosy! What's That? You Spell It How? *Economic Botany* 50: 3–9.

United States Pharmacopeial Convention. 1995. *The United States Pharmacopeia* XXIII. National Formulary 18. *USP* Convention, Rockville, Md.

Weaks, Mabel Clare. 1945. Medical Consultation on the Case of Daniel Vanslyke. *Bulletin of the History of Medicine* 18: 425–37.

Werbach, Melvyn R., and Michael T. Murray. 1994. *Botanical Influences on Illness: A Sourcebook of Clinical Research.* Tarzana, Calif.: Third Line Press.

Wichtl, Max. 1994. *Herbal Drugs and Phytopharmaceuticals: A Handbook for Practice on a Scientific Basis.* Trans. Norman Grainger Bisset. Boca Raton, Fla.: CRC Press.

Wilder, Alexander. 1904. *History of Medicine.* Augusta, Maine: Maine Farmer Pub. Co.

Wood, George B., and Franklin Bache. 1880. *The Dispensatory of the United States of America.* 14th ed. Philadelphia: Lippincott.

Rafinesque's Sentimental Botany: The School of Flora

BEVERLY SEATON

BETWEEN JANUARY 1827 and September 1832, Constantine Rafinesque published ninety-six short descriptions of plants, each one illustrated, in the pages of two magazines from the same Philadelphia publisher—*The Saturday Evening Post*, a weekly periodical, and *The Casket; or Flowers of Literature, Wit, & Sentiment*, which appeared monthly. He called these pieces "The School of Flora." In his introduction to the "School," he named his audience, "the youths of both sexes," and explained that he would be writing a kind of "floral history" of "pretty Flowers and useful Plants, such as are deemed ornamental and economical (both American and European)." Further, he promised that "every flower will be considered a moral emblem" (*Casket,* Jan. 1827, 32). These short articles constitute Rafinesque's main excursion into the field of sentimental botany, one aspect of the popularization of botany in the nineteenth century.

The study of botany as a pastime became popular in the late eighteenth century in Europe, and its appeal continued well into the late nineteenth century. Genteel Americans were quick to follow the lead of Europeans in this as in other cultural trends, and so the historian of popular culture in nineteenth-century America notes a proliferation of botany books for children and women (those perpetual amateurs). There was a general attitude that knowledge of botany was a necessary component of an accomplished person. A little poem published in *Punch* and reprinted in *The Garden* (Feb. 12, 1881) satirizes the concept of botany as one of a young lady's attractions to the opposite sex:

What reck I though she be fair
If the flowers are not her care;
If she ponder not upon
Many a Dicotyledon; . . .

She shall calmly learn to state,
Clover is tri-foliate;
And describe in words exact,
Awn and axis, blade and bract;
So shall I in her sweet presence,
Find my love hath inflorescence. (194)

Botany became sentimentalized as it broadened its appeal beyond that small circle of science-minded persons (of both sexes). The sentimentalization of botany in Europe has its origins in France, in the writings of Rousseau and Bernardin de St. Pierre, and in Germany, in the work of Novalis and Goethe. It was quickly capitalized on by lesser-known popularizers like Stephanie de Genlis and Charles Malo. One of the typical "botanical works" of the early century in France was Victorine de Chastenay-Lanty's *Calendrier de Flore* (1802–3), a mixture of botanical description, flower legends, and personal emotion which set the tone for countless other flower books of the period (Seaton 1983). These French books had an influence throughout the rest of Europe and in America as well.

Rafinesque naturally had a professional interest in the popularization of botany. At Transylvania University, much of his income depended on public lectures. In his introductory lecture *On Botany* (probably given there in the spring of 1820), he made every effort to relate the study of botany to practical concerns. He looked ahead to the breeding of better varieties of vegetables and made an eloquent plea for the conservation of maples (Rafinesque 1983: 11), similar to that made in Cooper's novel *The Pioneers*, published in 1823. While his appeal to his Lexington audience does not draw on sentiment at all, he knew the uses of sentimentalization. He wrote in a letter of December 1, 1819, to Zaccheus Collins, "I am patronized by the Ladies, and must endeavor to please them by telling them pretty things, rather than by displaying too much learning" (Dupre 1945: 16). He no doubt drew on this experience with female pupils in preparing the "School," which combines the appeal to utility with the appeal to the emotions.

The two magazines in which the "School of Flora" appeared are both famous in the history of Philadelphia publishing. Both magazines were published from the same office by Samuel C. Atkinson and Charles Alexander, who founded the *Saturday Evening Post* in 1821 and the *Casket* in 1826. Charles Alexander sold out his holdings in the *Post* to Samuel Atkinson in 1828, and Atkinson himself sold the magazine in 1839 (Mott, IV: 671–76). At the time Rafinesque published in the *Post* it had four pages and looked somewhat like a newspaper. [Its format can be seen here in fig. 15.3.] Most of its contents were lifted from other publications, especially the short items, as was the custom of the times. Rafinesque's little essays usually appeared on a page with advertisements for such products as farm equipment and patent medicines (including Pulmel, Rafinesque's own elixir), along with notices for lost horses, cows, and apprentices. Atkinson and Alexander's monthly, the *Casket*, was fairly successful under its founders, but when it was purchased in 1839 by George Graham and changed its name to *Graham's Magazine*, it became extremely popular (Mott, I: 544–46). As befitted a monthly magazine, the *Casket* was more pretentious than the *Post*. It had fashion notes from London, some original American short stories and poems, engravings of such important buildings as Congress Hall at Saratoga Springs, embroidery patterns, and in general more content aimed at the female reader. Sometimes Rafinesque's pieces in the *Casket* were followed by brief extracts on plants or gardening from other sources.

Fig. 17.1. Gilbert woodcut, used in the Casket *and* Saturday Evening Post, *June 1828.*

Beverly Seaton

One of these, from February 1828, explains that to "increase the odour of roses," one should plant a large onion just where it will touch the roots of the rose bush.

During most months of 1827, the first year of the "School," four pieces appeared in each *Casket* and, in roughly the same month, the same pieces appeared weekly in the *Post*.[1] For May and December, there were only three pieces, and there were none at all in October. The 1827 pieces were numbered consecutively, and there are forty-two. In January 1828 a "New Series" was announced, and thenceforth only two appeared each month in the *Casket*, also appearing in the *Post* in most cases (the exception is April).

There are twenty-four pieces in this Series, bringing the total of the "School" essays at the end of 1828 to sixty-six. That marks the end of the "School" pieces that have moral emblems for each flower. There was no "School" in 1829, and in February 1830 there appeared an item announcing another "New Series," which, however, was not new, but which was taken from the first volume of the already published *Medical Flora* (published by Atkinson & Alexander). From then on, each piece was announced as being from the *Medical Flora*, and indeed the material which appeared throughout 1831 and 1832 at the rate of one each month in each publication is simply taken right from the second volume of the *Medical Flora* (published now by Atkinson alone). Strictly speaking, then, the "School of Flora" as a separate entity in Rafinesque's work ceased in 1828, although his publishers continued to use the old title for the material. So while there are ninety-six pieces in the entire "School" (five of which are the same plant with a different description), it can easily be seen that as an endeavor distinct from the *Medical Flora*, the "School" contains only sixty-six plants.

One of the attractions of the "School of Flora" is its original plates published from woodcuts. Rafinesque got good mileage from his plates, using several of them in the "School," the *Medical Flora*, and an 1832 pamphlet titled *The American Florist*. In his day-book for January 1, 1832, he listed ownership of 200 woodcuts which he valued at $2,400. He had possession of 100, with the rest in the hands of S. C. Atkinson, as "belonging to Medical Flora." The present location (if any) of these woodcuts is unknown; some of those held by Atkinson were sold and used in later publications by others.[2] Generally speaking, the illustrations for the first sixty-six plants, those in the original "School of Flora," are less scientific in appearance than the ones used also in his *Medical Flora*. Some are very close to conventional flower pictures, those with no botanical pretensions (for example, the *Primula auricula* and the "pansey"). There seems to have been no attempt to make the drawings uniform in style even when three or four of the plants were pictured together.

Rafinesque's Sentimental Botany

EUPATORIUM PERFOLIATUM.

Fig. 17.2. Used in the Casket *and* Saturday Evening Post, *February 1830, and also in the first volume of the* Medical Flora.

In the *Medical Flora* (I: x), Rafinesque writes that "most of the figures have been drawn by the author, and a few reduced from Bigelow or Barton"; he does not state who cut the woodblocks, nor does the engraver's name appear on any of the prints.[3] Many were copied from several early sources, either exactly, with minor changes, or as mirror images. The last type suggests the use of a camera lucida. The sources named by Rafinesque were Jacob Bigelow's *American Medical Botany*, William P. C. Barton's *Vegetable Materia Medica*, and Barton's *Flora of North America*.

From Bigelow, vol. I (1817), Rafinesque made an exact copy of *Eupatorium perfoliatum;* a near copy of *Hyosciamus niger;* while his *Kalmia latifolia* is a partial copy and also a mirror image. From vol. II (1818), Rafinesque made an exact copy of *Asclepias tuberosa* and *Gentiana catesb[a]ei;* a near copy of *Statice caroliniana, Panax quinquefolium,* and *Apocynum androsemifolium;* and a rendering of *Cassia marilandica.* From vol. III (1820), he copied *Ilicium floridanum* with a very slight change; *Nymphea odorata* reversed and with a slight change; *Xanthoxylon fraxineum* much simplified; and the bottom part only of *Aristolochia serpentaria.*

From W. P. C. Barton's *Vegetable Materia Medica,* vol. I (1817), he copied *Gillenia stipulacea* with some changes, and he copied *Sanguinaria canadensis* as a mirror image with a few minor changes. From the same work, vol. II (1818), he borrowed the top portion of *Aristolochia serpentaria* to go with the bottom portion from Bigelow, as well as making simplified versions from Barton's *Cunila mariana* and *Spigelia marilandica*.

From Barton's *Flora of North America*, vol. I (1821), he made an exact copy of *Hibiscus speciosus*, *Oxycoccus macrocarpon*, *Scutellaria lateriflora*, *Aquilegia canadensis*, and *Pyrola maculata*, with a simplified mirror image of *Andromeda arborea*. There do not appear to be any borrowings from vol. II. From vol. III (1823), the exact copies are his *Gentiana saponaria*, *Nuttall[i]a grandiflora* (called *Bartonia ornata* in Barton), *Cypripedium luteum*, and *Ipomea quamoclit*, while his copies with slight changes are *Hepatica trilobata*, *Cephalanthus occidentalis*, and *Chelone glabra*.[4] There are numerous differences in nomenclature between Rafinesque and these sources, which I have not tried to sort out at all. For use in the "School," the book plates had to be reduced in size, of course, and sometimes the orientation of the plant had to be changed somewhat to fit. He used borrowed material for the early "School" pieces as well as for his *Medical Flora* series.[5]

While all of the plates after the end of 1828 also appeared in the *Medical Flora*, along with the complete accompanying exposition, Rafinesque also used the

Fig. 17.3. Used in the December 1830 Casket, *the 1 January 1831 issue of the* Saturday Evening Post, *and the first volume of the* Medical Flora.

Rafinesque's Sentimental Botany

Fig. 17.4. Auction woodcut, used only in the 1832 broadside titled American Florist.

plates in a less well-known work, the *American Florist,* which is simply a printed collection of some of the plates he had in his possession. It appeared in two editions in 1832, numbers 606 and 607 in Fitzpatrick, one with eighteen figures, another with thirty-six. I have studied the edition showing thirty-six plates, of which only ten are not found in the "School." In some cases Rafinesque shows different names with the illustrations, though, having changed the names of some plants between the first publication of the plate and 1832. Plants in the *American Florist* which do not appear also in the "School" are *Achillea ptarmica, Arctium latifolium, Viola canina, Saxifraga granularis* [i.e., *granulara*], *Barbarea alliaria, Chrysanthemum leucanthemum, Stellaria holostea, Melissa officianalis, Buplevrum* [i.e., *Bupleurum*] *rotundifolium,* and *Cardamine pratensis.*

As far as the contents of the "School" expository articles are concerned, there is a great difference between the early pieces and those from the *Medical Flora.* With few exceptions, every entry in the *Medical Flora* follows this standard pattern: Names and Classification; Description; History; Properties. Some of them have another section, Qualities. In most articles, the Properties part takes up the most space, although this varies. Sometimes Rafinesque speaks about only one species. Other times he describes various other species, as in his piece on *Nymph[a]ea odorata* he describes also *N. rosea, N. maculata,* and *N. spiralis.* The other species are usually considered part of the history. These *Medical Flora* pieces have no moral emblems, nor do they contain floral legends or other such material. While the appeal of the *Medical Flora* essays is to usefulness, and thus they would have some popular attraction, their audience is clearly the medical profession. This is a great change from the early pieces, which were addressed to young people of both sexes.

Because fully one-third of the "School" pieces are specifically from the *Medical Flora,* one would expect to find that plants of a useful nature far outnumber other sorts of plants in the entire series. And that is the case. The "School" contains both American and European plants, and Rafinesque often discusses both American and European species in the same article. It also includes a number of plants noted mainly for their garden use or their beauty in the wild. Dividing the plants very generally into American plants of mainly economic interest, European plants the same, American plants of an ornamental nature, and ornamental European plants, we can classify Rafinesque's plants in the "School" as follows:

American, economic interest	39
European, economic interest	12
American, ornamental	12
European, ornamental	24

This adds up to ninety-one different plants (there are five duplicates in the "School"). Plainly, Rafinesque's main interest was in the economic uses of plants, a regard that should not surprise us. However, he was not blind to their beauty or their uses in gardens, and he even recommended the planting of various wild flowers in gardens. Among the popular nineteenth-century plants included in the "School" are the tulip, pansy, periwinkle, moss rose, primrose, water-lily, dandelion, and ivy. There are seven trees and a few shrubs included, also. Among the pretty and familiar American wildflowers, he chose the columbine, yellow ladies' slipper, blue gentian, hepatica, mountain laurel, buttercup, bloodroot, and a lesser-known species of trillium. But of course many of the great flowers of the world are missing from the "School," especially the lily and the violet. And there are no daisies, another popular Victorian flower (no matter exactly what plant they meant by the name). Most popular writers about flowers of the period chose flowers for their associations and their popularity, but obviously Rafinesque had other criteria. Side by side with the moss rose and the pansy are comfrey and agrimony. One reader wrote in to the magazine suggesting the inclusion of the sensitive plant *(Mimosa pudica),* but although in Rafinesque's response to the letter he discussed various plants which might be called by that popular name, he did not include any of them in the series (*Casket,* May 1827, 195).

Rafinesque's Sentimental Botany

When we look at the contents of his early, sentimental "School" pieces, we find an interesting mixture of material, reminding us of Rafinesque's exotic background and his social accomplishments. Each little essay begins by giving the botanic name, the English name, and the French name of the plant. The plant is described and perhaps the meanings of its Latin names are explained. In keeping with his interest in the medical uses of plants, properties and qualities of the plants often appear. Sometimes—very seldom in fact—he gives bits of floral history, such as a brief account of the tulip craze of the seventeenth century. More frequently he has suggestions for garden uses of the plant. In his piece on the strawberry, he not only recommends a strawberry bed in every garden but also advises that more people ought to grow "monthly" strawberries, perhaps the ancestors of what we call "everbearing" strawberries today. He usually gives the emblematic meaning of the plant in a brief sentence, although sometimes he ventures on a short explanation of the reason for the emblem.[6]

Perhaps the most charming aspect of these essays is his emphasis on picturing the plant habitat. Sometimes we are given information on where the plant can be found growing, such as "on the Ohio and all the western streams" as well as "easterly on the banks of the Potomoc" (this is *Jeffersonia binata*, American twinleaf). Such descriptions give us a mental image of Rafinesque tramping the woods

Fig. 17.5. Auction woodcut, used in the Casket *and* Saturday Evening Post, *March 1832, and also in the second volume of the* Medical Flora.

of eastern North America, perhaps on a fine April afternoon. Other descriptions call to mind the younger Rafinesque, the cosmopolitan European, as when he explains that betony "grows all over Europe" and is "an essential border plant, like hyssop and sage" in European gardens. There are occasional little descriptions of scenes, such as of children picking buttercups or blowing dandelion seeds into the spring breezes or lovers exchanging a few pansy blossoms or a bloom of the herb truelove *(Paris quadrifolia)*. He suggests burnet for cattle feed and warns children against eating buttercups or green apples.

There is not much similarity between his material and that of the usual sentimental botany. The average writer of such books was either a journalist seeking to capitalize on a popular genre or a gentlewoman finding a way to combine her botanical and literary interests. In these "School" pieces, slight as they are, we see Rafinesque the scientist, the field naturalist, writing the kinds of remarks about plants that make sense to him. In contrast, in the average sentimental botany there would be brief botanical description, then much material on floral legends or associations and much quoting of poetry. Rafinesque never quotes poetry in the "School" series, but he does so in a little essay on *Melissa officinalis,* clearly intended to be part of the "School" (it has the moral emblem) that never appeared in either the *Post* or the *Casket,* but in his own *Atlantic Journal & Friend of Knowledge* (Fitzpatrick 1982: No. 621). Nineteenth-century readers seemed to value the quoting of little snippets of poetry arranged around a theme. For instance, in a piece about the violet, the typical editor might quote snatches of various poems mentioning the violet. Another common way to work in the poetry quotations was to assign the flower a meaning, then quote poetry concerning the meaning. While some of the writers described habitats so as to picture a mid-century rural ramble in the woods, the average flower books of this kind make flowers into a passive, sedentary interest.

The moral emblems in Rafinesque's little essays accompanying the woodcuts represent an interesting aspect of sentimental botany, the language of flowers. First of all, we need to realize that, in the nineteenth century, the word "moral" was applied far more widely than it is today. It referred to any sort of quality or characteristic of people, rather than ethical matters alone. In fact, it is the attachment of these emblems to his flower essays in 1827 and 1828 that make "The School of Flora" so obviously a work in the sentimental mode. In linking flowers to various qualities, situations, and states of mind, he was following one of the major forms of nineteenth-century sentimentalization of flowers. Given the history of the language of flowers, it seems entirely appropriate that Rafinesque should be its first American source—and I think he was that.

Rafinesque's Sentimental Botany

The language of flowers has a fairly complicated history (Seaton 1980, 1995). Of course flowers have been used as symbols or personified in other ways to represent human concerns as far back as we can go in western literary history. Sometime late in the eighteenth century, flowers began to play a more important role in nature description in poetry and became poetic subjects themselves in a new and different way. Along with that, flowers began to be personified in all sorts of moral and didactic poems, from John Langhorne's *Fables of Flora* (1771) to Maria Montolieu's *The Enchanted Plants* (1800). Also in the eighteenth century, European travelers to Turkey came back with stories of the *sélam*, a communication system in which objects, tied up in a scarf or handkerchief, formed a message. Romantic Europeans misunderstood this concept, which is described in two popular travel books, Aubrey de la Mottraye's *Travels through Europe, Asia, and into Parts of Africa* (1723) and Mary Wortley Montagu's *Turkish Embassy Letters* (1763, written about 1717), thinking of it finally as a language of flowers and imagining it to be a secret code language of lovers. The famous Austrian orientalist Joseph Hammer-Purgstall set them straight, or tried to, pointing out that the *sélam* is simply a game. After all, if everyone knows the significance of the objects and a young man has to convey a rather bulky package consisting of such items as fruit, flowers, stones, vegetables, and pieces of charcoal to a forbidden young woman, isn't he running a much greater risk of being caught than if he were carrying a simple note? Hammer-Purgstall reminded his readers that the *sélam* uses flowers among many other things, and the meanings are sound-related rather than symbolic (Hammer-Purgstall 1809, I: 32–42). But in many minds, a connection between the mysterious East and a secret language of flowers had somehow been made.

The symbolic language of flowers as we see it develop in the nineteenth century has its origins in France, during the first two decades of the century. In 1819 it was put into the form that made it so popular, in *Le Langage des Fleurs* by "Charlotte de Latour" (a pseudonym). After that year, it spread rapidly into Germany, England, and ultimately America. There were also many French imitations. There is no single language of flowers. The meanings compiled in the Latour book had the most influence, but every country had its own most popular version, and there were many variations. What is fairly consistent is that the language of flowers is the language of the love affair.

Simply defined, the language of flowers is a list of flowers (often including flowering trees, shrubs, and sometimes fruits and vegetables) attached to a list of meanings. The vocabulary, as it was called, enabled lovers to communicate with one another through bouquets, nosegays, or just single blossoms. There was an elaborate grammar of the language created, so that one could make "written" messages as well as direct ones. As far as I can tell, however, the language of flowers was just as much a game in the nineteenth century as the *sélam* was in the Turkish harems, for there was not much actual use of it. It was mainly a genteel amusement for women, developing into various actual games, such as a floral fortune-telling game (there were different versions).

The first two American language-of-flower books both appeared in 1829: Elizabeth Gamble Wirt's *Flora's Dictionary* was the first, followed by Dorothea Dix's *Garland of Flora*. *Flora's Dictionary* was much more popular and influential than Dix's short work, going through numerous editions. But while honors for the first American language-of-flower book go to Mrs. Wirt, Rafinesque's emblems in the "School" began fully two years before either of these books appeared, making them, to my present knowledge, the first set of flower emblems published in America and written by an American (or one soon to become an American). Rafinesque's Turkish birth, his French-German parentage, and his professional identity as a botanist make him almost an emblem, himself, of the language of flowers.

Floral emblems were not scorned by all botanists of the times. In those early days of its popularity, the language of flowers was viewed as a harmless, wholesome, and amusing aspect of botanical and horticultural interests. Mrs. Lincoln Phelps added a language of flowers vocabulary to her *Familiar Lectures on Botany* (1829) after the publication of the two American books, while the Rev. John Stevens Henslow, the distinguished British botanist, professor of botany at Cambridge and teacher of Charles Darwin, wrote parts of a sentimental botanical emblem book called *Le Bouquet des Souvenirs* (1840).

When we look at Rafinesque's emblems themselves, we see that they are a bit different from an ordinary vocabulary list. The main difference is that they are very negative, while most language-of-flower vocabularies stress positive meanings, allowing some negatives. In fact, Rafinesque's emblem set rather parallels his other writings on women, and perhaps even his own rather sad experiences with women. While in Sicily, he formed an alliance with a woman whom he could not marry because he was Protestant; and when he lost his fortune in a shipwreck off Long Island, she left him for another man, taking their child with her. His long poem *The World* has an entire section (187–200) on women, revealing a chivalrous yet detached attitude. He pictures a few bad women, those who create unhappy homes:

> But there also, if wicked temper comes
> In mood perverse, to drive the happy scene;
> There may arise a hell of doleful woes.
> If rare, it happens still, and men made sore,
> Reject the ties of love, to hatred changed. (195)

Rafinesque's Sentimental Botany

In his drawings of friends, published by Harry B. Weiss in 1936, Rafinesque also included some unpleasant women. In the pictures many of the women are surrounded by flowers, some of them emblematic. Weeping willow and bittersweet, carefully labeled, surround the image of "Juliet," who is further described in a verse,

> I knew her in the prime of her beauty and youth:
> When she was the chaste emblem of candor and truth.
> But alas! what a change! . . .

The set of floral emblems in "The School of Flora" contains a number of flowers meaning bad temper, including bittersweet, "inconstancy and bad temper." In all, forty-two of the meanings are negative, while twenty-four are positive. And of the positive ones, many are qualities not especially related to love: industry, gratefulness, strength, wealth, vernal delight, safety, practical philosophy, utility, thrifty and wary state, cordiality, good sense, favorable opportunity. Many of these in fact seem very commercial in meaning. The positive romantic meanings are few—only one dozen in all. Perhaps most telling of all is Rafinesque's emblem for

Fig. 17.6. One of Rafinesque's Kentucky friends.

true love—which traditionally is the myrtle, symbol of Venus. For this important meaning, Rafinesque chose the herb paris or herb truelove *(Paris quadrifolia),* a plant of the trillium family. He assigns this meaning on the strength of the European folk custom of lovers presenting it to their beloved. However, as he also explains, we know the herb paris of folklore in quite another association; it is the traditional symbol of good luck, in our culture usually represented by the four-leafed clover.

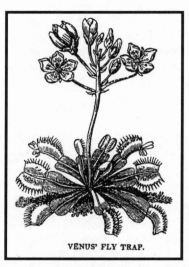

VENUS' FLY TRAP.

Fig. 17.7. Gilbert woodcut, one of the four illustrations that launched "The School of Flora" in the January 1827 number of the Casket. *This woodcut also appeared in the 6 January 1827 issue of the* Saturday Evening Post, *but nowhere else.*

Rafinesque often makes an attempt to explain his choice of meaning for a particular flower. Of the sixty-six emblems, only two are drawn from tradition, a variety of the rose (moss) representing female beauty and the ivy meaning friendship. In other words, he ignored popular meanings, such as the common equation between the pansy and thought. Some meanings relate to plant appearance, as *Andromeda arborea* (the sorrel tree) means tall nymph or lass of the hills; while others relate to the plant's effect, for instance, *Arum dracunculus* means ferocity, apparently because its leaves "have been compared to the claws of a dragon." Other meanings relate to the uses of the plant, as *Crat[a]egus ocycantha* (hedge hawthorn) represents vigilance and security. Some meanings come from properties of the plant, as *Dion[a]ea muscipula* (Venus's fly-trap) is the emblem of caution—understandable

Rafinesque's Sentimental Botany

from the perspective of the fly. *Erysimum alliaria,* or stinking gilliflower, represents hypocrisy and disguised bad temper, obviously because of its odor.

To give a specific example of how Rafinesque's emblem for one popular flower compares with those of other writers, let us look at the tulip. Rafinesque makes it the emblem of pride. In Latour, Wirt, and other writers, the tulip means declaration of love, probably the most popular meaning for the tulip throughout the century. In the work of these writers, other flowers represent pride—amaryllis, tall sunflower, crown imperial, and carnation, among others. But Rafinesque has no flower to mean declaration of love.

His "School of Flora" essays show Rafinesque attempting yet another common pursuit of the early-nineteenth-century botanist, the popularization of botany, imitating in many ways, in others going his own way, as he is so famous for doing in his more scientific projects. They tell us something about Rafinesque the botanist, Rafinesque the illustrator, and Rafinesque the man. As a historian of the language of flowers, I am gratified to find that such a suitable person seems to be the first American to popularize this "gentle science"—even if, in his hands, the love affair takes on the coloring of a Sicilian tragedy.

Notes

1. "The School of Flora" is completely documented in *Fitzpatrick's Rafinesque* (Fitzpatrick 1982), where, however, item 564 should be *Frasera verticillata,* rather than *Cunila mariana.* [See also n. 38 in the next chapter. **Ed.**]

2. Thomas Cooke published twenty in his periodical, *Botanic Medical Reformer* (1840–41), and Wooster Beach, also a Botanico-medical practitioner, published some in the 1847 edition of his *American Practice of Medicine.*

3. [A four-page manuscript in Rafinesque's hand at the American Philosophical Society lists generic names of plants under such titles as "Cuts of School of Flora not used in the Medical Flora" where 29 are attributed to "Auction" and 7 to "Gilbert." Another page lists 16 Auction cuts used mostly in the second volume of the *Medical Flora,* and 14 Gilbert cuts used in both volumes (all of which also appeared in the "School"). Auction has not been identified, but the other engraver was George S. Gilbert, named in the Philadelphia directory as an "engraver on wood," living at 13 South Fourth Street. Mott mentions (I: 545) that early numbers of *The Casket* contained "some unimpressive woodcuts of flowers and of American scenery by G. S. Gilbert." Of the woodcuts appearing here, those of Chrysanthemum (fig. 17.4) and Leontondon (fig. 17.5) are by Auction, those of Ipomea (fig. 17.1) and Dionaea (fig. 17.7) are by Gilbert. A comparison of these four shows that Gilbert was able to execute a more delicate line than Auction. **Ed.**]

Beverly Seaton

4. [When printed sources and, probably, his own herbarium failed to provide images, Rafinesque called on friends. An extant letter (Oct. 30, 1827) to George W. Clinton requests a drawing, four inches high, of *Pterospora*. Clinton did not reply until January 22, 1828, when he sketched in the margin of his letter the bell-shaped flower of the plant and its clawlike roots. In the illustration for *Pterospora andromedea* (*Medical Flora*, II: 68–69) only Clinton's sketch of the flower is used—almost verbatim. **Ed.**]

5. [Rafinesque's Dionaea, shown here as fig. 17.7, was a much borrowed image. He took it from the 1803 edition of Benjamin Smith Barton's *Elements of Botany*, where it had been engraved by Joseph H. Seymour. From there its pedigree traces back to the 1768 engraving by James Roberts for a pamphlet published in 1770 by John Ellis on how to bring viable seeds and plants from the New World, "To which is added, the figure and botanical description of a new sensitive plant, called Dionaea muscipula." It is believed that Ellis drew his representation of Dionaea from a living plant brought to England in 1768 by William Young after seeds sent by John Bartram failed to germinate. **Ed.**]

6. [Left unexplained is why he chose *Nuttal[i]a grandiflora* as the emblem of "presumption." To appreciate the witticism one has to know that fellow botanist Thomas Nuttall would not respond to Rafinesque's letters, had enjoyed financial assistance from Benjamin Smith Barton that Rafinesque envied, had traveled farther and published more discoveries—in short, was too uppity. **Ed.**]

References

Barton, William P. C. 1817–19. *Vegetable Materia Medica of the United States; or Medical Botany: Containing a Botanical, General, and Medical History, of Medicinal Plants Indigenous to the United States. Illustrated by Coloured Engravings, Made after Original Drawings from Nature.* 2 vols. Philadelphia: M. Carey & Son.

———. 1821–23. *A Flora of North America: Illustrated by Coloured Figures Drawn from Nature.* Philadelphia: M. Carey & Sons (vol. I); H. C. Carey & I. Lea (vols. II and III).

Bigelow, Jacob. 1817–20. *American Medical Botany, Being a Collection of the Native Medicinal Plants of the United States: Containing Their Botanical History and Chemical Analysis, and Properties and Uses in Medicine, Diet and the Arts.* 3 vols. Boston: Cummings & Hilliard.

Chastenay-Lanty, Victorine de. 1802–3. *Calendrier de flore; ou, etudes de fleurs d'après nature.* 3 vols. Paris: Crapelet.

De la Tour, Charlotte [pseud.]. 1819. *Le langage des fleurs.* Paris: Audot.

Dix, Dorothea Lynde. 1829. *The Garland of Flora.* Boston: S. G. Goodrich and Co.

Dupre, Huntley. 1945. *Rafinesque in Lexington.* Lexington: Bur Press.

Fitzpatrick, T. J. 1982. *Rafinesque: A Sketch of His Life with Bibliography. Revised and enlarged by Charles Boewe.* Weston, Mass.: M&S Press.

Rafinesque's Sentimental Botany

Hammer-Purgstall, Joseph von. 1809. *Fundgruben des Orients.* 6 vols. Wien: A. Schmid.

Henslow, J. S. 1840. *Le bouquet des souvenirs; a wreath of friendship.* London: R. Tyas.

La Mottraye, Aubry de. 1723. *Travels through Europe, Asia, and into parts of Africa; with Proper Cutts and Maps.* 2 vols. London: The Author.

Langhorne, John. 1771. *The Fables of Flora.* London: J. Murray.

Montague, Lady Mary Pierrepont Wortley. 1763. *Letters . . . Written during Her Travels in Europe, Asia and Africa . . . Which Contain, among Other Curious Relations, Accounts of the Policy and Manners of the Turks.* 3 vols. London: T. Becket and P. A. de Hondt.

Montolieu, Maria Henrietta. 1800. *The Enchanted Plants: Fables in Verse.* London: Thomas Bensley.

Mott, Frank Luther. 1938–68. *A History of American Magazines.* 5 vols. Cambridge, Mass.: Harvard Univ. Press.

Phelps, Mrs. Lincoln. 1829. *Familiar Lectures on Botany. Including Practical and Elementary Botany, with Generic and Specific Descriptions of the Most Common Native and Foreign Plants, and a Vocabulary of Botanical Terms. For the Use of Higher Schools and Academies.* Hartford, Conn.: H. and F. J. Huntington.

Rafinesque, C. S. 1828–30. *Medical Flora, or Manual of Medical Botany of the United States of North America.* 2 vols. Philadelphia: Atkinson & Alexander (1828); Samual C. Atkinson (1830).

———. 1832. *American Florist.* Philadelphia: The Author.

———. 1956 [1836]. *The World or Instability.* Ed. Charles Boewe. Gainesville, Fla.: Scholars' Facsimiles and Reprints.

———. 1983 [1820]. *On Botany.* Ed. Charles Boewe. Frankfort, Ky.: Whippoorwill Press.

Seaton, Beverly. 1980. The Flower Language Books of the Nineteenth Century. *Morton Arboretum Quarterly* 16: 1–11.

———. 1983. French Flower Books of the Early Nineteenth Century. *Nineteenth-Century French Studies* 11: 60–72.

———. 1995. *The Language of Flowers: A History.* Charlottesville, Va.: Univ. Press of Virginia.

Weiss, Harry B. 1936. *Rafinesque's Kentucky Friends.* Highland Park, N.J.: The Author.

Wirt, Elizabeth Washington Gamble. 1829. *Flora's Dictionary, by a Lady.* Baltimore, Md.: Lucas.

The Fugitive Publications of Rafinesque

CHARLES BOEWE

IT IS CONVENTIONAL WISDOM that C. S. Rafinesque set his name to approximately a thousand published books and articles—some of them mere snippets, to be sure; others reprints and translations. Yet, in the last year of his life, this naturalist, who was so immodest that he named three different plant genera in his own honor, listed himself candidly on one of his title pages as merely the "author of 220 works, pamphlets, essays, and tracts."[1] The glaring discrepancy of nearly 800 titles is cleared up by attention to how the Rafinesque bibliography has evolved; such scrutiny also highlights several dozen Rafinesque publications that probably were issued but, more than a century later, have never been found. These fugitive publications, understandably, do not appear in the Rafinesque bibliography; however, there are also a few titles that do appear there which never have been found either.

I

Richard Ellsworth Call compiled the first Rafinesque bibliography in pp. 135–214 of his *Life and Writings of Rafinesque* (1895).[2] His bibliographic work was hampered by the same impediments as his biographic work: a paucity of materials at hand and the need to rely on the assistance of others. Working in Louisville, Call had access to the library of Col. Reuben T. Durrett, which was substantial,[3] but he also had to depend on the willingness of people like G. Brown Goode to transcribe for him titles from the incomplete copy of *Speccio delle Scienze* (1814) at the Smithsonian Institution library, and he listed on the authority of Asa Gray's obituary article

The Fugitive Publications of Rafinesque

some titles of other works he could not see for himself. To his credit, he refused to endorse Gray's allegation that Rafinesque had classified in natural history style "twelve new species of thunder and lightning." "I have read the paper with the greatest care," Call wrote, "more than once." And he concluded that "It is time this fiction was destroyed"; yet, he could not resist adding that Rafinesque's article had no "scientific merit or value." As possessor of an M.D. degree, Call presumably was qualified to declare that *The Pulmist* (1829) "has no scientific medical value"; but he was equally willing, if less qualified, to decree that in the *Ancient Annals of Kentucky* (1824) is "a curious, though useless, ethnological and philological table of the primitive nations and languages." Swimming even farther beyond his depth, when he came to mention the *Genius and Spirit of the Hebrew Bible* (1838), the book in which Rafinesque explores Cabalistic meanings of Hebrew roots, Call could only sputter that "this work is very clearly that of a man who has lost the power of acute perception and correct ratiocination," a book "without the least value from any possible standpoint." Call had so many misgivings about all of Rafinesque's writings outside his own narrow field of interest that one wonders why he ever bothered with them at all.

Call also began the collection of secondary articles about Rafinesque and his works. He disclaimed any attempt at completeness and labeled this short section in his book a "Bibliotheca Rafinesquiana." The bookish term bibliotheca has continued in use by subsequent Rafinesque bibliographers, but not Call's equally pedantic Latinism, "Bibliographia Incerta," to head two items he had overlooked.

Next in succession was Thomas Jefferson Fitzpatrick, a perennial assistant professor of botany at a number of colleges in the Middle West who wound up, still an assistant professor because he never quite managed to complete a doctorate, at the University of Nebraska. Knowing the pension of only seventy dollars a month that he had to look forward to in retirement would not put much food on the table for him and his wife, Fitzpatrick had early decided the best way to prepare for old age was to capitalize on his skill in seeking out rare books likely to appreciate in value. He had whetted this talent working as a collector of local history for the State Historical Society of Iowa, a job that came to an abrupt end when the society discovered their collector was paying more attention to his personal acquisitions list than to theirs. While it lasted, though, the job took Fitzpatrick as far afield as the eastern seaboard, thus making it possible for him actually to examine in libraries Rafinesque texts he might never acquire for himself, books not even seen by R. E. Call. As Fitzpatrick lovingly turned them over, he was meticulous in recording precise bibliographic details: full transcription of titles, size, paper fold, pagination, publication data, type ornaments, etc. He did of course pay more attention to titles of botanical interest, and that was a very narrow interest, as his occasional annotations, focused mostly on taxonomy, do attest.

Like Call, who wrote near the end of his life that he "was down and out,"[4] Fitzpatrick died money poor (his savings account showed a balance of $11.86) but he was book rich. His widow realized $53,000 for the most valuable part of the collection that remained after 65 tons of dross had been cleared away from the accumulation of books and papers filling three houses and part of a barn that had left the Fitzpatricks barely enough room for two cots beside a pot-belly stove in the kitchen.[5]

The third in succession was Elmer Drew Merrill, whose academic career was more circumspect. Before reaching Harvard, where he spent his last years, Merrill had lived twenty-two years in the Philippines. During this period he left his family in Washington for health reasons, and returned after five years to find himself the father of a daughter aged five, whom he had never met. In 1949, when he published his *Index Rafinesquianus,* he created something of a crisis in the world of plant taxonomists by proposing that most of the sixty-seven hundred Latin binomials published by Rafinesque (and first collected in Merrill's *Index*) ought to be credited to their author under the provisions of the International Code of Botanical Nomenclature legislated by the world's botanists themselves. In the prefatory matter of this *Index* he also brought together his additions to the Rafinesque bibliography that had been printed piecemeal in his earlier articles.[6]

And, having revised and enlarged the listings of all the foregoing, I suppose I must include myself as fourth in succession, even if I lack my predecessors' colorful eccentricities. Already cited earlier in this book, my *Fitzpatrick's Rafinesque* (1982), added about a hundred pages of new material (including everything Merrill had discovered) to the complete 1911 text Fitzpatrick had published, which in turn had embodied all of Call's 1895 bibliography. Call and Fitzpatrick had both included what they called a bibliotheca of secondary items about Rafinesque, though neither made any special effort to locate such material. Merrill added to the secondary bibliography—especially items dealing with botany, including early European ones that would have been difficult for his predecessors to locate with their scanty library resources—and, in 1974, Ronald L. Stuckey and Marvin L. Roberts made a concerted effort to bring the secondary bibliography up-to-date.[7]

All the items collected in the foregoing publications—both primary and secondary—as well as additional ones I had turned up, were included in my 1982 revision of Fitzpatrick's book; yet, since then, 207 additional secondary items have been logged, 26 more of Rafinesque's short, original publications have been detected, 3 new Rafinesque items have been printed from his extant manuscripts,

The Fugitive Publications of Rafinesque

and 33 hitherto unlisted reprints or translations have been found—all of which were entered in *Mantissa: a Supplement to Fitzpatrick's Rafinesque* (2001).[8] Since Rafinesque is one of the most written-about American naturalists of the nineteenth century, the secondary bibliography should continue to swell, especially when it is recognized that he had important things to say in many fields unrelated to natural history.

II

To focus on Rafinesque's own published writings, though, the question naturally arises, why has it taken so long to find out what he actually published, with the end still not in sight? Both Call and Fitzpatrick were seriously hampered in gaining access to materials; in addition, both reflected their own scientific interests to the occasional neglect of other areas. Call alluded to, but did not list, three lengthy Rafinesque letters published in a Lexington newspaper.[9] One of these gives minute descriptions of prehistoric earthworks since obscured by the plow or lost under construction; but that had nothing to do with natural history, as Call understood it. Perhaps worst of all, Call decided that Rafinesque's own references to his published works were untrustworthy. This first bibliographer said that it appears to have been a custom with Rafinesque

> to prepare formal papers and forward them to journals, magazines, and societies; this constituted "publication" in his conception, and some of these memoirs and essays are quoted by him as having appeared. It is often quite difficult indeed to separate these titles from those which really were printed; but where specific journals are mentioned by him it is quite easy to make the discrimination. A case in point is a paper on "The Chinese Nations," said by Rafinesque to have been published in the *Knickerbocker Magazine* for 1834. The article never appeared. Similar instances might be multiplied. . . .[10]

As a matter of fact, the article *did* appear, exactly where Rafinesque said it was printed, though a year later than he remembered; and, despite the title, it is an important article for anyone interested in Rafinesque's pioneering studies of Indian languages. Call's dogmatic decree discouraged successors—including Merrill—from continuing the search when a title Rafinesque said he had published did not readily turn up. I shall have more to say later about what Rafinesque meant by "publication"; meanwhile, Call had, perhaps unwittingly, planted the idea that Rafinesque was capable of padding his own bibliography, and that suspicion has not enhanced his reputation among earnest scholars. Merrill added that

"Rafinesque's habit of listing his actually published works as well as his projected ones, . . . while it has helped bibliographers in their search for very obscurely published papers, has at the same time resulted in a considerable loss of time and effort on the part of those who have searched in vain for items that Rafinesque listed as published, or at least as having been submitted . . . for publication."[11] I, too, have endured many of these wild-goose chases, but in the process I have become convinced that in every instance—barring, perhaps, Rafinesque's occasional lapse of memory—there will be an explanation for why the paper was not discovered, when we learn fully to understand the author's signposts. In short, Rafinesque did not willfully mislead us.

On the other hand, Fitzpatrick's work, the most extensive of that of the Rafinesque bibliographers, has been misleading in several respects. A self-trained bibliographer, Fitzpatrick wrote elaborate descriptive details of Rafinesque books, which in turn convey the impression of utmost reliability. At the same time, he listed titles reported to him by others, which he had never seen, and which have never been seen by anyone since. A botanist, he reported motes of taxonomic significance and overlooked beams in other areas of interest. Like all of us, he must sometimes have pushed himself beyond the limits of his attention span, because he occasionally missed an article while examining a long run of a magazine.[12] He missed others because his resources—and their indexes—were not as comprehensive then as now. He was also extremely cautious about attributing to Rafinesque articles published over pseudonyms—for which Fitzpatrick ought to be forgiven, since he lacked any ancillary means to identify the authorship of such writings.

Both Call and Fitzpatrick chose to annotate some, but not all, of the items they listed. Annotation is always a perilous business, since no one can anticipate what information readers will wish to know. Call's annotations are generally innocuous, and occasionally helpful; but some of Fitzpatrick's are woefully inadequate. He could, for instance, append the laconic note "a popular article" to a short piece in which Rafinesque made important observations on botanical geography—one of the areas where he pioneered[13]—and in which he also gave some of the European sources for his extensive herbarium, much of which had been built up by such international exchanges.[14] One would think this kind of information interesting to the botanist Fitzpatrick, but a careful study of his annotations soon reveals that his concern was almost exclusively taxonomic.

Fitzpatrick's greatest disservice to Rafinesque studies results from the numbering system he adopted for his entries. Here, however, the fault was not his but that of all subsequent writers who have turned to the last item in his 1911 book, number 939, have generously rounded up the total to the next hundred, and have then grandly proclaimed that Rafinesque published nearly a thousand books and articles. All Fitzpatrick intended to do was give a serial number to each item, and, even then, he sometimes got the numbers mixed up. Every reprint and every translation (and there were many of these, in both French and German) got a separate number in Fitzpatrick's book. Each volume of a two-volume work got a separate number, even when both volumes were published the same year; volumes published in parts have separate numbers for each part, even when these, too, were all published in the same year. Rafinesque's serial publications, such as the ill-fated *Western Minerva*, got a separate number for the title of the magazine, then each of the individual articles within the magazine also got a number, including those attributed to someone other than Rafinesque! I continued the Fitzpatrick numbering system in the 1982 revision for the convenience of cross-referencing— emphasizing, however, that these are *serial numbers*, not totals. I ended with number 1,001 and would not be surprised to find Rafinesque classed one day with Scheherazade.[15] For reasons developed later, we probably never will know exactly how many titles Rafinesque published; meanwhile, his own estimate of 220 seems a reasonable guess.

III

As Merrill first observed, Rafinesque printed several lists of the titles of his own publications,[16] especially later in life, when he was making a futile attempt to persuade other naturalists to acknowledge his discoveries. Merrill, a botanist himself, was little dismayed that these lists emphasize botanical publications. They give short shrift to Rafinesque's zoological titles, and even shorter shrift to his publications that fall entirely outside the field of natural science.

While still in Sicily, Rafinesque also began the practice of printing lists in the endpapers of his books, both of his earlier publications and of his projected ones. Like most young authors, he must have found it beyond belief that an editor would reject one of his contributions. Thus, among the authorities cited in his 1813 *Chloris [A]Etnensis* [F 234] was Rafinesque's own "Monographie inserite nelle Transazioni della Società Linneana di Londra." The monograph, a thirty-four-page manuscript on the botanical genus *Callitriche*, had indeed been received by the Linnean Society of London in 1812; it is still there, and it has never been published.[17]

Thereafter, Rafinesque became more cautious. His usual practice, as one finds in the list of "Ouvrages et Essais déjà publiés par le même Auteur" in the endpapers of *Précis des Découvertes et Travaux Somiologiques* [F 230], was to include the word *envoyé*—or *sent*, when he was writing in English—before titles of papers he had dispatched but had not actually seen in print. In the *Précis* (1814), he exercises this caution about an essay on European plants naturalized in the United States that he had sent to Samuel L. Mitchill's *Medical Repository* and that had, in fact, been in print almost three years [F 21], though Rafinesque, still in Sicily, did not know this. Because he had learned so well the lesson of waiting to see the published article before he listed it in his bibliography, it appears reasonable to give him the benefit of the doubt when he cites a publication no one else has been able to discover.[18]

On the other hand, a serious bibliographer cannot record a publication unless he himself has seen it. One subject which remained unlisted until *Mantissa* appeared is Rafinesque's writing on a new species of a whalelike creature he named *Balena gastritis,* an accomplishment he must have been proud of, for he mentioned it in at least four different places as one of his published articles, and precisely stated that his account had appeared in the January 1815 number of the Sicilian *Portafoglio*.[19] No such periodical has been found in this country or at the British Museum; but there is a unique copy in Palermo, where, in 1987, Rosario Lentini found not one but two articles on the subject [F 1001, 1004] right where Rafinesque said they would be. On the other side of the coin, Rafinesque never mentioned anywhere his publication in 1818 of a new salamander and of two genera of aquatic worms in a short-lived magazine called the *Scientific Journal* [F 945], though these brief articles were discovered and republished in 1960.[20]

Rafinesque printed three short bibliographies of his own publications. In 1816, after his return to the United States, he distributed a *Circular Address on Botany and Zoology* [F 236]. Here he carefully cited his publications to date, including, in English translation, those appearing in *Specchio delle Scienze* [F 26–229], the periodical which went through two volumes that he had edited, had been its chief contributor, and had published in Sicily. He also added the titles of fifteen unpublished essays lost when he was shipwrecked on his return. The main purpose of the *Circular,* though, was to seek subscribers for two new periodicals he intended to issue. One of these was called *Annals of Nature,* which he continued to try to launch, without success, the rest of his life. In 1820, in Lexington, the first—and only—number actually was published as a sixteen-page pamphlet [F 370]; its lack of continuity has not prevented several libraries from listing the title as a periodical of which they have a broken set.

The other periodical proposed in the *Circular* was to be titled *Somiology of North America* and was intended to include nothing less that the complete flora and fauna of the United States and "the adjacent countries," with every plant and

The Fugitive Publications of Rafinesque

animal drawn by C. S. Rafinesque, "or under his direction, from a living or well preserved specimen . . . and engraved on wood in a plate of 8vo. size" (21). The author believed a total of 5,000 fascicles could be produced in eight to ten years. If the scheme had materialized the fascicles would have been a bargain. For instance, though Wilson's *Ornithology* cost $100, the birds alone of Rafinesque's *Somiology* were going to be made available for "about $20, or $40 if coloured." The beauty of the enterprise was that naturalists could subscribe to whatever sections most interested them. Then Rafinesque made the astonishing suggestion of 115 different ways the material might be divvied up—including, what surely would have been a very thin fascicle, the flora and fauna of "Groenlandica." Needless to say, the project never got off the ground.

This experience, however, did not dampen Rafinesque's enthusiasm for self-publication. Five years later and now settled in Lexington, he issued a flyer titled *Proposals to Publish by Subscription a Selection of the Miscellaneous Works and Essays of C. S. Rafinesque* [F 435]. This leaflet includes twenty-two categories of "selected unpublished tracts," and eighteen titles of tracts "already published," listed because of "being mostly printed in Europe, in the French and Italian languages, or inserted in periodical works of an ephemeral nature; hence they have not been generally known in the United States" (2). A comparison of these titles with those seen by Fitzpatrick shows all the ones actually published during the preceding two or three years in Brussels, Paris, and London are missing from Rafinesque's list. The lack of any notice of these foreign publications enhances again Rafinesque's reliability as a bibliographer of his own writings; he had not seen them, he did not list them. The list does, however, include one item that remains a ghost. This is the supplement said to be part of his "Survey of the Progress and Actual State of Natural Sciences in the United States" [F 251]. The "Survey" is a very creditable performance, one of the earliest such accounts, and was written by a participant personally acquainted with many of the persons discussed. What we have treats of "Collective Improvements and Labours," by which the author meant scientific and learned societies and spoke of the kinds of contributions made by members of different professions, such as physicians and clergymen. The second part, which promised to be "a particular notice of the works of each author" (85) has never been found and probably was never published.

The third major list of his own publications printed by Rafinesque himself causes some problems, problems sufficiently grave that we may question whether he was able to keep to his resolve to cite only articles he had seen in print. This list appears in the concluding number, number 8 [F 777], of another short-lived Rafinesque magazine, the *Atlantic Journal* [F 610–820], published in Philadelphia and dated Winter of 1833. There he listed titles year by year under the heading "Chronological Index of the Principal Botanical Works and Discoveries Published by C. S. Rafinesque" [F 779].

At the very least, entries such as "Botanical letters to Muhlenberg, Brickell, &c" for the year 1804 and "Botanical letters to Decandolle" for the year 1830 are misleading. Nine extant manuscript letters (all dated in 1803) to Henry Muhlenberg are known. Also known are one to John Brickell (dated in 1805), and a series of ten botanical essays written in the form of letters addressed to A. P. de Candolle during 1830 and 1831 as well as several more personal letters to Candolle both before and after this period. Since Candolle did occasionally mention a Rafinesque binomial as a synonym when he was compiling his *Prodromus*, a charitable explanation of this episode is to assume that Rafinesque believed Muhlenberg had also sent notice of Rafinesque's discoveries to his correspondent Carl Willdenow for inclusion in the latter's revision of Linnaeus's *Species Plantarum*, for this is the means by which Muhlenberg's own discoveries had become known. But even this generosity does not justify considering manuscript information sent to the obscure John Brickell as a form of publication.

Perhaps similar tortured reasoning caused Rafinesque to list this work: "1805. Discoveries in North America, Leghorn." No title of his has been found with Livorno as the place of publication. However, on his return to Europe, Rafinesque had to remain aboard ship during a forty-day quarantine period in the port of Livorno and thereafter he stayed in that city from March until he left for Palermo in May of 1805. He tells us in his autobiography that during this period of enforced idleness he sent specimens of American plants to Gaetano Savi (to whom there is an extant letter) in Pisa and to Giuseppe Raddi in Florence; and that he wrote other letters and brought his travel journals up to date. Unmentioned by Rafinesque himself are his extant letters written during the same period to Biagio Bartalini in Siena, the one to John Brickell earlier alluded to, and one to Manasseh Cutler, with whom he had carried on an extensive correspondence while still in the United States—all datelined from "Leghorn." As all of these letters discuss American discoveries, perhaps this flurry of letter writing is what Rafinesque meant by publication; it also is just possible that he did print a pamphlet in Livorno, but, if so, he never mentioned it again and it has not been found.

The 1833 list is further exasperating because Rafinesque sometimes gives there very precise data for a title, then adds a following entry that is hopelessly vague. For instance, he cites his *Chloris [A]Etnensis* [F 234] as published in 1815 in Catania as an appendix to the *Natural History of Etna* by Giuseppe Recupero—which is exactly correct—and follows it immediately by this: "Prodromus of New

The Fugitive Publications of Rafinesque

Genera. Pamphlet" under the same year. He sailed from Palermo in July of that year for the United States and never returned. Perhaps he left a pamphlet in press, somewhere in Sicily; perhaps it was published, perhaps not. We do not know.

An almost parallel situation occurs with this citation: "1825. Neogenyton or 66 N[ew] G[enera of] North Am[erican] plants, pamphlet, Lexington," which is immediately followed by the entry "Neocloris or N[ew] Sp[ecies] of Western America." The year 1825 was a thin one for Rafinesque publications as presently shown by the bibliographies, but the four-page *Neogenyton* [F 474] was indeed issued that year and two other Rafinesque contributions also appeared in Cincinnati's *Literary Gazette* [F 472, 473], though they are not botanical papers. Toward the end of June he traveled east, visiting Washington, Baltimore, and Philadelphia before returning to Lexington late in October, when he stopped to lecture in Cincinnati. He might have published "Neocloris" anywhere along the way, but such a title has not been found. Since the title is not called a pamphlet, an even better hypothesis is that it appeared—if indeed "Neocloris" was ever published—in some periodical, possibly a newspaper. New botanical species of western America were of interest to western newspaper editors. The title does not appear in the well-known Lexington papers, but only the period 1814–15 of Lexington's *Western Monitor* has ever been filmed and that was before Rafinesque came to Lexington. Rafinesque's Harvard-educated friend William G. Hunt became its editor in 1815, and the newspaper ceased publication in April 1825, but the one substantial file, at the Filson Club, has no issues for that final year. The University of Chicago reported scattered issues for 1825, but presently can find none of them. Issues at the Library of Congress reveal nothing. Two January 1825 issues at the American Antiquarian Society contain nothing by or about Rafinesque. At the moment, this clue cannot be pursued further.

Most of the items in the 1833 list can be identified without much trouble, sometimes under a slightly variant title. Occasionally, as was common with Rafinesque, his date was inaccurate by a year—which suggests that in 1833 he was writing from memory, not with the articles before him. Lapse of memory may explain why he now listed the Linnean Society MSS on botany as "pamphlets"; he always thought they ought to be printed, and after two decades perhaps he had persuaded himself that they had been. Yet there are also other remaining enigmas in the list.

For the year 1808, Rafinesque cites two of his articles on American plants published in New York in the *Medical Repository* [F 9, 10] while he was in Sicily. Many European botanists had published in Europe on American plants collected for them by others, but this may be the first time a European botanist had collected

Charles Boewe

his own specimens in America—and published them there. This innovation was quickly noted in Europe, and in less than a year the articles were translated into French by the American consul, David Bailie Warden, and published in Paris in Desvaux's *Journal de Botanique* [F 13, 14]—which Rafinesque also notes. Then he adds that they were additionally reprinted "in Archives of Discoveries," where they have not been found. And after inserting a dash he concludes the year 1808 with: "Icones Nov[um] Plantarum Americ[anum] 40 pl[ates]."

Fitzpatrick included these plates as a Rafinesque publication [F 11], but he cited them only from Rafinesque's own reference in *Specchio delle Scienze*, where they are called "cinquanta figure"—a discrepancy probably explained by the fact that 50 figures appeared on 40 plates. Fitzpatrick also cross-referenced to a unique set of "29 Plates & 46 figures" [F 232] at the New York Botanical Garden, supposing them a fragment of the larger Sicilian publication. One of the extant plates carries a handwritten legend which says, in part: "These Plates never published—only Proofs of Plates lost in 1815. . . . Deposited in the Lyceum [of Natural History] at the foundation in 1817, by the author."[21] Aside from the awkwardness—both for Rafinesque and Fitzpatrick—of listing as a publication an item printed but never distributed, the question arises of how the author could have deposited the proofs if, in fact, he lost everything in his 1815 shipwreck, including the copper plates the proofs were pulled from. The question is worth pursuing because it also casts light on how, later in life, Rafinesque could offer for sale in America books he had published in Sicily. It was just such contradictions that caused many to doubt, according to Call, that Rafinesque had ever been shipwrecked at all.[22]

For one thing, after his return to the United States in 1815, Rafinesque conducted more than one trading venture with his younger brother, Antoine, who had taken the elder brother's position at the American consulate in Palermo and remained in Sicily for several years. The brigantine *Indian Chief,* under the command of Capt. Simeon Price, which brought a Rossier, Roulet, and Rafinesque cargo to Palermo in 1816, could easily have included Rafinesque books when she left on March 17 to return to her home port, New Bedford.[23] For another, while he himself was employed by the consulate, Rafinesque had available communication facilities hitherto unsuspected. Though signed by the consul, Abraham Gibbs, there is an extant letter in Rafinesque's hand dated January 6, 1808—the same year the plates were made—and addressed to the Hon. James Madison, Secretary of State. At the close of this official despatch—still in Rafinesque's hand—is a postscript: "I Beg leave to enclose a Letter for the hon[ora]ble Sam[ue]l Mitchill, Senator."[24] Editor of the *Medical Repository* in New York and one of the founders

The Fugitive Publications of Rafinesque

of the Lyceum of Natural History there, Mitchill is little remembered for his senatorial career in Washington. His receipt of Rafinesque contributions for his magazine through the diplomatic pouch may be among his more notable public acts. It is plausible to believe that, having received proofs of the twenty-nine plates also through this channel, Mitchill returned them to Rafinesque after the latter came back to the United States.

Two other alleged botanical publications recorded in the 1833 list have entailed more than a little fruitless search on the part of Rafinesque's bibliographers. One is the title *Neophyton Botanikon, or N[ew] Plants of N[orth] America,* 1828, which Fitzpatrick included [F 553] on Asa Gray's towering authority. Gray, however, had merely mentioned it as one of several Rafinesque titles he himself had not seen.[25] Rafinesque had published four short pieces under the general title *Neophyton* in the *Cincinnati Literary Gazette* [F 452, 453, 463, 464]—but in 1824, not 1828—and he used "Neophyton" as a subtitle for the second part of his *New Flora of North America* [F 865], but that was published four years later, in 1837 though dated 1836 on the title page. Yet, preserved at the American Philosophical Society are three small notebooks written in the copperplate script Rafinesque used when preparing manuscripts for the press. The title of the first is "Neophyton / or New Plants, Genera & Species from / North America / By C. S. Rafinesque / Book first—Philad[elphi]a 1830—300 numbers"; it consists of 36 pages and is continued as "Book 2d from N. 301–640" (40 pages), which in turn is continued as "Book 3d from N. 641– 1058" (44 pages). It is entirely possible that this MS was published somewhere, though printed copies have never been found. Since many of Rafinesque's later works were issued in editions of a hundred copies or less, and all were ignored by his contemporaries, such a book could literally have disappeared from the face of the earth. Whether printed in 1828 or not, it deserves to be published now because among its entries are some of Rafinesque's best technical descriptions of taxa he either claimed to have discovered or segregated from other genera.

The other equivocal title is *Florula Miss[o]urica, Mandanensis and Oregonensis,* 1817, which Rafinesque flatly calls a pamphlet. Unlike several botanical works he proposed but never wrote, this one must have existed on paper, for in 1817 he was seeking financial assistance to publish a "Florula of Missouri . . . one volume 12mo 40 pages."[26] In his autobiography *A Life of Travels* [F 863] (64), he mentions having later sent the MS "to [William] Swainson in England," adding, "whom I fear never received it"—which probably is true, for it is not among the Swainson papers at the Linnean Society of London. But Rafinesque also kept a copy, because, in 1821, he sent it again, in translation—along with six other essays (none of which has been found)—to Bory de Saint-Vincent for publication in the *Annales Générales des Sciences Physiques* (where it also never appeared).[27] All we can ascertain now of its contents comes from a four-page list of seventy-eight species published in *Herbarium Rafinesquianum* [F 786] under the surprising title "Index of the Florula Mandanensis of Bradbury and Rafinesque, published in 1817 and in 1820, with Notes and additions" [F 801]—unless, of course, copies of the pamphlet were printed but not found since.[28]

Finally, the 1833 list alludes to 20 numbers of "The Cosmonist" published in Lexington in 1822. Numbered serially, these short contributions appeared in the *Kentucky Gazette,* and Fitzpatrick found eleven of them in the file of that newspaper at the Library of Congress. It happens that these were copies of the newspaper Henry Clay received while he was in Washington and never took back to Lexington with him. It is unlikely that the "Cosmonist" series extended beyond number XV (which is in fact the total Rafinesque mentioned elsewhere in a letter to Jefferson),[29] but the last Fitzpatrick could find was one numbered XIV, and the file also was deficient in the issues for February 7, March 7, May 2, and November 7—which should contain, respectively, numbers II, VI, XI, and XV. The newspaper has been filmed three times, but these films also lack the issues for February 7, March 7, and May 2. One, however, filmed at West Virginia University, contains the issue for November 7, and in 1984 Ronald L. Stuckey found there "Cosmonist No. XV"—an important discovery, for this article is Rafinesque's first published account of the Kentucky Yellowwood *(Cladrastis lutea),* since become a popular ornamental tree, which he encountered on the banks of the Kentucky River and for which he devised the generic name that still stands. "Cosmonist" II, VI, and XI are still fugitive.[30]

IV

Among Rafinesque's unpublished papers, including letters, are numerous clues to unlisted titles. For example, at the American Philosophical Society is a four-page handwritten list of his "Original Tracts," several of which are known to remain in manuscript. However, before the titles of several of these the plus sign has been inserted, apparently indicating that these were published after the list was made, for three of the titles so marked have been found and listed by bibliographers. Yet two others marked with a plus have never been found: one is a "Dissertation on the Modern Languages & Dialects of Europe"; the other is on "Poetical Bravura." The same repository has a six-page MS review in Rafinesque's hand of W. P. C. Barton's *Flora of North-America* (1820–23), which is unknown as a Rafinesque

The Fugitive Publications of Rafinesque

publication. This document is puzzling because it is dated "Nashville January 1824"—a city Rafinesque is not known to have visited—but this may be explained by the fact that it also is signed "C. S. Rafinesque & Co." His friend William G. Hunt, who had edited in Lexington the *Western Monitor* (where Rafinesque's writing appeared), moved to Tennessee in 1824 to edit the *Nashville Banner* newspaper. Perhaps he took a copy of the review with him and, adding to its content himself, perhaps he published it. At present we do not know. Equally enigmatic is a two-page review in Rafinesque's hand of his own *Medical Flora* [F 554, 557], also at the American Philosophical Society. Signed with the pseudonym "Linneus," it may have been published somewhere—though such a review has never been listed. We can hope it was published, for every author knows in his heart that the only way to get the review his book deserves is to write it himself.

While it has been in print since 1944, another document full of clues was not intended by Rafinesque to be printed at all. This is his "Catalogue of the Principal Works, Essays & Manuscripts published or written by C. S. Rafinesque" and enclosed with his letter of February 15, 1824, to Thomas Jefferson when he was seeking a position at the new University of Virginia.[31] Several titles given there as published—especially when European publication is alleged—have never been found. However, one of them, listed as "Printed in 1822," is given as "Psyche or the Immortality of the Soul, Lexington," and it very much exists, though it concerns an area of interest seldom associated with Rafinesque. Each prefaced by a few lines of original verse, the five "Psyche" papers were signed "Constantine" and they develop a sustained naturalistic argument for the immortality of the soul. They were answered anonymously by a fundamentalist who signed himself "Philo-Constantine" and contended that Constantine [Rafinesque] had ignored the only consequential evidence for immortality, that of divine revelation. All of these articles appeared in the *Western Monitor* newspaper during the period from February to April 1822, and were detected after the publication of the 1982 Revised Fitzpatrick.[32]

The list of titles prepared for Jefferson also hints at a distinction Rafinesque may have made between printing and publishing. Though not all of them have been found, 49 titles are listed as printed between 1818 and 1823; five others are listed as "Manuscripts read to the Kentucky Institute" (one of which, in fact, was printed later); eight others as "Manuscripts sent for publication" (at least two of which also were printed later); and four others as "Manuscripts already published." This last category consists of courses of lectures, some of which are extant in manuscript, and, as far as we know, never were printed.[33] It appears that in Rafinesque's view they were "published" to fee-paying Lexington audiences made up of persons from both town and gown, at public lectures open to all. If manuscripts read before the Kentucky Institute did not seem to him equally a form of "publication," it must be because the institute was a small, tightly controlled group of Lexington intellectuals, whose meetings were not open to the public. In other words, he may have understood publication to mean literally "to make public," and printing was only one means of doing that.

V

Many of Rafinesque's personal letters are full of publication plans, some of which reached fruition; most did not. Mentioned here are two that seem so positive it is hard to believe they represent mere hope. Writing to John Torrey in 1831, Rafinesque announced that "I have issued just now the first number of a Monthly Journal connected with another attempt for my hist[ory] of America. You will receive it by mail"[34]—as perhaps Torrey did. We have no record of such a journal however, and no other hint has appeared to suggest even its title. In connection with the same area of interest—one little exploited by the naturalist bibliographers—Rafinesque wrote to the Boston bookseller O. L. Perkins in 1834 offering for sale a number of his manuscripts (most of which have since disappeared) as well as several of his printed works. One of the latter was titled *Antiquities of Somerset County,* for which Rafinesque says he had but two copies remaining.[35] No such title has ever found its way into the Rafinesque bibliography.

The earliest epistolary reference to an unknown Rafinesque publication appears in a letter he wrote in Sicily in 1810 to William Swainson, where he tells his English friend that he is enclosing "a Number of our Journal which contains an Article of mine on a New fish."[36] Rafinesque continued to send copies of his publications to Swainson, and a decade later, now settled in Lexington, he enclosed in a letter "one of my Odes, lately published in the newspapers." No clipping has been retained with the letter, nor has the poem been found in any Kentucky newspaper.

Rafinesque's French friend the geologist Alexandre Brongniart kept cryptic notes on his correspondence, though his letters to Rafinesque are no longer extant. In one note where he mentioned his receipt of three known Rafinesque publications he also recorded having received Rafinesque's "Memoir on the trilobites, cyclorites, ditoxasus," which has never been found.[37] Also in French, this time in a letter to his sister, Rafinesque mentioned that he had "undertaken to compile 2 or 3 small works for school, such as a *School of Flora* for children and ladies" and "the *Elements of the Chronology of America* for schools."[38] Nor is the *Elements of Chronology*

The Fugitive Publications of Rafinesque

the only pedagogical title by Rafinesque that may be fugitive. He remarked to Charles Wilkins Short in an 1822 letter that he was enclosing his "plan of a Class of Pupils on General knowledge." If the plan was a printed document it is unknown; there is no record that he ever taught such a class in Lexington.

VI

In the endpapers of Rafinesque's pamphlet *American Manual of the Mulberry Trees* (1839) appears the statement that five numbers of his *New Aurora* were published in 1837. "Aurora" refers to a company he and Charles Wetherill founded to develop a settlement in Illinois which was intended to include a university that remarkably anticipated the land-grant universities of a generation later. Most of the little we know about it comes from the "Second Edition, 1st June, 1832," of the *Plan of the Philadelphia Land Company of Aurora*. No other Aurora pamphlets have been found. This same *American Manual of the Mulberry Trees* itself poses a bibliographical enigma. The only state known [F 894] is dated 1839, yet Rafinesque told the governor of New York in a letter dated March 15, 1836, that he had "published pamphlets on the American Vines & Mulberry trees." Perhaps the pamphlet we know is an unmarked second edition or a reprint of the first.

Equally mysterious are two titles among the "Other Works of Prof. Rafinesque for Sale by him and his Agents in Europe and America" listed in No. 7 of his *Bulletin of the Historical and Natural Sciences* [F 886]. These titles are: "Early history of Pennsylvania, 50 cts., printing" and "The true names in the Bible, D[itt]o." *Bulletin* No. 8 is fugitive and *Bulletin* No. 9—the last known—makes no further mention of these titles.[39] For that matter, Numbers 2 and 6 of the *Bulletin* also are fugitive. While the content of these late publications is intrinsically trivial, the *Bulletin*s do help us understand Rafinesque's efforts to dispose of books and manuscripts during the last five years of his life.

These years saw a flurry of publication on the part of Rafinesque: books, pamphlets, and various broadsides. Since all were published privately, it is hard to square their existence with the widely held belief that Rafinesque died in a garret, in abject poverty, and was buried in Philadelphia's potter's field.[40] The printing bill was paid by Charles Wetherill. One of the earliest of these publications was a reprint with preface (4–8) and notes (147–58) by Prof. Rafinesque of *An Original Theory or New Hypothesis of the Universe* (1750) by Thomas Wright of Durham, issued in Philadelphia in 1837 under the title *The Universe and the Stars* [F 874], where the title page states that it is printed for Charles Wetherill. The existence of such a reprint completely nonplussed the modern editor of Thomas Wright of Durham,[41] and the Philadelphia edition has been passed over in embarrassed silence by all commentators on Rafinesque.

Another of their schemes was the Philadelphia Land Company of Aurora, which planned to take up 66,640 acres of western land; Charles Wetherill was president and C. S. Rafinesque secretary of the company. This resulted in the printing of an eight-page pamphlet titled *Plan* [F 876], of which only the second edition, dated June 1, 1837, is known. Elsewhere[42] Rafinesque seems to be alluding to the same scheme when he says that five numbers of the *New Aurora* were printed in 1837. The same source of funds may have subsidized Rafinesque's printing of a four-page broadside titled *First Scientific Circular for 1837* [F 995], where he informed all the governors, lieutenant governors, and speakers of state legislatures about his ideas on natural history surveys and of the need to build state museums and state libraries; he mentioned[43] a second such a *Circular* for 1838, which has not been found. If the Panic of 1837 had not dampened these development schemes, Wetherill's death the next year probably would have ended them anyway; yet, in the spring of 1838 Rafinesque was planning to travel to Illinois to begin the organization of what he now called the Central University of North America.[44] This accounts for the series of publications—some of them books—"Published for the Central University of Illinois," and probably paid for by the president of the Aurora Land Company, Charles Wetherill. Of course Rafinesque never went, and the university, like the utopian community it was intended to serve, remained a dream. It is interesting, though, that according to the *Plan*, the shining new city where the university would be sited was going to be named Agathopolis. This is bogus Greek in about the same way that Urbana, where the University of Illinois actually was founded in 1864, is spurious Latin.

In addition to the hints and clues found in Rafinesque's own published and unpublished writings, one final avenue to the discovery of his fugitive titles has yielded some small successes. This is by searching the journals published or edited by any of his many associates who might have been persuaded by friendship to give him a forum. By this means it was possible to turn up, before the Revised Fitzpatrick appeared, three Rafinesque contributions [F 988, 990, 993] to the *Bulletin de la Société de Géographie* because of his known friendship with Edme François Jomard, president of the society. Since the publication of the revised bibliography, two additional contributions [F 1020, 1021] were uncovered in the *Bulletin Universel des Sciences et de l'Industrie, publié sous la direction de M. le B[ar]on de Férussac*, whose editor, the French zoologist André Férussac, was also a Rafinesque acquaintance.

The Fugitive Publications of Rafinesque

But if we are still in the dark about the total output of Rafinesque's pen, he himself was equally in doubt. A decade before his death he complained to Candolle (as here translated):

> Finally I have discovered by accident in the Library of the Lyceum [of Natural History, in New York] my Botanical & Zoological memoirs sent in 1819 to Messrs. Cuvier & Blainville, & inserted [in the *Journal de Physique, de Chemie et d'Histoire Naturelle*] the same year. [Their] having neglected to send me a Copy, it has taken *10 years!* before I acquired certitude of their reception and publication; it is very annoying, because thinking these Memoirs lost or neglected I often repeated myself since. . . . This negligence of Mr Blainville has had several regrettable consequences for me, believing my Mem[oirs] lost like so many others, I did not send him any longer my Zoological Discoveries. I accepted in 1820 & 21 the offer of Mr Bory de St Vincent to publish in the *Annales [Générales des Sciences Physiques]* of Brussels. This work was suspended in 1822, and Mr Bory still has in his hands a dozen Memoirs intended for it, of which I do not know what has been done. Can you tell me if any other Zoolog[ical] or Botanical Mem[oirs] have been published in France by me between 1822 & 1825? & which. Not having had any more News of these last Memoirs, I have sent two Memoirs to Mr Thiebaut de Berneaud Secr[etary] of the Linnean Soc[iety] who in 1824 nominated me his correspond[en]t. I have heard nothing![45]

Notes

1. C. S. Rafinesque, *The Good Book, and Amenities of Nature*. Dated January 1840, this is the first—and only known—number of a journal Rafinesque intended to issue quarterly or semi-annually. He died in September of that year.

2. Richard Ellsworth Call, *The Life and Writings of Rafinesque* (Louisville, 1895). This is Filson Club Publication No. 10. The only recent biographical notice of Call is my own sketch in Keir B. Sterling, et al., *Biographical Dictionary of American and Canadian Naturalists and Environmentalists* (Westport, Conn., 1997), 132–33. See also the obituary article by Charles Keyes, "R. Ellsworth Call and Iowa Geology," *Proceedings of the Iowa Academy of Science* 27 (1920): 35–40.

3. For its time and place, this was a notable library; the Filson Club met there, and Theodore Roosevelt used it when researching his *Winning of the West*. When sold to the University of Chicago after Durrett's death in 1913, it amounted to 20,000 volumes, 250 boxes of pamphlets, 200 volumes of atlases and maps, hundreds of runs of newspapers, and

Charles Boewe

thousands of manuscripts. Call mentions that Durrett had the first number of Rafinesque's *Atlantic Journal,* but titles from the remaining seven numbers were transcribed for him by some unnamed person from copies at Harvard. He also remarked that the only copy in Kentucky of *Florula Ludoviciana* was one in Lexington.

4. R. E. Call to Daniel B. Fearing, Apr. 25, 1905. This was Call's explanation for offering for sale his own leatherbound copy of his Rafinesque book—grangerized into two volumes by the addition of tipped-in engravings and a number of manuscript letters—for $135, what it had cost him to make it. Call's letter is itself tipped into the first volume of the two-volume set, purchased by Fearing and presented by him to Harvard University, where it is now preserved at the Houghton Library. The only MS letter by Rafinesque included in the volumes is one Call published in his 1899 reprint of Rafinesque's *Ichthyologia Ohiensis.*

5. I dealt more extensively with Fitzpatrick's virulent bibliomania in the introductory pages (1–18) of *Fitzpatrick's Rafinesque* (Weston, Mass., 1982). For a vivid word picture of the squalor resulting from the bibliographer's addiction, see Robert Vosper, who visited the site and wrote about it in "A Pair of Bibliomanes for Kansas: Ralph Ellis and Thomas Jefferson Fitzpatrick," *Papers of the Bibliographical Society of America* 55 (1961): 207–25. Vosper was librarian at the University of Kansas library, where the most valuable scientific portions of Fitzpatrick's collections wound up.

6. Elmer D. Merrill, *Index Rafinesquianus: The Plant Names Published by C. S. Rafinesque with Reductions, and a Consideration of His Methods, Objectives, and Attainments* (Jamaica Plain, Mass., 1949).

7. Ronald L. Stuckey and Marvin L. Roberts, "Additions to the Bibliotheca Rafinesquiana," *Taxon* 23 (1974): 356–72.

8. I used an earlier version of this essay as an introduction to *Mantissa,* which was issued by the same publisher as *Fitzpatrick's Rafinesque.* One purpose of the essay was to advise any successors about productive avenues as well as dead ends I had found in seeking out Rafinesque's elusive publications, since I expect never again to publish another supplement. And having completed after many years the editing of all the known Rafinesque correspondence but aware that I may never live to see it in print, I introduced in *Mantissa* as an innovation in Rafinesque bibliography a list of these letters showing the repositories that hold each.

9. Addressed to Thomas Jefferson, these are listed as item 957 in the revised (1982) *Fitzpatrick's Rafinesque.* Hereafter, serial numbers of that edition preceded by the letter F are used to identify Rafinesque titles and provide a ready reference to additional bibliographical information about them. Serial numbers in *Mantissa* follow sequentially those of *Fitzpatrick's Rafinesque.*

10. Call, *Life and Writings of Rafinesque,* 136. "The Chinese Nations and Languages" is F 992.

The Fugitive Publications of Rafinesque

11. Merrill, *Index*, 41.

12. A notable example is Rafinesque's review [F 946] of Maclure's *Observations on the Geology of the United States* (1817) in a magazine Fitzpatrick otherwise covered pretty well. In the review Fitzpatrick missed, Rafinesque clearly expressed the concept of index fossils, an understanding which placed him in the vanguard of those who recognized the need for this means of dating geological strata.

13. See Joseph Ewan, *A Short History of Botany in the United States* (New York, 1969), 115–16.

14. The article is "The Cosmonist, No. VIII" [F 441].

15. Those who, from time to time, have sought to place all of Rafinesque's publications on the *Index Librorum Prohibitorum* of science think that he created an Arabian Nights entertainment anyway. Other writers have pronounced learnedly on Rafinesque titles without reading them; Bernard Jaffe, for instance (*Men of Science in America* [New York, 1946], 123), was so carried away by the lilting Latin of Rafinesque's botanical titles *Alsographia Americana* [F 883] and *Sylva Telluriana* [F 885] that he alluded to the books as poetry. With *Mantissa*, the last numbered Rafinesque item becomes 1030.

16. Merrill, *Index*, 40–43. Merrill also conveniently summarized many of Rafinesque's lists of proposed but presumably unpublished works, most of which probably were never written. Although possibly published, eleven of Fitzpatrick's entries were not examined by him and most of them remain doubtful; these are: F 7, 11, 12, 23, 309, 465, 475, 553, 851e, 851f, and 933.

17. The monograph on *Callitriche,* along with four other Rafinesque contributions submitted between 1810 and 1812, are entered in the Linnean Society's "Register of Papers Read," though it appears that Rafinesque did not learn this until a note appeared in the London "Proceedings of Philosophical Societies," *Annals of Philosophy* 4 (1814): 448–49. Later, in New York, he adopted such quasi-publication as a means of announcing some of his discoveries. Among "Tracts of C. S. Rafinesque" in the endpapers of *Florula Ludoviciana* [F 257] are several which were orally "communicated to the Lyceum" of Natural History though never printed. One of the Linnean Society manuscripts has been edited and published by L. B. Holthuis as "Rafinesque's Crustacean Genera *Heterelos* and *Yalomus,*" *Zoologische Medelingen* [Leiden] 59 (1985): 133–47.

18. There is no reason to doubt that Rafinesque actually published the "opuscolo" mentioned in the endpapers of *Analyse de la Nature* [F 235] as *Relazione del concorso per la Cattedra di professore de Botanica nella R[eále] Università di Palermo* when he was seeking an academic chair in Sicily; the title is repeated in French (apparently the pamphlet was bilingual), which adds that it was a "brochure avec une vue du Jardin des Plantes de Palerme et une Notice de plusieurs N[ouvelles] Esp[èce] de Plantes siciliennes." Quite properly, Fitzpatrick did not list the pamphlet, and it has never been found.

19. In a printed letter dated January 25, 1816, addressed to the trustees of the University of Pennsylvania [F 942], where Rafinesque was seeking a position. He had also mentioned it as in press in the endpapers of his *Analyse de la Nature* [F 235], where the specimen of the new species was reported "échouée," in all likelihood on the beach. It was cited again in Rafinesque's *Circular Address on Botany and Zoology* [F 236]. His fourth reference was in an unpublished list (Library of Congress) of his titles dated 1821 and probably prepared when he was seeking a position at the University of Virginia. Strangely enough, "Description of Balena gastrytis" is one of two titles for 1815 listed by S. S. Haldeman in his obituary "Notice of the Zoological Writings of the Late C. S. Rafinesque," *American Journal of Science* 42 (1842): 280–91. This suggests that Haldeman may actually have had a copy of the article, for he also listed several unpublished Rafinesque MSS then in his possession that have not been seen since.

20. By George H. Goodwin Jr., "Unrecorded Papers of Rafinesque and Jacob Green," *Systematic Zoology* 9 (1960): 35–36. Nor did Rafinesque even list all of his botanical contributions. Merrill found one, in the *Actes de la Société Linnéenne de Bordeaux* (1834), and reprinted its substance in "A Generally Overlooked Rafinesque Paper," *Proceedings of the American Philosophical Society* 86 (1942): 72–90.

21. As stated by W. R. G[erard], "Reliquiae Rafinesquianae," *Bulletin of the Torrey Botanical Club* 12 (Apr. 1885): 37–38. These are the plates that excited Croizat's wrath, p. 181.

22. An American newspaper account of the shipwreck verifying Rafinesque's own description of the event is reprinted in *Précis ou Abrégé des Voyages, Travaux, et Recherches de C. S. Rafinesque* (1833), ed. Charles Boewe, Georges Reynaud, and Beverly Seaton (Amsterdam, 1987), 56.

23. "Shipping Register / of the/American Consulate / of / Palermo / or Statement of the Arrivals and / Departures of American Vessels / to & from the Port of Palermo / Begun March 1806," a bound MS volume at the National Archives (R.G. 84), with the title page in Rafinesque's hand.

24. Abraham Gibbs to James Madison; Palermo, Jan. 6, 1808, National Archives (Record Group 59).

25. Asa Gray, "Notice of the Botanical Writings of the Late C. S. Rafinesque," *American Journal of Science* 40 (1841): 221–41. Call saved himself the trouble of puzzling over the bibliographical problem of *Neophyton Botanikon* because he considered Rafinesque's *Atlantic Journal*, where it is listed, a publication that "has absolutely no scientific value" (*Life and Writings of Rafinesque*, 192).

26. Rafinesque to Reuben Haines, October 1, 1817, Univ. of Kansas.

27. Mentioned in a letter [F 413] dated January 7, 1821, that appears in *Western Minerva*. The *Western Minerva*, another abortive attempt by Rafinesque to publish a magazine, is

The Fugitive Publications of Rafinesque

itself a bibliographical oddity. Rafinesque wrote that he saved but three copies when the magazine was suppressed by his Lexington foes before its publication. The only one of these known is a set of proof sheets corrected in his hand and preserved at the Academy of Natural Sciences of Philadelphia. The *Western Minerva* could not be inspected by readers until these proof sheets were printed by photo-offset in 1949; nevertheless, in his 1911 bibliography Fitzpatrick gave the full contents of the magazine [F 375–430], thirty-eight years before they could be read unless one visited Philadelphia. In the letter to Bory, Rafinesque also listed the titles of twenty-one contributions he had sent to Paris in 1820; most of these were published in the *Annales Générales des Sciences Physiques*.

28. Presumably most of the plant specimens came from John Bradbury, who had traveled in the region of their origin as Rafinesque had not. It is believed that Bradbury was so distraught he gave up botany entirely when Frederick Pursh got hold of a duplicate set of his plant specimens in England and "published the most interesting" of them in the appendix of his *Flora Americae Septentrionalis* (1814). In the only book he published, *Travels in the Interior of America* (1817), Bradbury gave a bare list of the names of ninety-eight "rare or valuable plants" of the region, only a few of which, Call discovered (*Life and Writings of Rafinesque*, 190), are identical with the names in Rafinesque's "Index." At the American Philosophical Society is a four-page Rafinesque MS titled "Florula Mandanensis . . . by Mr. Bradbury," where seventy-eight numbered species are mentioned and, unnumbered, ten new genera, which are briefly described. Another four-page Rafinesque MS at the same repository also is titled "Florula Mandanensis"; it lists bare names without descriptions, some of the plants being attributed to Meriwether Lewis and others to Frederick Pursh.

29. Edwin M. Betts, "The Correspondence between Constantine Samuel Rafinesque and Thomas Jefferson," *Proceedings of the American Philosophical Society* 87 (1944): 368–80.

30. The only hope for finding them seems to lie in turning up scattered issues of the newspaper. I have searched for these as far afield as the Österreichische Nationalbibliothek without success. (An unpublished letter reveals that Rafinesque sent some of the "Cosmonist" articles to Karl von Schreibers in Vienna.) The "Cosmonist" articles, which deal with several of Rafinesque's varied interests in natural history, deserve to be collected and reprinted. Each is prefaced by a few lines of original verse, and they were written as one of his several attempts to make natural science popular. See also n. 37.

31. Betts, "Correspondence," 378–79.

32. This newspaper contains several other short Rafinesque publications, now listed in *Mantissa*, as well as scattered notes about his activities. Among the latter is the surest evidence so far found that he actually was awarded an honorary doctorate by a German university— a claim long regarded as evidence of Rafinesque's megalomania. Other Rafinesque contributions unknown to both Call and Fitzpatrick were discovered in Hunt's magazine, the *Western Review and Miscellaneous Magazine*, and were included in the 1982 Revised Fitzpatrick.

33. An exception is *On Botany* (1820), which the Whippoorwill Press (the hobby press of a Transylvania professor) printed in 1983. Since this pamphlet had almost no distribution, it may be questioned whether the title was "published" in any but the most literal sense. Its text was transcribed from Rafinesque's MS introductory lecture on the subject at Transylvania University. The manuscript of *On Botany* belongs to the Academy of Natural Sciences of Philadelphia and, like many of the academy's Rafinesque MSS, is deposited at the American Philosophical Society.

34. Rafinesque to John Torrey, Mar. 6, 1831, Trent Collection, Medical Center Library, Duke Univ.

35. Rafinesque to O. L. Perkins; Dec. 5, 1834, Gray Herbarium Library, Harvard Univ. A further puzzle about this title is that no Kentucky county is named Somerset, and all the other Rafinesque studies of Indian antiquities were based on his own explorations of prehistoric mounds in Kentucky. One such investigation took place in the vicinity of Somerset Creek, but Rafinesque knew very well that that is in Montgomery County.

36. "Our Journal" had to be a periodical published in Palermo, and it was not Rafinesque's own *Specchio delle Scienze*, which had not yet begun in 1810. The fish may have been included also in Rafinesque's *Indice d'Ittiologia Siciliana* being printed in Messina at that time, but there is no way of knowing.

37. Actually he wrote "Mem[oir] sur le trilobites, cyclorite[s] . . . ditoxasus," where the ellipsis sign is in the MS. Rafinesque's "Cosmonist I" in the *Kentucky Gazette* was "On a large Fossil Trilobite of Kentucky." The issue of the newspaper where "Cosmonist II" should have appeared has not been found. It is a reasonable guess that "Cosmonist II" dealt with both Cyclorites and Ditoxasus.

38. His actual words are: "j'ai entrepris de rediger 2 ou 3 petits Ouvrages d'Ecole, teles que une *Ecole de Flore* pour les enfans & Dames—Les *Elemens de la Chronologie d'Amérique* pour les Ecoles." If the chronology book—or more likely, a pamphlet—was printed it has never been found.

The "School of Flora" had a long and complicated publishing history. Short articles under this title, illustrated by woodcuts, ran in the two magazines published by Samuel Atkinson, the *Saturday Evening Post* and the *Casket*, from January 1827 to mid-1832. Some of them were gathered as a broadside and published as *American Florist*; only the "Second Series" (1832), consisting of 18 woodcuts, is known [F 606], but in March of that year Rafinesque wrote to John Torrey that he had published the "first series" of this title in January and that the edition was nearly sold out. He went on to say that he wanted to issue the *American Florist* as a book, a plan he had entertained at least since 1828, when he advised Torrey that he was "making another popular Work[,] School of Flora with 200 figures for $2"—which, if published, and that seems unlikely, has never been found. However, there did appear, also in 1832, a pamphlet version of the *American Florist* [F 607], consisting of 36

The Fugitive Publications of Rafinesque

figures printed one side only on 36 leaves, and which may be series one and series two combined. Rafinesque continued to get good mileage out of the woodcuts, using many of them in his two-volume *Medical Flora* (1828–30). See Beverly Seaton, pp. 363–29, for further details on the publishing history of these woodcuts of plants.

39. It is just possible that "The true names in the Bible" refers to Rafinesque's *Genius and Spirit of the Hebrew Bible* [F 882], a book issued the same year (1838) as *Bulletin* No. 7. However, F 882 is a substantial volume of 264 pp., priced at a dollar. A variant title for the book on Pennsylvania was also mentioned as "under the press" in *Bulletin* No. 5 [F 997]. Called there *Enquiries on the Early History of Pennsylvania*, it is described as "one volume 12mo." It has not been found.

40. See "The Last Days of Rafinesque," chap. 3.

41. Michael A. Hoskin (London, 1971), who could only identify Rafinesque as a person "who describes himself as a Professor of historical and natural sciences" and "has some inkling of Wright's conception of our star system" (29). The reason for the unusual reprint was explained by Wetherill himself in a three-page introduction "To the American Public," where he said that "In as much as I have been for the last two years engaged in the production of a work on the electric theory of the solar system. . . . I do issue this work to demonstrate that I am not singular in my views"; and the back cover stated that "Another work supplemental to this will soon be printed for Charles Wetherill, on the Electric theory of the solar system," but Wetherill did not live to complete it. Perhaps Rafinesque tried to remedy the lack by his publication the following year of *Celestial Wonders and Philosophy, or the Structure of the Visible Heavens* [F 884], where he signed himself "Founder of the Central University of Illinois." *Celestial Wonders* also has had no attention at all from Rafinesque scholars.

42. *American Manual of the Mulberry Trees* [F 894], 97, where the *New Aurora* is listed among the "Publications of the Eleutherium of Knowledge." The Eleutherium was a "free school of useful knowledge" planned as a kind of extension division of the Central University of Illinois.

43. *Bulletin No. 7 of the Historical and Natural Sciences* [F 886], 44. In the 1837 *Circular* [F 995] Rafinesque also said that in 1833 he published a pamphlet on incombustible architecture—which exists perhaps in a unique copy at the Library of Congress that is a four-page reprint of pp. 183–86 of the *Atlantic Journal*.

44. The trip to Illinois is mentioned in a letter (Rafinesque to A. P. de Candolle, June 1, 1838) that Call reproduced in facsimile between pp. 48–49 of his *Life and Writings of Rafinesque*, but made no allusion to in the text.

45. C. S. Rafinesque to A. P. de Candolle; New York, July 1830, Conservatoire Botanique de Genève. The French original: Enfin j'ai découvert par hazard dans la Bibl[iothèque]. du Lycée mes mémoires Botaniques & Zoologiques en voyés en 1819 à Mess Cuvier & Blainville, & inserés la même année. Ayant négligé de m'envoyer une Copie, il a

fallu 10 ans! avant que j'aie a[c]quis la certitude de leur reception & insertion; ce là est bien facheux, car croyant ces Memoires perdus ou négligés je me suis souvent repeté depuis. . . . Cette négligence de Mr Blainville a eu plusieurs consequences facheuses pour moi, croyt mes Mem~ perdus comme tant d'autres, je ne lui <ai> plus envoyé mes Découvertes Zoologiques. Je me suis prévalu en 1820 & 21 de l'offre de Mr Bory de St Vincent pour publier dans les Annales de Bruxelles. Cet ouvrage a été suspendu en 1822, et Mr Bory a encore entre les mains une Douzaine de Mémoires destinés pour lui, dont je ne sais pas ce qu'il a fait. Pourriez vous me dire si aucun autre Mem. Zoolog[.] ou Botanique a été publié en france &c de moi entre 1822 & 1825? & lequels. N'ayant plus des Nouv~ de ces derniers Memoires, j'ai envoyé deux Mem~ à Mr Thiebaut de Berneaud Secr. de la Soc. Lin[n]eene qui en 1824 me nomma son correspondt. Je n'en ai aucune Nouvelle!

Part V

The Legend

XIX

The Eccentric Naturalist

JOHN JAMES AUDUBON
(1785–1851)

"WHAT AN ODD-LOOKING FELLOW!" said I to myself, as, while walking by the river, I observed a man landing from a boat, with what I thought a bundle of dried clover on his back; "how the boatmen stare at him! sure he must be an original!" He ascended with a rapid step, and approaching me asked if I could point out the house in which Mr. Audubon resided. "Why, I am the man," said I, "and will gladly lead you to my dwelling."

The traveller rubbed his hands together with delight, and drawing a letter from his pocket handed it to me without any remark. I broke the seal and read as follows: "My dear Audubon, I send you an odd fish, which you may prove to be undescribed, and hope you will do so in your next letter. Believe me, always your friend B." With all the simplicity of a woodsman I asked the bearer where the odd fish was, when M. de T. (for, kind reader, the individual in my presence was none else than that renowned naturalist) smiled, rubbed his hands, and with the greatest good-humor said, "I am that odd fish I presume, Mr. Audubon." I felt confounded and blushed, but contrived to stammer an apology.

We soon reached the house, when I presented my learned guest to my family, and was ordering a servant to go to the boat for M. de T.'s luggage, when he told me he had none but what he brought on his back. He then loosened the pack of weeds which had first drawn my attention. The ladies were a little surprised, but I checked their critical glances for the moment. The naturalist pulled off his shoes, and while engaged in drawing his stockings, not up, but down, in order to cover the holes about the heels, told us in the gayest mood imaginable that he had walked a great distance, and had only taken a passage on board the *ark*, to be put on this shore, and that he was sorry his apparel had suffered so much from his late journey. Clean clothes were offered, but he would not accept them, and it was with evident reluctance that he performed the lavations usual on such occasions before he sat down to dinner.

At table, however, his agreeable conversation made us all forget his singular appearance; and, indeed, it was only as we strolled together in the garden that his attire struck me as exceedingly remarkable. A long loose coat of yellow nankeen, much the worse for the many rubs it had got in its time, and stained all over with the juice of plants, hung loosely about him like a sac. A waistcoat of the same, with enormous pockets, and buttoned up to his chin, reached below over a pair of tight pantaloons, the lower parts of which were buttoned down to the ankles. His beard was as long as I have known my own to be during some of my peregrinations, and his lank black hair hung loosely over his shoulders. His forehead was so broad and prominent that any tyro in phrenology would instantly have pronounced it the residence of a mind of strong powers. His words impressed an assurance of rigid truth, and as he directed the conversation to the study of the natural sciences, I listened to him with as much delight as Telemachus could have listened to Mentor. He had come to visit me, he said, expressly for the purpose of seeing my drawings, having been told that my representations of birds were accompanied with those of shrubs and plants, and he was desirous of knowing whether I might chance to have in my collection any with which he was unacquainted. I observed some degree of impatience in his request to be allowed at once to see what I had. We returned to the house, when I opened my portfolios and laid them before him.

He chanced to turn over the drawing of a plant quite new to him. After inspecting it closely, he shook his head, and told me no such plant existed in nature; for, kind reader, M. de T., although a highly scientific man, was suspicious to a fault, and believed such plants only to exist as he had himself seen, or such as, having been discovered of old, had, according to Father Malebranche's expression, acquired a " venerable beard." I told my guest that the plant was common in the immediate neighborhood, and that I should show it him on the morrow. "And why to-morrow, Mr. Audubon? Let us go now." We did so, and on reaching the bank of the river I pointed to the plant. M. de T., I thought, had gone mad. He plucked the plants one after another, danced, hugged me in his arms, and exultingly told me that he had got not merely a new species, but a new genus. When we returned home, the naturalist opened the bundle which he had brought

The Eccentric Naturalist

on his back, and took out a journal rendered water-proof by means of a leather case, together with a small parcel of linen, examined the new plant, and wrote its description. The examination of my drawings then went on. You would be pleased, kind reader, to hear his criticisms, which were of the greatest advantage to me, for, being well acquainted with books as well as with nature, he was well fitted to give me advice.

It was summer, and the heat was so great that the windows were all open. The light of the candles attracted many insects, among which was observed a large species of Scarabæus. I caught one, and, aware of his inclination to believe only what he should himself see, I showed him the insect, and assured him it was so strong that it would crawl on the table with the candlestick on its back. "I should like to see the experiment made, Mr. Audubon" he replied. It was accordingly made, and the insect moved about, dragging its burden so as to make the candlestick change its position as if by magic, until coming upon the edge of the table, it dropped on the floor, took to wing, and made its escape.

When it waxed late, I showed him to the apartment intended for him during his stay, and endeavored to render him comfortable, leaving him writing materials in abundance. I was indeed heartily glad to have a naturalist under my roof. We had all retired to rest. Every person I imagined was in deep slumber save myself, when of a sudden I heard a great uproar in the naturalist's room. I got up, reached the place in a few moments, and opened the door, when to my astonishment, I saw my guest running about the room naked, holding the handle of my favorite violin, the body of which he had battered to pieces against the walls in attempting to kill the bats which had entered by the open window, probably attracted by the insects flying around his candle. I stood amazed, but he continued jumping and running round and round, until he was fairly exhausted, when he begged me to procure one of the animals for him, as he felt convinced they belonged to "a new species." Although I was convinced of the contrary, I took up the bow of my demolished Cremona, and administering a smart tap to each of the bats as it came up, soon got specimens enough. The war ended, I again bade him good-night, but could not help observing the state of the room. It was strewed with plants, which it would seem he had arranged into groups, but which were now scattered about in confusion. "Never mind, Mr. Audubon," quoth the eccentric naturalist, "never mind, I'll soon arrange them again. I have the bats, and that's enough."

Some days passed, during which we followed our several occupations. M. de T. searched the woods for plants, and I for birds. He also followed the margins of the Ohio, and picked up many shells, which he greatly extolled. With us, I told him, they were gathered into heaps to be converted into lime. "Lime! Mr. Audubon; why, they are worth a guinea apiece in any part of Europe." One day, as I was returning from a hunt in a cane-brake, he observed that I was wet and spattered with mud, and desired me to show him the interior of one of these places, which he said he had never visited.

The cane, kind reader, formerly grew spontaneously over the greater portions of the State of Kentucky and other western districts of our Union, as well as in many farther south. Now, however, cultivation, the introduction of cattle and horses, and other circumstances connected with the progress of civilization, have greatly altered the face of the country, and reduced the cane within comparatively small limits. It attains a height of from twelve to thirty feet, and a diameter of from one to two inches, and grows in great patches resembling osier-holts, in which occur plants of all sizes. The plants frequently grow so close together, and in course of time become so tangled, as to present an almost impenetrable thicket. A portion of ground thus covered with canes is called a *cane-brake*.

If you picture to yourself one of these cane-brakes growing beneath the gigantic trees that form our western forests, interspersed with vines of many species, and numberless plants of every description, you may conceive how difficult it is for one to make his way through it, especially after a heavy shower of rain or a fall of sleet, when the traveller, in forcing his way through, shakes down upon himself such quantities of water as soon reduce him to a state of the utmost discomfort. The hunters often cut little paths through the thickets with their knives, but the usual mode of passing through them is by pushing one's self backward, and wedging a way between the stems. To follow a Bear or a Cougar pursued by dogs through these brakes is a task the accomplishment of which may be imagined, but of the difficulties and dangers accompanying which I cannot easily give an adequate representation.

The canes generally grow on the richest soil, and are particularly plentiful along the margins of the great western rivers. Many of our new settlers are fond of forming farms in their immediate vicinity, as the plant is much relished by all kinds of cattle and horses, which feed upon it at all seasons, and again because these brakes are plentifully stocked with game of various kinds. It sometimes happens that the farmer clears a portion of the brake. This is done by cutting the stems—which are fistular and knotted, like those of other grasses—with a large knife or cutlass. They are afterwards placed in heaps, and when partially dried set fire to. The moisture contained between the joints is converted into steam, which causes the cane to burst with a smart report, and when a whole mass is crackling, the sounds resemble discharges of musketry. Indeed, I have been told that travellers floating down the rivers, and unacquainted with these circumstances, have been induced to pull their oars with redoubled vigor, apprehending the attack of a host of savages, ready to scalp every one of the party.

A day being fixed, we left home after an early breakfast, crossed the Ohio, and entered the woods. I had determined that my companion should view a cane-brake in all its perfection, and after leading him several miles in a direct course, came upon as fine a sample as existed in that part of the country. We entered, and for some time proceeded without much difficulty, as I led the way, and cut down the canes which were most likely to incommode him. The difficulties gradually increased, so that we were presently obliged to turn our backs to the foe, and push ourselves on the best way we could. My companion stopped here and there to pick up a plant and examine it. After a while we chanced to come upon the top of a fallen tree, which so obstructed our passage that we were on the eve of going round, instead of thrusting ourselves through amongst the branches, when, from its bed in the centre of the tangled mass, forth rushed a Bear, with such force, and snuffing the air in so frightful a manner, that M. de T. became suddenly terror-struck, and, in his haste to escape, made a desperate attempt to run, but fell amongst the canes in such a way that he looked as if pinioned. Perceiving him jammed in between the stalks, and thoroughly frightened, I could not refrain from laughing at the ridiculous exhibition which he made. My gayety, however, was not very pleasing to the *savant*, who called out for aid, which was at once administered. Gladly would he have retraced his steps, but I was desirous that he should be able to describe a cane-brake, and enticed him to follow me by telling him that our worst difficulties were nearly over. We proceeded, for by this time the Bear was out of hearing.

The way became more and more tangled. I saw with delight that a heavy cloud, portentous of a thunder gust, was approaching. In the mean time, I kept my companion in such constant difficulties that he now panted, perspired, and seemed almost overcome by fatigue. The thunder began to rumble, and soon after a dash of heavy rain drenched us in a few minutes. The withered particles of leaves and bark attached to the canes stuck to our clothes. We received many scratches from briers, and now and then a switch from a nettle. M. de T. seriously inquired if we should ever get alive out of the horrible situation in which we were. I spoke of courage and patience, and told him I hoped we should soon get to the margin of the brake, which, however, I knew to be two miles distant. I made him rest, and gave him a mouthful of brandy from my flask; after which, we proceeded on our slow and painful march. He threw away all his plants, emptied his pockets of the fungi, lichens, and mosses which he had thrust into them, and finding himself much lightened, went on for thirty or forty yards with a better grace. But, kind reader, enough—I led the naturalist first one way, then another, until I had nearly lost myself in the brake, although I was well acquainted with it, kept him tumbling and crawling on his hands and knees until long after mid-day, when we at length reached the edge of the river. I blew my horn, and soon showed my companion a boat coming to our rescue. We were ferried over, and on reaching the house, found more agreeable occupation in replenishing our empty coffers.

M. de T. remained with us for three weeks, and collected multitudes of plants, shells, bats, and fishes, but never again expressed a desire of visiting a cane-brake. We were perfectly reconciled to his oddities, and, finding him a most agreeable and intelligent companion, hoped that his sojourn might be of long duration. But, one evening when tea was prepared, and we expected him to join the family, he was nowhere to be found. His grasses and other valuables were all removed from his room. The night was spent in searching for him in the neighborhood. No eccentric naturalist could be discovered. Whether he had perished in a swamp, or had been devoured by a Bear or a Gar-fish, or had taken to his heels, were matters of conjecture; nor was it until some weeks after that a letter from him, thanking us for our attention, assured me of his safety.

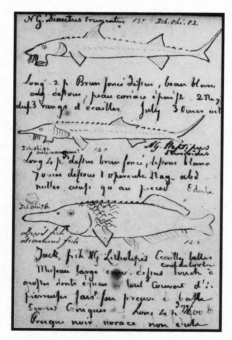

Fig. 19.1. From Rafinesque's field notes, the bottom figure even depicts a diamond-like scale above the nose of the Litholepis *monster.*

Who's Buried in Rafinesque's Tomb?

CHARLES BOEWE

Un voyageur dès le berceau,
Je le serais jusqu'au tombeau. . . .

I

WHEN RAFINESQUE PUT this doggerel French couplet—A traveler from the cradle, / I'll be one to the grave—on the title page of his autobiography he could not have envisioned the putative posthumous travel that will be narrated here. Nor would I have imagined the bizarre events that would need to be explored before an answer to the title of this essay emerged from documents still extant in two of the cities where Rafinesque lived: Lexington in Kentucky and Philadelphia in Pennsylvania. The quest was prompted by a visitor I had in 1956, when I had just settled into my first full-time teaching job at Lehigh University, in Bethlehem, Pennsylvania. The search could not be completed, however, until I myself also lived in Philadelphia for a spell, and thirty years later in Lexington as well, and had had leisure to rummage through archives in those two cities.

My Bethlehem visitor, up from Philadelphia in the interest of Appleton-Century-Crofts, the book publisher he represented, was Robert K. Spencer. Low-key book salesmen in those days spent as much time getting to know the interests of their clients as they did pushing books. Somehow Spencer had learned that during my recent past as a graduate student at the University of Wisconsin I had developed an extracurricular curiosity about C. S. Rafinesque, in part I suppose the serendipitous result of there being so nearly a complete collection of his published writings at the Wisconsin Historical Society library, also on the university campus. These publications were there because the society's resourceful collector, the intrepid Lyman C. Draper, had galloped on horseback throughout much of Kentucky around the turn of the century, collecting for Wisconsin more primary historical documents relating to Kentucky than Kentucky itself has today. Such is the caprice of historical documentation.

Early in our acquaintance, the canny Robert K. Spencer made the bold assertion that he was the only man alive "who has set eyes on C. S. Rafinesque." Spencer also had ancestral ties with Kentucky, but he was living with his family in Philadelphia and going to high school there when Rafinesque's grave in Ronaldson's Cemetery was opened in 1924. This act had been carried out under the direction of his father, James A. Spencer, an executive of the Chas. M. Stieff piano company. Young Robert had skipped school that day; as the others present were long dead, there was a certain truth in his assertion that no one else then alive had seen Rafinesque.

Spencer's memory of what he had witnessed proved somewhat disappointing. He recalled only a few long bones, and he was not sure whether they were from an arm or a leg. There was a "kneecap bone"—he remembered that. But especially, he said, he was impressed by the size of the skull, a skull which suggested a man with a bulging forehead—appropriate, he thought, to the towering intellect it once had housed. He could not remember seeing any vertebrae or even a pelvis. "To tell you the truth," he remarked, "it was a cold, raw day, and we all wanted to hurry up the work and get back inside. I've wished since I had taken notes at the time." He had forgotten—and I only learned long after he himself was dead—that he had in fact written about what he saw. His account, which is still extant, was put on paper within two weeks of the event.

Spencer and I talked of Rafinesque on more than one of his visits. His memory never recovered more details than have been given, but he always came back to that massive, bulging skull which, we both agreed, was consonant with the one picture we knew to be authentic, the frontispiece of *Analyse de la Nature*. Neither of us knew then that there had been an autopsy, and when I came across this in Pennell's 1949 report Spencer was no longer alive.

I said earlier (p. 89) that I would reserve the anatomical findings of the autopsy report for this place. Those significant here are quickly stated. The report includes, among other anatomical and physiological details, a minute description of Rafinesque's brain and its arachnoid membrane. It remarks on the thickness of the brain's cortex; it gives the depth of the fissures in its convolutions; and it

Who's Buried in Rafinesque's Tomb?

confirms a cerebellum "of normal size, measuring four inches transversely."[1] Whatever all this may or may not imply about the mental capacity of the living man, it is information that could be obtained by no other means than by opening the skull. I have often wondered what Spencer would have made of this fact.

In 1924, when Rafinesque's grave in Philadelphia was opened, the autopsy report was unknown. It was received opinion then—and continues to be widely believed today—that Rafinesque "died in a garret, in abject poverty"; that his frail body was let down from a window by ropes, by a few kind-hearted friends who spirited away his remains to save their being sold by a rapacious landlord to a medical school for dissection; and that he was buried in Philadelphia's "potter's field." Yet, eighty-four years later, Robert K. Spencer saw an intact skull of impressive size brought out of Rafinesque's grave.[2] What he remembered thirty-odd years after that was a skull distinguished by its swelling brow line—not a skull in two parts, with the cranium sawed through, as indeed it had to be if the skull were Rafinesque's. Faulty as memory might prove for anyone, surely the one thing a person would remember—when he remembered even a patella bone—would be a skull sawed open.

II

To follow this enigmatic skull the scene now shifts to Lexington, Kentucky. We already know from Pennell's account how Rafinesque came to teach there as a result of the efforts of his friend, trustee John D. Clifford. Famous in its time as the only university west of the Appalachian mountains, its origin dating back to a charter of 1780, Transylvania has educated scores of congressmen, senators, governors, ambassadors, and historical celebrities ranging from Stephen F. Austin to Jefferson Davis, as well as hundreds of western physicians and lawyers. Over the years it has had is ups and downs like other institutions of some antiquity. One of its brightest periods was that presided over by its president the Rev. Horace Holley, which also was roughly the period of Rafinesque's tenure there. Not that there was a causal connection. A trustee had thrust Rafinesque upon Holley, who deeply resented the "Constantinopolitan," as he called the naturalist in private letters, and more than once tried to discharge him.

It was Holley who turned a backwoods academy into a real university, but his conception of university education was the conventional one of his age. It was based on the students attaining sufficient mastery of Greek and Latin to read some of the masterpieces in those languages. There was little or no room in the curriculum for natural science, while the learned professions were represented by the university's faculties of medicine and of law. After Holley's presidency, Transylvania lost its medical school to Louisville in 1859 and its law program became defunct in 1912. For thirteen years it had submerged its identity with that of Bacon College in a hybrid called Kentucky University, and during this period, 1865–78, its identity was further confused when it provided space on its campus to house a separate College of the Bible. In 1915 its status as an undergraduate facility was recognized in the name change to Transylvania College. But in 1969 it readopted the name Transylvania University in celebration of its heritage, though, in fact, it continues as a regional liberal arts college today.[3] Such a capsule history of Transylvania is necessary to understand the deep sense of historic pride and ineluctable destiny that may characterize those associated with it.

Alas, the view of outsiders is likely to be quite different. Out-of-state tourists, driving down Lexington's North Broadway through the two block campus and, ignorant of early American history and equally innocent of Latin, invariably associate the word Transylvania with Bram Stoker's *Dracula,* a novel they probably have not read but may have seen in one of its several film versions. This distressing association may be underscored if they stop to visit the campus. Transy, as the students call it, is a hospitable southern institution, so a student will be assigned to guide them around the handful of buildings. After passing the campus watering-hole called "The Rafskeller"—a pun appealing most to those whose pronunciation of German is shaky—the tour will conclude with Transylvania's greatest wonder: the Tomb of Rafinesque.

After unlocking an unmarked door set flush in the corridor wall of a building called Old Morrison, the guide slowly pushes back two creaking iron grills to reveal a cubical room, with no windows, dimly lit by indirect lights. The dusty concrete floor and peeling plaster of this room—in sharp contrast with the fresh paint, polished brass, and gleaming pegged floors elsewhere in the building—hint at unspeakable mysteries concealed from all but the initiated. The building itself, named for an early benefactor, is a registered National Historical Landmark. One of Gideon Shryock's earliest designs, it is Greek Revival in style, the pediment of its facade supported on six Doric columns, and it has what Lexington intellectuals delight in calling "antepodia"—boxlike banisters on either side of the wide steps that sweep up to the second floor. These steps are seldom used, except for ceremonial occasions such as group photos. One actually enters the building by a side door on the first floor. By entering on the west side of the building and turning right immediately after entry, one faces the door to Rafinesque's tomb. The crypt housing it is the interior of the western "antepodium."

Who's Buried in Rafinesque's Tomb?

Fig. 20.1. Old Morrison. The western "antepodium" is on the left.

While a visitor begins to take in the above-ground tomb, which is completely covered by what appears to be a stone slab, the guide recounts that Rafinesque was a great unappreciated scientist, but he neglected his classes to botanize in the woods; that he "discovered the theory of evolution before Darwin"; that he had a torrid love affair with Mary Holley, the president's wife; and, of course, that he died in a garret and was buried in a potter's field.[4] The guide may tell you that Rafinesque Day, occurring about Halloween time, has become a tradition at Transylvania, and that on that night, four lucky students, two boys and two girls, are selected to pass the night together in the crypt, "right here on these graves!" At this point the visitor's attention will be called to a second tomb, its stone identifying it as that of "Prof. S. F. Bonfils, / a native of France who in / consequence of political con- / vulsions emigrated to the / United States. He devoted / the last thirty years of / his life to the education of the / youth of his adopted country. / Died in Lexington / July 6, 1849 / Aged 54." The more flippant guides say they are "tombmates."[5]

The guide is less likely to explain that part of the Rafinesque Day festivities consists of a bonfire that night, and that boys dressed as undertakers solemnly carry a black coffin around the fire, while, in some years, students have been known to scream obscenities to vent whatever primordial urges bedevil students in the autumn when the moon is full. Nor are you likely to be told it is believed that some years the grave itself has been opened (the cover is not cemented down; three or four strong athletes could lift it off), and a moldy leather bag containing the actual bones has been stuffed inside the symbolic coffin.[6] The guide is sure to dwell long on the infamous "Curse of Rafinesque." If it were not kept alive by Transylvania students themselves, the curse story would go on being resurrected annually by Kentucky newspapers needing filler for their Halloween issues.[7]

A few newspaper writers have gone to the trouble of seeking out the source of the story, which is a single sentence in Rafinesque's autobiography. Returning from a trip to Philadelphia in 1825 (for which, incidentally, he punctiliously requested leave of absence, though he did remain away longer than President Holley expected), he found his rooms in the college had been broken open during his absence; one had been given to students; and his collections and books were in disarray. Rafinesque blamed the violation on Holley's "hatred against sciences," and he wrote: "I took lodgings in town and carried there all my effects: thus leaving the College with curses on it and Holley; who were both reached by them soon after, since he died next year at sea of the Yellow fever, caught at New Orleans, having been driven from Lexington by public opinion: and the College has been burnt in 1828 with all its contents."[8]

It happens, however, that *A Life of Travels* is Rafinesque's own English translation of words he wrote earlier, in 1833, in French. In his native language all he had said was: "Je pris logement en Ville & y transportai tous mes effets, maudissant de bon coeur l'injustice de mon expulsion prématurée, et dévouant le Collège à la justice divine. Qui a atteint l'un et l'autre, car Holley mourut bientôt après a la N. Orleans de la fièvre jaune, ayant été obligé de quitter Lexington par l'Opinion publique, & le Collège a été brûlé en 1828 avec ce qu'il renfermait!"[9] One might translate this as follows: "I took lodgings in Town & carried there all my effects, cursing heartily the injustice of my untimely eviction, and consigning the College to divine justice. Which overtook both, because Holley died soon afterwards in N. Orleans of the yellow fever, having been forced to leave Lexington by public Opinion, & the College burned down in 1828 with everything in it!" Cursing injustice and relinquishing the college to the impartial judgment of the divine will are rather different from a direct and baleful malediction.

Whether knowledge of Rafinesque's original words would have forestalled the legend of the curse is anybody's guess. Probably not. Having pondered his ill treatment for three years, when he came to translate his own words, Rafinesque himself hinted at causality; to minds less clear than his, subsequent disasters at Transylvania came to be regarded almost as proof. In 1961 a woman student was found strangled to death in her car, parked directly outside the Rafinesque tomb—

Who's Buried in Rafinesque's Tomb?

a crime never solved, people will tell you, shuddering over the thought that she must have been the victim of an unnatural force. On a cold January night in 1969 Old Morrison was gutted by fire, with only Rafinesque's tomb, it is now said, escaping unscathed. And more recently, when a workman fell to his death while repairing the gymnasium roof, people again recalled the curse of Rafinesque with nervous tittering.

A more salutary side of Transylvania's infatuation with Rafinesque has been its attempt, in the twentieth century, to make his genuine accomplishments better known. At appropriate intervals, it has sponsored two symposia on his life and works—the first, in 1940, the centennial of his death; the second, in 1983, the bicentennial of his birth. And its library, which already had some manuscripts in his hand and several of his rare printed works, tries to purchase all significant books about him, and now possesses the three portraits said to be of Rafinesque.

At the same time, the university's administration has been reluctant to let specialists have a close look at the notorious Rafinesque bones. What can be said about their authenticity? It happens that a few years ago a highly experienced

Fig. 20.2. Rafinesque's gravestone covers the bones of Mary Passimore in the tomb on the visitor's right.

physical anthropologist who is certified by the American Academy of Forensic Sciences offered to come to Lexington and examine the contents of the tomb. He proposed to make precise measurements of all osseous material, to submit his scientific report to the university, and to offer it for publication only to one of the staid journals of forensic science. Above all, there was to be no sensationalism, and the investigator himself said it was not within his competence to render a judgment on whether the bones were those of Rafinesque or of someone else. That would be up to historians and biographers, who might use his findings as evidence they would have to interpret. The tomb, which is built up of bricks from the concrete floor, would be undamaged, and its massive cover, which is not cemented down, would only have to be lifted off and replaced.

After some time, the proposal was rejected by the university's president, who wrote that the project would be too disruptive to Transylvania. Within a year it was possible to present a similar proposal again, to the successor of that president. This time it had the enthusiastic endorsement of Dr. Jean Rafinesque, the naturalist's great-great grandnephew, who, himself a physician, had come to Lexington to participate in the bicentennial celebration.[10] The new president curtly rejected the proposal, saying the matter had already been settled. And there the question remains, with Transylvania tolerating student high jinks bordering on the occult while refusing to permit a scientific investigation which might dampen them.

Since it is unlikely the tomb will soon be opened, it occurred to me one day that another approach to the mystery of its contents might be tried. This perception can be accomplished by bringing together documents at either end of the Pennsylvania–Kentucky axis, papers never before brought into juxtaposition with each other. As in Charles Willson Peale's celebrated self-portrait in which he draws back a velvet curtain to display the wonders in his Philadelphia museum, it is time now to draw back the curtain on this cabinet of curiosities and reveal who *is* buried in Rafinesque's tomb.

III

As Pennell pointed out, the small wooden marker that may once have indicated the location of Rafinesque's grave in Ronaldson's Cemetery in Philadelphia soon rotted away, and the exact position was unknown for more than seventy years.[11] Then, about 1914, a man named Anthony M. Hance took an interest in trying to find the grave and, though he "came away disappointed," his efforts did lead to the success of others. Samuel N. Rhoads,[12] proprietor of the Franklin Bookshop in Philadelphia, read Hance's article[13] and undertook to search the cemetery records. In 1919 he wrote to Henry C. Mercer,[14] who had encouraged Hance's efforts, that he would be happy to contribute to the cost of the memorial stone Mercer planned to place on the grave. He told Mercer that, with the aid of the sexton, he had "located the exact spot easily enough, for the plots and lots all have their exact locations marked by stone corner monuments with registry Nos and letters corresponding to the registers in the sexton's office. These I verified personally and *stood* on the *site* of Rafinesque's grave, which is flat, no mound[,] and thickly set with grass and weeds, (which the man was mowing off adjoining lots) and by now, no doubt is cleared."[15] He gave the coordinates of the site he identified as 16 south, 11 west, 3rd grave.

Ronaldson's Cemetery, at the southwest corner of Ninth and Bainbridge Streets, had fallen on bad times by 1919. For one thing, it was full—one might say, in the light of further discoveries, it was running over. For another, this was no longer a fashionable part of the city. In such a place, strewn with broken bottles and covered by weeds, it is little wonder everyone accepted without question that Rafinesque, wretched in his last days, had been shoveled without ceremony into a pauper's grave. However, founded in 1827 by the philanthropic James Ronaldson, the burial ground known officially as the Philadelphia Cemetery after its incorporation in 1833 was, for a time, a show place. Its plantings and graveled walks made it a park for the neighborhood, where people strolled of an evening for pleasure. Its most notable facility was a bell house. There a person found dead could be laid out for a day or more, with a string in his hand that would ring the bell should he happen to revive. Protection against premature burial was as highly prized by the nineteenth century as "perpetual" care is by the twentieth.

Moreover, Ronaldson, first president of the Franklin Institute, who had grown wealthy as a type founder, specifically laid out the burying ground as an alternative to the potter's field. To keep out the riffraff, Ronaldson specified in his deed of trust that no coroner might ever own a lot there. The cemetery was especially intended for "strangers"—that is free-thinkers and those, whether of modest means or not, who were denied burial in a churchyard because they lacked church membership. As a consequence, many of the early burials were of visitors to the city, including actors. A Protestant by family background, Rafinesque seems never to have been much of a churchgoer; his friend and executor, the socially prominent Dr. James Mease, made a logical choice in selecting his resting place. Lot 16 south, 11 west, was one of several reserved by Ronaldson himself, who lived another year beyond Rafinesque's death. Ronaldson probably had known Rafinesque, and may have donated the burial space.

Charles Boewe

Despite all this, Rhoads—who had no reason to go into the history of the desolate cemetery—accepted the sexton's opinion that Rafinesque's remains had been located in a pauper's lot, for this helped to explain the startling information he had further to report to Mercer: "Strangely enough, the records show that two persons were buried in the same space, 7 ½ x 4 ft. in 1831; then in 1840 a man named '*Passimore*' and Rafinesque were interred in the same space! In 1847 Mary Passimore's bones (etc) were confided to that identical spot, and in 1848 Julia Steen's remains were added to the conglomeration!!!" According to the sexton, some graves had been dug thirty feet deep, "to allow of successive superimposed layers of bodies to be placed in the same vertical space"; but in the case of Rafinesque and Passimore, the sexton speculated that two bodies had been "closely laid side by side to occupy one grave space normally devoted to *one* person of means."[16] And there the matter rested as far as Rhoads and Mercer were concerned. Their interest was only to identify and mark the correct 7 ½ x 4 foot spot, not to sort out the remains of those who lay beneath its surface.

The sexton's speculation would convey more certainty if it were known that both men interred in 1840 were buried only in winding sheets. However, the bill presented to Dr. Mease by the Bringhurst firm is extant, and it shows that a walnut coffin was used for Rafinesque. (See Pennell, p. 50) It appears highly unlikely that both a coffin and another body, however prepared for burial, could be placed in a space only four feet wide, especially when allowance is made for the improbability of removing earth to that exact width. Moreover, there were intervals, however brief, between burials made in grave 3 which, though of no interest to Rhoads, do give credence to the sexton's hypothesis of a grave thirty feet deep to begin with. And, probably also of little interest to Rhoads, the fact is that serial burial in Ronaldson's Cemetery was the norm, not the exception. The records show that in the lot where Rafinesque lay, grave 1 had seven burials, grave 2 had five, and grave 4 had nine; elsewhere, too, in the cemetery, including in family plots, serial burial was common, as it still is in many European cemeteries and is practiced in some military cemeteries in this country. Here is the complete record for Rafinesque's grave site:

1831–Dec 15	Martha Williams, 52 years, found dead
1831–Dec 18	Robt Grey, 38 years, Infl[amation of the] Brain
1840–July 1	Saml Passimore, 68 years, cholera
1840–Sep 19	C. S. Rafinesque, disease[d] spleen [Marginal note: "Rafinesque—Removed 1924"]
1847–May 28	Mary Passimore, 62 years, Consumption
1848–June 21	Julia A. Steen, 37 years, Pneumonia[17]

Who's Buried in Rafinesque's Tomb?

Without further regard for the other five occupants of grave 3, Mercer completed his arrangements to cover the entire grave site with a memorial to Rafinesque. In correspondence with Rhoads, he worked out a suitable inscription and cast a flat rectangular marker in his own Moravian Pottery and Tile Works in Doylestown. It was made of fine grain concrete, with this inscription, in intaglio, on the surface:

> HONOR TO WHOM HONOR IS OVERDUE
>
> CONSTANTINE S. RAFINESQUE
> BORN CONSTANTINOPLE 1783
> DIED, PHILADELPHIA, SEPTEMBER 18, 1840
> TO DO GOOD TO MANKIND HAS
> EVER BEEN AN UNGRATEFUL TASK
> THE WORKS OF GOD TO STUDY
> AND EXPLAIN
> IS HAPPY TOIL AND NOT TO
> LIVE IN VAIN
> THIS TABLET PLACED HERE SEPTEMBER 1919.[18]

Mercer was too ill to accompany the stone to Philadelphia, but he sent it to Rhoads along with the letter of authorization he had obtained by correspondence with the secretary of the cemetery. Rhoads hastened to write to Mercer as soon as the job was completed, mentioning that the slab had been placed just as Mercer planned, on concrete posts at each corner to raise it about an inch above the surface of the ground. He assured Mercer he had cautioned all those present to keep quiet about what they had done, for all along Mercer had been writing to him that if the newspapers got wind of the project they would "slime" it. Among those present had been four members of the Academy of Natural Sciences representing various branches of botany and zoology, as well as Prof. John Harshberger of the University of Pennsylvania, the five of them making up no doubt a more imposing group than the mourners in 1840. And Rhoads added that moving the slab from the truck to the grave site presented a problem solved only when eight men—of whom he was "proud to have been one"—carried it there by brute force.[19] In his reply, Mercer mentioned hearing from the truck men that additional help for moving the slab had providentially appeared. "As I am slightly superstitious I regard the appearance of the gigantic Italian and his friends in the nick of time to carry the slab either as that of ghosts from the haunted house at Tenth and Bainbridge Streets or of early Sicilian friends of Rafinesque himself."[20]

Charles Boewe

IV

The newspapers did learn about the event, of course, for the next month Mercer had a letter from Mrs. Charles F. Norton in Lexington, saying she had read about his "noble" act and wanted to thank him on behalf of the friends of Transylvania.[21] Because she was Rafinesque's eventual successor as Transylvania's librarian, she felt it appropriate to write a thank-you note; and, as would appear later, she had a personal interest in Rafinesque, believing he had sketched a portrait of one of her female ancestors.[22] Thus began, through librarian Elizabeth Norton, motivation for returning Rafinesque's remains to Lexington when Ronaldson's Cemetery was threatened with destruction.

Even as the memorial stone had been put in place, Mercer learned from a Philadelphia friend that Ronaldson's Cemetery was likely to be turned to other uses. Though, in fact, this did not happen for another thirty years, the same fear was evident in Lexington within five years, when the forthright Mrs. Norton resolved to salvage the bones of Rafinesque. She promptly got in touch with her brother in Philadelphia—James A. Spencer.

Out of filial regard for his native state and with fraternal affection for his sister, Spencer agreed to take on the assignment and thought the necessary permissions had been obtained when Transylvania's dean, Thomas Macartney, journeyed to Philadelphia to ceremonially escort back to Lexington what were respectfully being referred to as "the ashes." Then the trouble began—trouble that produced the documentation from which we can today reconstruct pretty much what happened.

Spencer had made arrangements with the Oliver H. Bair Company to open the grave; he had the good wishes of the secretary of the mayor; and he had a permit from the city's Board of Health—a permit valid only for seventy-two hours when he wrote to his sister on February 14, 1924. Dean Macartney arrived, but the job was finished only on February 27, when Spencer wrote again, nearly two weeks later. Meanwhile they had been chased from pillar to post.

First of all, the sexton balked at letting them into the cemetery, since it was private land, not under city control. Trying to be helpful, the undertaking firm advised them to telephone a Mr. Taylor. Mr. Taylor in turn referred them to the secretary of the cemetery. The secretary proved too feeble to talk with them in person but, through his daughter, advised them to seek the approval of the heirs of James Ronaldson. James Ronaldson's heirs turned out to be two elderly women living somewhere in New Jersey. Along the way there they picked up the suggestion that having permission from the man who placed the stone would be helpful, so they also drove to Doylestown and returned with Mercer's agreement. Armed with all these approvals, they went back to the cemetery, only to be told that they absolutely had to have the permission of Mr. Taylor—the same Mr. Taylor they had earlier telephoned—who was acting manager of the cemetery. Driving some distance outside the city, to wherever Mr. Taylor lived, they at last confronted him in person.

Mr. Taylor agreed that their undertaking was a magnanimous one, and he assured them that he personally had no objection to their mission. But you see, he pointed out, there was a dollar a year ground rent due on every grave, which the relatives and friends of the deceased had been remiss in keeping paid up. Licking the stub of a pencil, he quickly calculated that for a burial dating from 1840 the sum of eighty-four dollars was due.

As reported by Spencer, the visitors said they understood that this demand would have to be met. "We would agree to pay the charge in order that there be no further delay, but"—and Spencer now sprang his own surprise—"we of course would only pay $\frac{1}{9}$ of the assessment as there were 8 other bodies buried in this 1 grave"! All consideration of ground rent was then dropped, and Spencer remarked to his sister that "the suggestion that we make public the fact that there were 9 bodies in this one grave I think was the reason he agreed to let us proceed." Spencer went on to say that "Taylor seemed to be a nice gentleman, and it is possible he too was glad to know that the body of this great man, who had lain so long in a pauper's grave was to be removed, & taken to Lexington, where it would be placed in the keeping of that grand old institution whose board of trustees were so anxious to do it honor."[23]

Even if his belief included three too many bodies, Spencer knew the grave contained multiple burials—a fact he might have learned through his conversation with Mercer in Doylestown, for there is no evidence that he or anyone else in 1924 actually inspected the cemetery records. Perhaps there seemed to be no need to do so, because the site was already marked by Mercer's slab. Nor is there any way to ascertain how he determined the serial order of the burials; this question is not addressed in any of the extant letters by him or by Dean Macartney. The important point is that they—or their workmen—made a slight miscalculation.

We can detect the error now only because the inquisitive John Harshberger, author of *The Botanists of Philadelphia* (1899), who had been present at the laying of the memorial stone in 1919, read about the exhumation in the *Philadelphia Inquirer* (Mar. 4, 1924) and wrote to ask the dean whether he could "give me some details of the condition of [Rafinesque's] remains"; "that is, as to the condition of the coffin and the bones accompanying it."[24]

Charles Boewe

Probably wearied by the whole affair, the dean was unable to provide many concrete details, but his reply deserves extensive quotation since his appears to be another eyewitness account of what was removed from the grave. He wrote:

> In reply to your inquiries I would say that the coffin in which Rafinesque was buried was found to be intact except that the top had caved in under the weight of earth. The wood of which the coffin was constructed was perhaps not over half an inch in thickness. A portion of the top of the coffin was removed with the bones, but the bottom of the casket was so well preserved and so firmly embedded in the earth that it resisted attempts to remove it and so was left in the grave, as we did not wish to disturb a body that the records said was buried beneath Rafinesque.
>
> I am sorry not to be able to tell you the exact nature of the bones found. To corroborate and extend my own knowledge I write to a man who was present at the exhuming, but his knowledge of anatomy is even more inadequate than my own. I am writing now to another man who I am sure can give more definite information, and I shall be glad to communicate this to you later. The skull was very well preserved, there were two long bones which were evidently those of the arms, some of the ribs, one knee cap, a part of the hip bone and many small bones and pieces of bones. The undertaker said that considering the age of the man and the length of time since the burial, the body had been very well preserved.[25]

The first man, whose knowledge of anatomy was inadequate, could have been James A. Spencer. His March 21 response to his sister (who had requested the "full details concerning the removal of the ashes of Rafinesque") told how he and Macartney had been beset by obstructions at every turn and how they finally had overcome them. There is not a word about what was removed from the grave. If the second, more knowledgeable man ever replied, no record of his response has been found. What remains is the dean's observation, like that of the young Robert Spencer, that the "skull was very well preserved."

In fact, the dean must have borrowed the observation from Robert. In his response to Harshberger he used Robert's language, whose letters lay before him. In reply to a letter from the dean, Robert had written:

> Last Wednesday morning Dad and I were at the graveyard at promptly eight o'clock. We were the only ones that were tho. A short time after Mr. McNab and Mr. Taylor came and finally the truck men. The men who were to move the stone had not come so the truck men picked it up and carried it

away. It was lying on top of five supporters that were sunk into the grave. Dad left with the truck men and Mr. Taylor did soon after. When the men first started to dig they found it hard work because the ground was frozen and the supporters were in the way. After these were removed I found they were made of terra cotta piping filled with concrete and weighted with heavy lumps of cement. About ten the men from Costow's came to remove the stone. It was funny to hear Mr. McNab send them about their business.

By this point Robert has revealed why his father could not report on what came out of the grave—he had not been present—and, since Robert wrote this for the dean's information, the dean likely had not been present either. Left now as probably that of the only eyewitness who wrote about what he saw, young Robert's testimony takes on even more weight.

> The first body was found after digging perhaps five or six feet. The only remains were a few bones and a piece of the skull. These were put in a small box before they were replaced in the grave.
>
> Just here some water flowed into the grave making it very muddy and hard to work in. About a foot deeper were found the remains for which we were looking. Mr. McNab said they were very well preserved considering the age of the man at his death and the length of internment [*sic*]. The coffin was perfectly intact except that the top had caved in with the weight of the earth. The wood of which the coffin was constructed was not over a half inch in thickness. The shape was like that of a mummy case and was long and narrow. Part of this was removed with the remains but the floor of the casket was so firmly embedded in the earth that it resisted all attempts to move it and so was left in the grave. The box in which the first body was placed was then put back into the grave and it was refilled.[26]

Wishing to know more, the dean must have queried Robert again. A week later Robert dispatched another letter to Macartney, first explaining that McNab was the man from the undertaking firm—another reason for believing the dean was not present—("I tried to see him before writing this but he is out of town"), then providing most of the language Macartney used without attribution in his letter to Harshberger about the condition of the bones. "I am not sure that I can tell you exactly how many and which of Rafinesque's bones were found as I did not notice particularly. Anyhow there was the skull, two long bones which were evidently the upper arms, the ribs, one knee cap (I remember this especially as the men spoke of it), part of the hip bone and many smaller bones and pieces of bones. I am sorry that this is not more accurate but I hope it will answer."[27]

We may never know why the men from the Bair undertaking firm thought Rafinesque's remains lay at the second level from the surface, or whether the excavation would have gone deeper, to the third and correct level, had the elder Spencer been present. However, the few bones and partial skull they found five or six feet below the surface were those of Julia (or Juliet) A. Steen, who died of pneumonia in 1848 at the age of thirty-seven. Those found about a foot lower at the second level, including the well-preserved skull Robert K. Spencer remembered for three decades, were the remains of Mary Passimore (or Passmore), who died of consumption in 1847 at the age of sixty-two. Since the bottom of Mary Passimore's coffin stuck fast and the excavation went no deeper, lying below and left untouched were the remains of C. S. Rafinesque and three others.[28]

With a shipping label still preserved in the Transylvania archives, Mary Passimore's consumptive bones were dispatched by rail to Kentucky the same day, on February 27, 1924, under Rafinesque's name—and even that was misspelled. The shipping label fallaciously certified that "the body contained in this case died of a non-contagious disease," and it made the further implausible assertion that it had been "arterially embalmed." Later, Mercer's stone also was shipped to Lexington. It rests on the tomb today, identifying the remains of Mary Passimore as those of C. S. Rafinesque.

V

If the troubled spirit of Mary Passimore has sometimes disrupted the tranquility of Transylvania because of the unseemly hoopla performed annually over her mortal remains, the bizarre error that brought them to Kentucky might have been suspected earlier. One journalist went far enough into the Transylvania archives to come across James A. Spencer's letter, and to write that the Philadelphia grave contained nine bodies, as Spencer thought. Written before recent calamities began striking the university, this Halloween article, however, expressed in its heading the pious hope that "Rescue from Grave May Have Lifted Rafinesque Curse on Transy."[29]

Finally, in 1950, the land occupied by Ronaldson's Cemetery at Ninth and Bainbridge was turned into a public playground, and it continues in that use today. At that time the remains known to be of Revolutionary War veterans were reinterred in the burial ground of Old Swedes Church. The others were trucked eighteen miles northeast to Forest Hills Cemetery, Somerton, which is still within the present-day boundaries of the City of Philadelphia.[30] If the bulldozers went

Who's Buried in Rafinesque's Tomb?

deep enough—that is, more than six feet—they scooped up all that was left of Rafinesque; if not, he now lies beneath the pounding feet of Society Hill soccer players. In either event, there is a kind of symmetry in the fact that, like his father, he lies in an unmarked grave, somewhere in Philadelphia.

Notes

1. Edward Hallowell, "Case of Cancer of the Stomach and Liver," *Medical Examiner* 3 (Sept. 19, 1840): 597–99. This weekly Philadelphia medical journal clearly was unable to keep to its publication schedule. Though the date of this number is the day the autopsy was performed, the article itself is dated October 1840. Each of two sets of cemetery records gives a different cause for Rafinesque's death, neither of which is cancer.

2. Several persons who knew Rafinesque personally have testified that his head struck observers as large; John James Audubon remarked on his "broad and prominent" brow if we accept the "Eccentric Naturalist" as intended for Rafinesque.

3. The best history is John D. Wright Jr., *Transylvania: Tutor to the West* (Lexington, 1975). Ash Gobar and J. Hill Hamon, *A Lamp in the Forest* (Lexington, 1982), focuses on the golden years of Horace Holley.

4. Most of this blather is conventional "wisdom," of course. Only the two misconceptions directly pertaining to Transylvania require comment. It would have been hard to neglect classes nobody was required to take anyway, since the academic year at that time was scheduled for those months during winter when the rewards of plant hunting are few. Mary Holley, as befits a college president's wife, invited Rafinesque to tea; she may have studied Spanish with him. The only evidence ever advanced for the love affair is a pastoral poem Rafinesque wrote. The title is "Lines to Maria, who asked me if I should like to love in a cottage," where *love* is a misprint for *live*.

5. Very little is known about Bonfils, except that during their lives he could have had no connection with Rafinesque. His full name was given by the Louisville *Courier-Journal* (June 6, 1939) as St. Sauveur François. It added that he was a native of Corsica, served with Napoleon before Waterloo, and died of cholera. Wright, *Transylvania*, 373, says that his remains were shifted from Lexington's Episcopal cemetery to the Transylvania crypt in May 1939, at the request of his great-granddaughter.

6. A countermyth surfaces at Transylvania now and then. This story has it that when the "body" arrived in Lexington it was put in storage for a time in the attic of the College of the Bible, while the tomb was being prepared. The College of the Bible, an independent training school for ministers of the Disciples of Christ (more commonly known in the South and Middle West as the Christian Church), stood at that time on the Transylvania

Charles Boewe

campus. Then the remains were confused with other "bodies" stored in the same place, with the result that the wrong remains were entombed. However, the notion of dead bodies stacked up like cord wood in a theological institute is so mind boggling it strains even the credulity of Kentucky Christians.

7. Lexington newspapers usually publish an anonymous squib, often quoting one or another Transylvania professor on how it feels to work in an institution burdened by such an imprecation. Louisville's *Courier-Journal*, however, has been diligent in keeping the legend alive by publishing signed pieces. The following, all from the *Courier-Journal*, came to hand without much searching: Gerald Griffin, "The Professor's Curse," Aug. 1, 1937; John S. Gilkinson, "Distinguished Oddity," Dec. 11, 1949; Bob Cooper, "Curse Endears Rafinesque," Oct. 25, 1964; Joe Ward, "The 'Curse' of Rafinesque—An Eerie Tale Has Inspired an Annual Ritual on Transylvania Campus," Oct. 26, 1974; Billy Reed, "Transy's Mad Botanist: Brilliant But Careless," Feb. 23, 1976; and Byron Crawford, "Pioneering Professor . . . He Worked on the Frontier of Knowledge," Sept. 15, 1982.

8. *A Life of Travels*, 78. It seem that Rafinesque's historical inaccuracy did not spoil the efficacy of his curse. Actually, the college burned May 9, 1829, and Holley died July 31, 1827.

9. *Précis ou Abrégé des Voyages, Travaux, et Recherches de C. S. Rafinesque (1833); the Original Version of A Life of Travels (1836)*, ed. Charles Boewe, Georges Reynaud, and Beverly Seaton (Amsterdam, 1987), 72. The manuscript of this earlier form of Rafinesque's autobiography was unknown until 1982.

10. My own interest, of course, was to ascertain whether the tomb contained an intact skull, such as Robert K. Spencer believed was removed from the grave in Philadelphia. I might as well confess, though, that I also hoped to leave behind a small memento mori. Doctor Rafinesque had brought me, as a personal gift, two bottles of a noble cognac from Paris, France (as it always is called in the Bluegrass, to distinguish the City of Light from the better-known village of the same name in Bourbon County). I wanted to place one of these bottles, in a lead-lined oaken box, in the tomb before it was sealed, with instructions on the box that it not be opened until a century had passed. Inside the box, on a slip of parchment, were to be lettered these words: "As we are now, so you will be: / Drink of this cognac merrily; / But let a tear fall in your glass, / For Constantine . . . and Jean . . . and Chas."

11. On the dubious authority of "H. H.," in his letter printed in the *Philadelphia Ledger*, May 5, 1877; the author had Rafinesque's dates wrong and even confused the order of his initials, which he said were painted on the board.

12. Though a bookseller, Rhoads published the most intelligent commentary yet to appear on Rafinesque's contributions to American ornithology; his three articles are "Rafinesque as an Ornithologist," *Cassinia* 15 (1911): 1–12; "Additions to the Known Ornithological Writings of C. S. Rafinesque," *The Auk* 29 (1912): 191–98; "Ornithological Notes of Rafinesque in the *Western Review and Miscellaneous Magazine*," *The Auk*, 29 (1912): 401.

Who's Buried in Rafinesque's Tomb?

13. Anthony M. Hance, "Grave of Rafinesque, the Great Naturalist," *Bucks County Historical Society Papers* 4 (1917): 510–29. This is badly mistitled, for it says little about the grave but rehearses the outlines of Rafinesque's life as it appears in his autobiography. After the exact location of the grave had been ascertained, Hance reprinted the article as a separately paged pamphlet and here included, inside a black box on the last page, the coordinates of the grave site discovered by Rhoads. The coordinates are correct, but Rafinesque's age at death is given as sixty-three, which is wrong. However, this error supplies a helpful hint about what may have happened during the later exhumation. There are *two sets of cemetery records;* the one which lists the names of burials alphabetically also makes this error in Rafinesque's age, while the other which lists burials by lot numbers gives no age. Though Rhoads used the lot record, Hance's repetition of the correct coordinates—which appear in both sets of records—along with this erroneous age suggests that he, and perhaps others later, consulted only the alphabetical record. Moreover, Hance also gives "visceral obstr[uction]s" as the cause of death, and this appears only in the alphabetical record. To identify other burials in the same grave from the alphabetical record alone would require examination of hundreds of entries on pages widely separated from each other, and could easily lead to an error in their serial order.

14. Henry C. Mercer's many accomplishments in archaeology have been less remembered than his one unfortunate book, *The Lenape Stone, or the Indian and the Mammoth* (New York, 1885), a study of an artifact since proved to be fraudulent. It was this study that brought Rafinesque to his attention because of Rafinesque's Lenape creation myth, the *Walam Olum.* Independently wealthy, Mercer had the means to honor Rafinesque in a fitting manner.

15. S. N. Rhoads to Henry C. Mercer, Aug. 11, 1919; Henry C. Mercer Correspondence, Fonthill Manuscript Collection, Bucks County Historical Society, Doylestown, Pa.

16. S. N. Rhoads to Henry C. Mercer, Aug. 11, 1919, Mercer Correspondence.

17. "Ronaldson Cemetery Interments: Lots," II: 249, Historical Society of Pennsylvania. Photostat of a handwritten ledger, this is one of two sets of burial records; the other is an alphabetical list by name. The alphabetical list records the name of the last burial as that of Juliet A. Steen, and that of Mary Passimore as Mary Passmore. Probably Passmore is the more common spelling of the name. The surname Passmore occurs with some frequency in Philadelphia history, a Thomas Passmore being mentioned as one of six city auctioneers in James Mease's *Picture of Philadelphia* (1811).

18. Most of the words of the inscription come from Rafinesque's own writing, polished up a bit by Mercer. Rhoads wanted to add "Time renders justice to all at last," from Rafinesque's autobiography. Rejecting this, Mercer did add the name of his pottery works on the side of the four-inch slab and, underneath, he incised these initials to commemorate those responsible: A. M. H., S. N. R., H. C. M.

19. S. N. Rhoads to Henry C. Mercer, Sept. 24, 1919, Mercer Correspondence.

20. Henry C. Mercer to S. N. Rhoads, Sept. 25, 1919, Mercer Correspondence.

21. Mrs. Charles F. Norton to Henry C. Mercer, Oct. 16, 1919, Mercer Correspondence.

22. Harry B. Weiss to Elizabeth Norton, July 4, 1931; Transylvania Univ. Archives (hereafter referred to as TU Archives). Planning to publish a group of portrait sketches by Rafinesque, Weiss was in touch with Mrs. Norton to identify some of the subjects. That titled "S. S. of Frankfort" in Weiss's book, *Rafinesque's Kentucky Friends* (Highland Park, 1936), may be the one she had in mind, but the sitter had to be an ancestor more remote than Mrs. Norton's mother, as Weiss believed.

23. James A. Spencer to Elizabeth Norton, Feb. 14, 1924, Feb. 27, 1924, Mar. 21, 1924, TU Archives.

24. John W. Harshberger to F. B. McCartney [*sic*], Mar. 6, 1924, TU Archives. The exhumation was also put on more permanent record than that of newspapers by David Starr Jordan, "The Bones of Rafinesque," *Science* 2nd series, 59 (1924): 553–554. Jordan was one of the revisionists who had tried to rehabilitate Rafinesque's scientific reputation— at least in ichthyology—while adding to misconceptions about his life.

25. Thomas Macartney to John W. Harshberger, Mar. 31, 1924, TU Archives.

26. Robert K. Spencer to Thomas Macartney, Mar. 8, 1924, TU Archives.

27. Robert K. Spencer to Thomas Macartney, Mar. 15, 1924, TU Archives.

28. To crosscheck the accuracy of the records, I have scrutinized the coordinates of the graves for every name listed in the "Alphabetical Record" and found that the only persons recorded for lot 16 south, 11 west, 3rd grave are identical with those listed for this grave in the second volume of "Ronaldson Cemetery Interments: Lots." At the end of the "Alphabetical Record" appear a few burials of unknown persons, but none of these is listed for lot 16 south, 11 west, 3rd grave.

29. Joan Hickerson, *Lexington Herald-Leader,* Oct. 11, 1959. Oddly enough—whether by guess or by error—she also states that Rafinesque's was the third body down from the surface; she only failed to read far enough to learn that the remains that were removed were at the second level.

30. Joseph T. Reichwein, "Old Cemetery in South Phila. Soon To Be a City Playground," *Philadelphia Evening Bulletin,* Mar. 12, 1950.

CONTRIBUTORS

Artist/writer **JOHN JAMES AUDUBON,** "a man who needs no introduction," will not receive a redundant one here.

VILEN V. BELYI was a professor of linguistics at the Technical University in Vinnitsa, when he wrote the essay included here. He is now with the Department of Information at the Institute of Advanced Studies in Arad, Israel.

After retiring from the academic world twenty years ago, **CHARLES BOEWE** has edited Rafinesque's correspondence and kept his bibliography up to date.

The biologist **ARTHUR J. CAIN** was curator of zoology and university lecturer at Oxford University until 1964. He ended his academic career in 1989 as Derby Professor of Zoology at the University of Liverpool but was active in retirement until his death in 1999.

Botanist **LEON CROIZAT** was E. D. Merrill's assistant at Harvard's Arnold Arboretum from 1936 to 1946. His last years were spent as director of the Jardin Botanico, which he and his wife had founded in Coro, Venezuela, where he died in 1982.

When **MICHAEL A. FLANNERY** wrote the article included here, he was director of the Lloyd Library and Museum in Cincinnati. Currently he is associate director for Historical Collections of the Lister Hill Library of the Health Sciences at the University of Alabama at Birmington.

After twenty-two years as a botanist in the Philippines, **ELMER D. MERRILL** was dean of the University of California College of Agriculture for six years, then chief executive officer of the New York Botanical Garden from 1930 to 1935, and, finally, from 1936 to 1946, director of the Arnold Arboretum. He died in 1956.

DAVID M. OSTREICHER is among the leading authorities on the language and culture of the Lenape Indians, with whom he has worked for twenty-five years. He holds a master's degree in anthropology and Hebraic studies and has a Ph.D. in anthropology from Rutgers University. He is currently preparing for publication a book-length treatment of the Walam Olum.

Contributors

FRANCIS W. PENNELL was an associate curator at the New York Botanical Garden from 1914 to 1922, and thereafter he served as curator of plants at the Academy of Natural Sciences in Philadelphia until his death in 1952.

Until his recent retirement, GEORGES REYNAUD was a researcher at the Laboratoire de Morphogénése Expérimentale & Caryologie of the Université de Provence in Marseille. His hobby is genealogy.

In addition to her work on the language of flowers, BEVERLY SEATON, associate professor emeritus of English at Ohio State University of Newark, has pubished many articles on various aspects of American popular culture of the nineteenth and early twentieth centuries.

A specialist on the ancient Maya, GEORGE E. STUART served as the staff archaeologist at the National Geographical Society for nearly forty years. He now directs the Center for Maya Research in Barnardsville, North Carolina.

RONALD L. STUCKEY was professor of botany for twenty-six years at Ohio State University, where he also was director from 1967 to 1976 of the university's herbarium. In retirement he writes on botanical history.

INDEX

Only those Latin scientific names used by Rafinesque are indexed. The titles of his own publications and manuscripts also are indexed.

Index

Index

Index

Index

Index

Index

Index

Index

Index

Index

Index

Index

Index

Index

Index

Index

Profiles of Rafinesque

was designed and typeset on a Macintosh computer system using QuarkXPress software. The text is set in Baskerville, and display type is set in Bodoni Poster Compressed. This book was designed and typeset by Bill Adams and manufactured by Thomson-Shore, Inc.